Foundations of
ELECTRONICS

J . R . C O G D E L L

University of Texas at Austin

ECE Department

Austin, Texas

PRENTICE HALL

Upper Saddle River, New Jersey 07458

Library of Congress Cataloging-in Publication Data

Cogdell, J. R.
    Foundations of electronics / J. R. Cogdell.
      p. cm.
    Includes bibliographical references and index.
    ISBN 0-13-907759-6
    1. Electronics.   I. Title.
TK7816.063 1999
621.381—dc21   98—44924
                     CIP

Acquisitions Editor: **Alice Dworkin**
Editorial/Production supervision: **Sharyn Vitrano**
Copy editor: **Barbara Danziger**
Managing editor: **Eileen Clark**
Editor-in-chief: **Marcia Horton**
Director of production and manufacturing: **David W. Riccardi**
Manufacturing buyer: **Pat Brown**
Editorial assistant: **Dan DePasquale**

© 1999 by Prentice Hall, Inc.
Upper Saddle River, New Jersey 07458

The author and publisher of this book have used their best efforts in preparing this book. These
efforts include the development, research and testing of the theories and programs to determine
their effectiveness. The author and publisher make no warranty of any kind, expressed or implied,
with regard to these programs or the documentation contained in this book. The author and
publisher shall not be liable in any event for incidental or consequential damages in connection
with, or arising out of, the furnishing, performance, or use of these programs.

Printed in the United States of America
10 9 8 7 6 5 4 3 2 1

ISBN 0-13-907759-6

Prentice Hall International (UK) Limited, *London*
Prentice Hall of Australia Pty. Limited, *Sydney*
Prentice Hall Canada Inc., *Toronto*
Prentice Hall Hispanoamericana, S.A., *Mexico*
Prentice Hall of India Private, Limited, *New Delhi*
Prentice Hall of Japan Inc., *Tokyo*
Simon & Schuster Asia Pte. Ltd, *Singapore*
Editora Prentice Hall do Brasil, Ltda., *Rio de Janeiro*

Contents

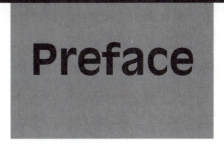

Preface

The need for this book.

About 10 years ago, I started teaching electrical engineering to nonmajors. We used a well-known text, but I found that my students had trouble with the book. The encyclopedic scope of this text of necessity forced it to be superficial. I wanted a text that presented the important ideas in depth and left many of the details for future learning in further study or professional practice. So I wrote *Foundations of Electrical Engineering,* choosing "Foundation" for the title to point to the few important principles that upheld the entire superstructure of electrical engineering. *Foundations of Electrical Engineering* was well received and went into a much-changed second edition in 1996.

The Foundations series. The *Foundations of Electrical Engineering* has now been divided into three stand-alone books: *Foundations of Electric Circuits, Foundations of Electronics,* and *Foundations of Electric Power.* The purpose was to reduce the student's cost and to target specific one-semester (one- or two-quarter) courses as follows:

One semester course	Foundation books required
Circuits for majors or nonmajors	*Foundations of Electric Circuits*
Circuits and electronics for nonmajors	*Foundations of Electric Circuits* (skipping Chaps. 5 and 7) and *Foundations of Electronics*
Electronics for nonmajors	*Foundations of Electronics*
Power for majors or nonmajors	*Foundations of Electric Power*

For a full-year survey of electrical engineering, including topics in circuits, electronics, and power, *Foundations of Electrical Engineering* is the best choice.

The present volume corresponds to Chaps. 7–12 of the second edition of *Foundations of Electrical Engineering.* We have added a chapter summarizing the needed material on circuits to allow the book to stand on its own. Our study of electronics begins with a chapter introducing solid-state phenomena and devices, leading to simple electronic circuits such as power supplies, digital circuits, and analog circuits. The last three chapters deal with electronic systems—instrumentation systems, communication systems, and linear systems.[1]

---

[1] The chapter on linear systems is very similar to Chap. 5 of *Foundations of Circuits.* If a course includes both books, we recommend that this material be covered as part of the study of electronic systems.

Prerequisites

Prerequisite material is a course in differential and integral calculus as well as a course in electrical physics. The exposure to electric circuits in such a physics course plus the study of Chap. 1, should give adequate background in electric circuits for the chapters on electronics.

Pedagogy of the book.

One way to view the structure of an engineering subject is shown in the following diagram:

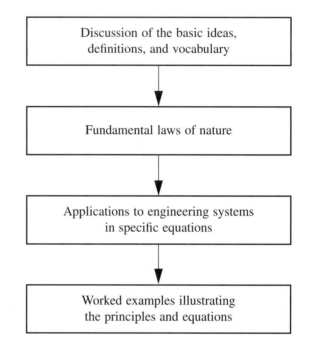

The instructor's emphasis goes from top to bottom. The structure in the instructor's (and author's) mind is primarily ideas and vocabulary, followed by laws, equations, then examples—from the general to the specific. But most students seem to learn in the opposite order: first examples, then equations, then laws, and finally ideas. One suspects some students never go beyond studying the examples, and certainly many believe the equations are what's important. When the instructor probes their understanding of the more general principles by giving a quiz problem that differs from previous examples, many students serve up memorized "solutions" or protest that "the quiz wasn't anything like the lectures and homework." The goals and needs of both instructors and students were paramount in the writing and design of this book.

Aids to learning

This text addresses the viewpoints of both students and instructors in a number of ways

Conservation of Energy

■ **Ideas.** We have identified eight basic ideas of electrical engineering and identified them with a light-bulb icon in the margin, as shown at left. All of these ideas are used in the present volume. These icons hopefully will convince students that electrical engineering is based upon a few foundational ideas that come up again and again.

- **Key terms** are italicized and emphasized by a note in the margin, as shown, when they are first introduced and defined. *Vocabulary* cannot be overemphasized because we communicate with others and, indeed, do our own thinking with words. Each chapter ends with a Glossary that defines many key terms and refers to the context where the words first appear in the text.

- **Causality diagrams**[2] that show cause/effect relationships are given in complicated situations. One reason students have trouble solving problems is that they look at equations and don't see what the important variables are. Understanding consists largely in knowing the causal connections between various factors in a problem so that equations are written with a purpose. Causality diagrams picture these cause/effect relationships.

- **Objectives.** Chapters begin with stated objectives and end with summaries that review how those objectives have been met. In between, we place marginal pointers that alert the student to where the material relates directly to one of the stated objectives. Our intention is to give road signs along the way to keep our travelers from losing their way.

EXAMPLE P.1 — **This is the title of an example**

The numerous examples are boxed, titled, and numbered.

SOLUTION:

Solutions are differentiated from problem statements, as shown. Students are lured beyond passively studying the examples by a WHAT IF? challenge at the end.

WHAT IF?

What if the student were asked to rework the example with a slight change?[3] The answer to the WHAT IF? challenge appears in a footnote for easy checking of results.

- **Check your understanding.** We have "Check Your Understanding" questions and problems, with answers, after major sections.

Problems

Three types of problems have different levels of difficulty and appear at three places in the development

1. WHAT IF? challenges follow most examples. These problems present a slight variation on the examples and are intended to involve the student actively in the principles illustrated by the examples. The answers are given in footnotes.

---

[2] See page 94 for an example.

[3] That would get them involved, wouldn't it?

2. Check Your Understanding problems follow most sections. These are intended as occasions for review and quick self-testing of the material in each section. The answers follow the problems.

3. The numerous end-of-chapter problems range from straightforward applications, similar to the examples, to quite challenging problems requiring insight and refined problem-solving skills. Answers are given to the odd-numbered problems. We are convinced that the only path to becoming a good problem solver passes through a forest of nontrivial problems.

Acknowledgments

I gratefully acknowledge the assistance of many colleagues at the University of Texas at Austin: Lee Baker, David Bourell, David Brown, John Davis, Mircea Driga, "Dusty" Duesterhoeft, Bill Hamilton, Om Mandhana, Charles Roth, Irwin Sandberg, Ben Streetman, Jon Valvano, Bill Weldon, Paul Wildi, Quanghan Xu, and no doubt others. My warmest thanks go to my friend, Jian-Dong Zhu, who worked side by side with me in checking the answers to the end-of-chapter problems. Special thanks go to my son-in-law, David Brydon, who introduced me to Mathematica and to my son, Thomas Cogdell, who produced most of the figures in the text. My heartfelt thanks for numerous corrections, improvements, and wise advice go to the reviewers of *Foundations of Electrical Engineering*: William E. Bennett, U.S. Naval Academy; Richard S. Marleau, University of Wisconsin-Madison; Phil Noe, Texas A&M University; Ed O'Hair, Texas Tech University; and Terry Sculley and Carl Wells, Washington State University, and to Dolon Williams for corrections and suggestions.

I wish to thank my wife for her support and encouragement. Finally, I thank the Giver of all good gifts for the joy I have in teaching and writing about electrical engineering.

John R. Cogdell

1

Electric Circuit Theory

1. To understand the physical variables used to describe electric circuits, particularly their relationship to energy
2. To understand how to apply Kirchhoff's Voltage Law and Kirchhoff's Current Law to electric circuits
3. To understand the properties of resistors, inductors, and capacitors
4. To understand basic methods for analyzing electric circuits
5. To understand the significance of impedance level in interactions between circuits
6. To be able to formulate and solve first-order transients
7. To be able to analyze AC circuits by frequency domain techniques
8. To understand the voltage, current, and impedance transforming properties of transformers

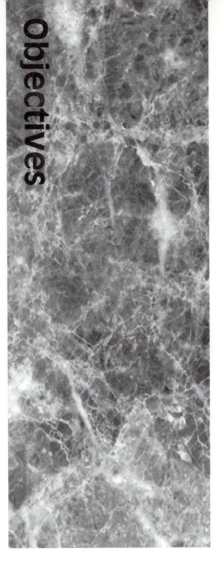

1.1 CHARGE, CURRENT, VOLTAGE, ENERGY, AND POWER

The scope of this chapter. This chapter reviews the fundamentals of electric circuit analysis. The presumption is that the reader has studied electric circuit theory previously, perhaps in a physics course, and a concise review is needed to refresh the memory, establish notations, and set forth a perspective.

In this first section we define current and voltage, present Kirchhoff's Current and Voltage Laws (KCL and KVL), and show how knowledge of voltage and current allows us to monitor the energy processes in an electric circuit. We conclude with Ohm's law relating voltage and current in an ideal resistor.

Energy and Charge

charge

Charge is a fundamental physical quantity. *Charge*, like mass, is a property of matter; indeed, charge joins mass, length, and time as one of the fundamental units from which all scientific units are derived. There are two types of charges, *positive* and *negative*. The names fit because the two types of charges produce opposite effects. Thus equations describing the effects of charges encompass both types of charges if we associate a positive number with one type and a negative number with the other. Traditionally, the electron has been assigned a negative sign and the proton a positive. The magnitude of the charge of the electron is the smallest possible charge; in the MKS system of units this is

$$e = -1.602 \times 10^{-19} \text{ coulombs (C)} \qquad (1.1)$$

Since the mass of the electron is 9.11×10^{-31} kg, the charge-to-mass ratio of the electron is 1.76×10^{11} in the MKS system. Since the charge-to-mass ratio of the fundamental bits of matter is so great, electrical effects usually dominate mechanical inertia effects. Hence we usually talk about charge as if it were not tied to mass—in fact as if it were massless.

energy

The importance of energy. *Energy* is the medium of exchange in a physical system, like money in an economic system. Energy is exchanged whenever one physical thing affects another. In mechanics it takes force and movement to do work (exchange energy), and in electricity it takes electrical force and movement of charges to do work (exchange energy). The electrical force is represented by the voltage and the movement of charge by the current in an electrical circuit. By knowing the voltage and current throughout a circuit, we can monitor energy exchanges.

Current

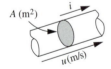

A (m²)

u(m/s)

Figure 1.1 Wire with current.

current

Current is charge in motion. Electrical conductors have mobile (conduction) electrons capable of moving in response to electric forces. Nonconductors have plenty of charges but charges that cannot move. The movement of charges is defined as electric current.

Consider a wire with a cross section of A m² with charges moving with a velocity u from left to right as pictured in Fig. 1.1. If in a period of time Δt, ΔQ coulombs cross A in the indicated direction, we define the *current* to be

$$i = \frac{\Delta Q}{\Delta t} \quad \text{C/s or ampere, A} \qquad (1.2)$$

If the electrons are moving with a velocity u, the number of electrons crossing A in Δt would be $\Delta n = n_e A u \Delta t$, where n_e is the electron density. Hence the current would be[1]

$$i = \frac{\Delta Q}{\Delta t} = \frac{e \Delta n}{\Delta t} = e n_e A u \quad A \tag{1.3}$$

For example, consider electrons traveling downward in a No. 12 wire (0.081 in. in diameter) at a snail's pace of 0.1 mm/s. A copper wire has a concentration of conduction electrons of $n_e = 1.13 \times 10^{29}$ electrons/m^3. From Eq. (1.3), the charges constitute a current of $i = -5.8$ A downward, as shown in Fig. 1.2. We could also express this result by saying that the current is $i = +5.8$ A upward.

physical current

Reference directions. The fact that we can express the current two ways reveals the importance of reference directions. We write "downward," to indicate the direction to which we are referring current flow. To specify a current, we require a reference direction plus a numerical value, which may be positive or negative. Current reference directions are indicated by arrowheads drawn on the lines representing the conductors as in Fig. 1.2, or with arrows beside the wires, or with subscripts. The relationship between the reference direction and the numerical sign of the current is shown in Fig. 1.2, which expresses the same current in all three ways. The direction of the *physical current* is opposite to the direction of electron motion, and hence the physical current is by definition numerically positive.

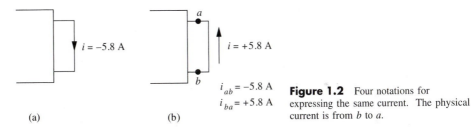

(a)   (b)

$i_{ab} = -5.8$ A
$i_{ba} = +5.8$ A

Figure 1.2 Four notations for expressing the same current. The physical current is from b to a.

Assigning reference directions. The engineer is responsible for assigning current reference directions as a beginning step in analyzing a circuit. These reference directions may be assigned without regard for the direction of the physical currents; they are assigned for bookkeeping purposes. This freedom in assigning reference directions will be clarified below when we state Kirchhoff's current law.

On the other hand, experienced engineers will usually define the current reference direction in the direction of the physical current if that direction is evident. They know that normally the physical current flows *out* of the + terminal on a battery, through the circuit, and returns into the − terminal. It's reasonable, therefore, to make your best guess as to which ways the physical currents will go, but you must write and solve the equations to find out for sure.

Summary. Moving charges constitute a current. To specify the current in a conductor, we need both a reference direction and a numerical value, which can be positive or negative.

---

[1] Note that e is negative, so the direction of numerically positive (physical) current is opposite to the direction of electron motion.

Relationship between charge and current. If we know the current in a wire, $i(t)$, the charge q passing a cross section of the wire during the period $t_1 < t < t_2$ would be

$$q = \int_{t_1}^{t_2} i(t)dt \quad \text{coulombs} \tag{1.4}$$

The charge q in Eq. (1.4) would be the charge that would accumulate on one side of a capacitor if $i(t)$ were the current through the capacitor.

Kirchhoff's Current Law

Conservation of charge and charge neutrality. All the evidence suggests that the universe was created charge-neutral: that is, there exists somewhere a positive charge for every negative charge. Positive and negative charges can be separated by natural causes (lightning) or man-made causes (TV tubes), but most matter is charge-neutral. Furthermore, charge is neither created nor destroyed in electrical circuits. This conservation principle leads directly to a constraint on the currents at a junction of wires.

Kirchhoff's Current Law (KCL) at a junction of wires. The junction of two or more wires is called a *node*. The constraint imposed by conservation of charge and charge neutrality is known as *Kirchhoff's current law* (KCL) and can be stated as follows:

node

Three forms of Kirchhoff's current law

1. *The sum of the currents leaving a node is zero at all times.*
2. *The sum of the currents entering a node is zero at all times.*
3. *The sum of the currents referenced into a node is equal at all times to the sum of the currents referenced out of the node.*

Figure 1.3 illustrates KCL at node a. Kirchhoff's Current Law (KCL) in the first form requires a $+$ sign for currents referenced departing from the node and a $-$ sign for currents referenced toward the node. In general form, KCL is

$$\sum_{\text{node}} \pm i_n = 0 \tag{1.5}$$

The signs ($+$ or $-$) come from the reference directions; the i's are counted $+$ if referenced departing from the node and $-$ if referenced entering the node.

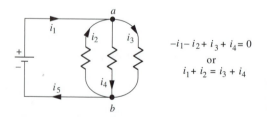

$$-i_1 - i_2 + i_3 + i_4 = 0$$
$$\text{or}$$
$$i_1 + i_2 = i_3 + i_4$$

Figure 1.3 Conservation of charge at nodes a and b is expressed by Kirchhoff's Current Law (KCL).

Definition of Voltage

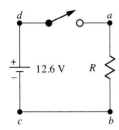

Figure 1.4 A simple circuit: battery, switch, and resistor.

Voltage and energy exchanges. Voltage expresses the potential of an electrical system for doing work. In Fig. 1.4 we show the circuit modeling a battery, switch, and resistor. We have identified points in the circuit with the letters a, b, c, and d. After the switch is closed, the motion of charges around the circuit effects the transfer of energy from the battery to the resistor. The work done by the electrical system in moving a charge from a to b is indicated by the voltage. Specifically, the voltage from a to b is defined to be[2]

$$v_{ab} = \frac{\text{work done by the electrical system in moving } q \text{ from } a \to b}{q} \tag{1.6}$$

In Fig. 1.4, the voltage between d and c is 12.6 because of the battery. With the switch open the voltage does not get to the resistor—with the switch closed the voltage is applied to the resistor. After we explore the concept of voltage more thoroughly, we will present the laws that describe how voltage distributes throughout a circuit.

Voltage as potential. Although our definition implies that charges must move for the voltage to exist, such motion is not required. With the switch open, voltage would still be present in the circuit. The voltage expresses the *potential* for doing work; that is, it measures how much work would be done if a charge were moved from a to b.

subscript notation

Voltage reference directions. We will use two conventions for defining the reference direction of voltages. The more explicit convention uses subscripts, v_{ab}, as in Eq. (1.6), to define the beginning point, a, the ending point, b, and thus the direction traveled, from a to b. Often in a circuit, however, we desire to express the voltage across a single element such as a resistor. In this case we can mark both ends of the circuit element with polarity symbols: a $+$ at one end and a $-$ at the other end. With this simpler notation, the convention is that the $+$ represents the first subscript and the $-$ the second subscript of the voltage.

physical voltage

Physical voltage. The physical polarity of the voltage is the reference direction markings that give a positive voltage. Thus the *physical voltage* is always positive and has polarity markings assigned accordingly. That is, if the voltage with arbitrary $+/-$ reference direction assignment is numerically negative, the physical voltage is polarized in the opposite direction. Most voltmeters have red and black leads, with the black marked "common." Such voltmeters will indicate a positive voltage when the common is connected to the minus polarity mark of the physical voltage and the red is connected to the plus polarity mark. On an auto battery, for example, the markings on the battery terminals are those of the physical voltage.

Kirchhoff's Voltage Law (KVL)

 Conservation of Energy

Conservation of energy. Kirchhoff's voltage law (KVL) expresses conservation of electrical energy in electrical circuits. The charges traveling around a circuit transfer energy from one circuit element to another but do not receive energy themselves on the average. This means that if you were to move a hypothetical test charge around a complete loop in a circuit, the total energy exchanged would add to zero.

---

[2] Some contexts require taking the limit as $q \to 0$.

Forms of KVL. Because the energy sum is zero, it follows from the definition of voltage that the voltage sum around a closed loop is zero also. This is *Kirchhoff's voltage law* (KVL):

$$\sum_{\text{loop}} \text{voltages} = 0 \tag{1.7}$$

We can apply KVL to the circuit in Fig. 1.4

$$v_{cd} + v_{da} + v_{ab} + v_{bc} = 0 \tag{1.8}$$

The direct connection between b and c implies an ideal connection and hence no voltage between these points: $v_{bc} = 0$ and Eq. (1.8) reduces to

$$v_{cd} + v_{da} + v_{ab} = 0 \tag{1.9}$$

We will solve this equation with the switch both open and closed after we reformulate it using the $+/-$ polarity convention.

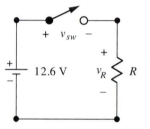

Figure 1.5 Circuit with voltages marked with the $+/-$ convention.

Use of +/– polarity convention. When we apply KVL to voltages whose reference directions are indicated with the $+$ and $-$ polarity convention, we find that we cannot guarantee, as in Eq. (1.9), that all the signs in the resulting equation will be positive. Thus we must rewrite KVL in the form:

$$\sum_{\text{loop}} \pm v\text{'s} = 0 \tag{1.10}$$

In writing KVL equations with $+$ and $-$ polarity symbols, we write the voltage with a positive sign if the $+$ is encountered before the $-$ and with a negative sign if the $-$ is encountered first as we move around the loop. Applying this rule to Fig. 1.5 we express KVL as

$$-(+12.6) + v_{sw} + v_R = 0 \tag{1.11}$$

where v_{sw} and v_R are the voltages across the switch and resistor, respectively, with the polarity markings shown in Fig. 1.5.

Equation (1.11) results from starting at the lower left corner and proceeding clockwise around the loop including the switch and resistor. The minus sign results because we encounter the $-$ first at the battery. The two plus signs occur because we encounter the $+$ reference marks first for both v_{sw} and v_R. All the signs in Eq. (1.11) result from the voltage reference directions marked in Fig. 1.5, not from the signs of the actual voltages.

Solving for the voltage. With the switch closed, $v_{sw} = 0$ and Eq. (1.11) reduces to

$$-(+12.6) + v_R = 0 \implies v_R = +12.6 \text{ V} \qquad (1.12)$$

which means simply that the battery voltage is applied to the resistor. With the switch open no current can flow and the voltage across the resistor must be zero. Thus Eq. (1.11) reduces to

$$-(+12.6) + v_{sw} = 0 \implies v_{sw} = +12.6 \text{ V} \qquad (1.13)$$

which means that the voltage is held off by the open switch, much as hydraulic pressure must be held off by a closed valve.

Two rules for subscripts. As a further application of KVL, we will demonstrate two properties of the subscript notation for voltages. For this purpose we refer to Fig. 1.6. One application of KVL is to go from a to b and then return to a. The resulting equation is

$$v_{ab} + v_{ba} = 0 \implies v_{ab} = -v_{ba} \qquad (1.14)$$

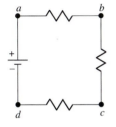

Figure 1.6 Circuit for proving rules of subscript notation.

Thus we see that reversing the subscripts changes the sign of a voltage or vice versa.

Let us now determine v_{ac} by moving from a to c to b and back to a:

$$v_{ac} + v_{cb} + v_{ba} = 0 \implies v_{ac} = -v_{ba} - v_{cb} = v_{ab} + v_{bc} \qquad (1.15)$$

We changed the signs in the last form and reversed the subscripts. Note the pattern in the subscripts in the last form of Eq. (1.15)—the second subscript (b) of the first term, v_{ab}, is identical to the first subscript of the second term, v_{bc}. The final form of v_{ac} suggests that the middle point always drops out. You can verify this by writing v_{ac} in terms of v_{ad} and v_{dc}. You should get

$$v_{ac} = v_{ad} + v_{dc} \qquad (1.16)$$

Again the middle point drops out.

Voltage, Current, and Power

power

Power from voltage and current. Electrical engineers design systems to control electrical energy. Usually, energy is exchanged between parts of a circuit, and the rate of energy flow, the power, must be calculated. We will now show that knowledge of voltage and current everywhere in an electrical circuit allows computation of energy flow throughout that circuit. Indeed, merely restating the definitions of voltage and current reveals their relationship to power flow. Voltage is the energy exchanged per charge and current is the rate of charge flow. *Power* is defined as the rate of energy exchange; hence the power is by definition the product of voltage and current

$$v\left(\frac{\text{work}}{\text{charge}}\right) \times i\left(\frac{\text{charge}}{\text{time}}\right) = \frac{\text{work}}{\text{time}} = \text{power} \qquad (1.17)$$

Thus if we know the voltage across a circuit component and the current through that element, we can determine the power into or out of that component by multiplying voltage and current (Volts × Amperes = Watts).

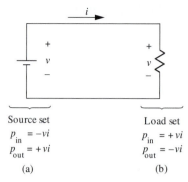

Source set

$p_{in} = -vi$

$p_{out} = +vi$

(a)

Load set

$p_{in} = +vi$

$p_{out} = -vi$

(b)

Figure 1.7 Sign convention for power calculations. A source set (a) is normally used with sources such as batteries, and a load set (b) is normally used with passive circuit elements such as resistors.

source set, load set

Load sets and source sets. We must consider reference directions when we write a power formula. The sign in a power formula depends on the *combination* of the voltage and current reference directions. We show both possibilities in Fig. 1.7. In a *source set*, Fig. 1.7(a), the current reference direction is directed *out of* the + polarity marking, or the first subscript, of the voltage. This combination is normally used for a source of energy to the circuit such as a battery. In a *load set,* Fig. 1.7(b), the current reference direction is directed *into* the + polarity marking (or the first subscript) of the voltage reference direction. This is the convention normally used with a resistance or some other element that receives power from the circuit. For both cases in Fig. 1.7 we can speak of the power out of the component and the power into the component. For example, although a battery normally gives power to a circuit ($p_{out} = +$), it can be receiving power ($p_{out} = -$ or $p_{in} = +$) when the battery is being charged.

We must stress that the formulas given in Fig. 1.7 relate solely to reference directions. Any of the numerical values of the voltages, currents, or powers can be positive or negative. The direction of energy flow at a component will be established only after the sign depending on the reference directions is combined with the numerical signs of the voltage and current. The use of load and source sets will be clarified in the example in Fig. 1.8.

IDEA 2

Conservation of Energy

EXAMPLE 1.1 **Calculating the power in and out of circuit elements**

Find the power out of the battery and the power into the two resistances. Show that energy is conserved.

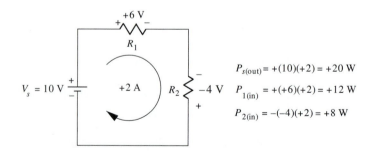

$P_{s(out)} = +(10)(+2) = +20$ W

$P_{1(in)} = +(+6)(+2) = +12$ W

$P_{2(in)} = -(-4)(+2) = +8$ W

Figure 1.8 The battery and R_2 have source sets, but R_1 has a load set. The voltages and current are given.

SOLUTION:

With the battery we have a source set since the current reference arrow is out of the + polarity marking. Hence the battery supplies a power of $+vi$ or $+(+10)(+2) = +20$ W. In the preceding sentence, the + of the "$+vi$" is from the source set of reference directions and the + of the "$+20$" comes from the combination of the reference directions and the numerical values of the voltage and current. The interpretation of the positive power out of the source is that the battery is delivering energy to the circuit. If the power out of the battery had been negative, we would learn that the battery is being recharged by some other, more powerful source in the circuit.

Resistance R_1 has a load set and proves to be receiving $+12$ watts from the circuit. Resistance R_2 is also receiving energy from the circuit because a positive sign is produced by the combination of the minus sign from the reference directions and the minus sign of the voltage. Thus in calculating power we must continue to distinguish between signs arising from the reference directions and signs arising from the numerical values of the voltages and currents. Only the meaning of the power variable ("into" or "out of") and the numerical sign of the result allow the final conclusion as to which way energy is flowing at a given instant. The power out of the source in the previous example, $+20$ W, is equal to the sum of the powers into the resistances, 12 W $+ 8$ W; hence electric energy is conserved in the circuit.

Energy and power. When a charge dq is moved from a to b, the energy given by the electrical circuit is

$$dW = v_{ab}dq \text{ joule (J)} \tag{1.18}$$

When continuous charge flow, a current, is involved, the power, p, would be

$$p = \frac{dW}{dt} = v_{ab}\frac{dq}{dt} = v_{ab}i \text{ watts (W)} \tag{1.19}$$

where i is the current referenced from a to b. Finally, the energy exchanged in a period of time $t_1 < t < t_2$ can be computed by integration

$$W = \int_{t_1}^{t_2} p \, dt = \int_{t_1}^{t_2} v_{ab}i \, dt \text{ J} \tag{1.20}$$

Summary. Voltage and current are easily measured. They allow us to monitor energy flow in an electrical circuit and also obey simple laws—KVL and KCL. For these reasons voltage and current are universally used by electrical engineers in describing the state of an electric network.

Resistances and Switches

LEARNING OBJECTIVE 3.

To understand the properties of resistors, inductors, and capacitors

In the previous sections we defined current and voltage and showed how these describe energy exchanges in an electrical circuit. We stated Kirchhoff's Current Law and Kirchhoff's Voltage Law, which express conservation of charge and electric energy in a circuit, respectively. For illustrative purposes we presented some *circuit elements*—sources, switches, and resistances—without careful definition or explanation. We now define and explain these circuit elements.

Ohm's law. Let us again think about what occurs physically in the circuit of Fig. 1.5 with the switch closed. Chemical action produces charge separation within the battery. This charge separation appears at the battery terminals, and electrical forces are experienced by all the charges in the vicinity of the battery, significantly by charges in the wire and the resistor. Because of these forces, electrons move in the wire and resistor, a current exists in the circuit, and the resistor gets hot.

For a large class of conductors, the current increases in direct proportion to the voltage. Physical experimentation leads to the following equation, known as *Ohm's law*:

$$i = \frac{v}{R} \quad \text{or} \quad v = Ri \tag{1.21}$$

where v is the voltage across a resistance in volts, i is the current through the resistance in amperes, and R is the *resistance* in ohms. Ohm's law relates voltage and current for resistors; later we will discover other voltage–current relationships for capacitors and inductors. The physical behavior of such resistors leads directly to the circuit theory definition of an ideal resistance.[3]

Circuit theory of resistance. The circuit symbol for a resistance, R, is given in Fig. 1.9. Note that the reference directions of the voltage and current form a load set. Ohm's law is presented in graphical form in Fig. 1.10. We have shown voltage to be the independent variable (cause) and current to be the dependent variable (effect). The slope of the line is the reciprocal of resistance. Customarily, Ohm's law is also stated in terms of reciprocal resistance, which is called *conductance*

$$i = Gv, \quad \text{where} \quad G = \frac{1}{R} \tag{1.22}$$

The unit of conductance is amperes per volt, but we use siemens (the official unit) or mho (the unofficial but more common term). The symbol for siemens is S, and the common symbol for the mho[4] is the upside-down omega ($\mho$).

Power into resistors. Because we have a load set, the power into the resistance is $+vi$. We can express the power into a resistance in several ways, as follows:

$$p = +vi = (Ri)i = i^2 R$$
$$= v\left(\frac{v}{R}\right) = \frac{v^2}{R} \quad \text{watts} \tag{1.23}$$

where v is the voltage across the resistance and i is the current through the resistance. Note that the power into a resistance is always positive. Resistance, therefore, always removes electrical energy from a circuit.

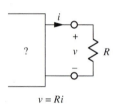

resistance, resistor

$v = Ri$

Figure 1.9 The circuit symbol for a resistance. Note that the voltage and current reference directions form a load set.

conductance

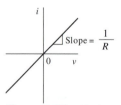

Slope $= \frac{1}{R}$

Figure 1.10 Ohm's law in graphical form. We show the voltage as the independent variable (cause) and the current as the dependent variable (effect).

---

[3] We should perhaps distinguish a *resistor* from a *resistance*. A resistor is a device, often cylindrical in shape, exhibiting the property of resistance, placed in an electrical circuit for the purpose of influencing some voltage or current. Wires, lightbulbs, heater elements, and resistors all exhibit resistance. However, in common parlance *resistor* and *resistance* are synonymous.

[4] "ohm" spelled backward.

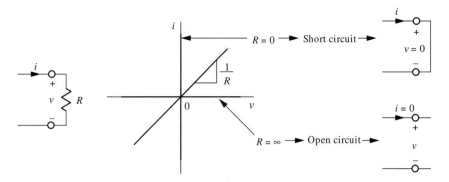

Figure 1.11 Circuit symbol and graphical characteristic of a resistance. The characteristic of a short circuit ($R = 0$) is vertical and that of an open circuit ($R = \infty$) is horizontal.

short circuit and open circuit

Open circuits and Short circuits. Figure 1.11 shows the characteristics of two special resistances. A short circuit ($R = 0$) permits current to flow ($i \neq 0$) without any resulting voltage ($v = 0$), and an open circuit ($R = \infty$) permits voltage ($v \neq 0$) with no current ($i = 0$). In both cases, Eq. (1.23) shows no power is required for the open or short circuit.

Switches. An ideal switch is a special resistance that can be changed from a short circuit to an open circuit to turn an electrical device ON or OFF. Ideal switches receive no electrical energy from the circuit.[5] Figure 1.12(a) shows a *single-pole, single-throw* switch in its open (OFF) state. Figure 1.12(b) shows a *single-pole, double-throw* switch, which switches one input line between two output lines. Figure 1.12(c) shows a *double-pole, single-throw* switch; the dashed line indicates mechanical coupling between the two components of the switch to cause simultaneous switching. Clearly switches can have any number of *poles* and *throws*.

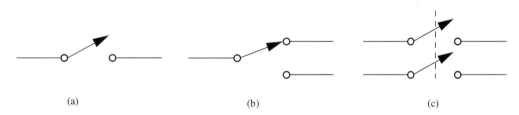

(a)   (b)   (c)

Figure 1.12 Circuit symbols for switches: (a) a single-pole, single-throw switch in the OFF state; (b) a single-pole, double-throw switch; and (c) a double-pole, single-throw switch in the OFF state. The dashed line in (c) indicates mechanical coupling to ensure identical states for both poles.

Voltage and Current Sources

ideal voltage source

Ideal voltage sources. Figure 1.13 shows the circuit symbol, mathematical, and graphical definitions for an ideal general and dc voltage source. The *ideal voltage source* maintains its prescribed voltage, independent of its output current. In general, an

---

[5] Of course some energy is required to "switch" the switch between states.

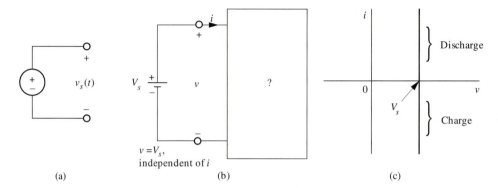

(a)   (b)   (c)

Figure 1.13 (a) General symbol for a voltage source; (b) The circuit symbol and mathematical definition for a dc voltage source; (c) graphical characteristic for a dc voltage source. The source determines the voltage, but the current is determined by the load and the source.

ideal voltage source may have positive or negative voltage, and it may be constant or time-varying. We use the battery symbol for a constant (dc) voltage source and always consider the battery voltage to be positive. We have used a source set for the voltage source. Normally, a voltage source would produce a physical current out of the + terminal and thus act as a source of energy for the circuit; but it is possible that some other, more powerful source might force the physical current *into* the + terminal of the voltage source, thus delivering energy to the source. When this happens for a battery, the current as defined in Fig. 1.13 is negative and we say that the battery is being charged.

ideal current source

Ideal current sources. Figure 1.14 shows the circuit theory symbol, mathematical definition, and graphical characteristic of a general and a dc current source. The *ideal current source* produces its prescribed current independent of its output voltage. Like the ideal voltage source, an ideal current source will deliver any required amount of energy. Unlike the voltage source, there is no physical device at your local convenience

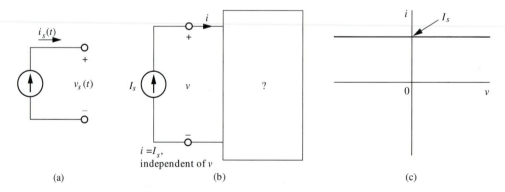

(a)   (b)   (c)

Figure 1.14 (a) General symbol for a current source; (b) the circuit symbol and mathematical definition for a dc current source; (c) graphical characteristic for a dc current source. The current is determined by the source, but the voltage is determined by the load and source.

store whose electrical properties resemble those of a current source; however, we can build electronic devices that act like current sources and the idea is useful in circuit analysis.

Check Your Understanding

1. Does the physical current come out of the plus or minus terminal of a battery? (Which is correct?)

2. A tape player requires three 1.5-V batteries. What voltage is developed if one battery is inserted backward?

3. A battery charger puts 5 A into a 12.6-V auto battery. What is the power into the battery?

4. Is Ohm's law written with a + sign when the reference directions of the voltage across and current through the resistance are related by a load set or source set?

Answers. **(1)** plus, unless the battery is being charged; **(2)** 1.5 V; **(3)** 63.0 W; **(4)** Load set.

1.2 CIRCUIT ANALYSIS TECHNIQUES

In this section we present basic methods for analyzing circuits. We consider only constant (dc) sources, but the techniques are general for dc, ac, or other sources.

LEARNING OBJECTIVE 4.

To understand basic methods for analyzing electric circuits

Series Resistances and Voltage Dividers

Resistances in series. Circuit elements are connected in *series* when the same current flows through them. Figure 1.15 shows a series connection of three resistances and a battery. The method of voltage dividers requires the engineer to anticipate the direction of the physical current, so we have defined the current reference direction such that the current will be positive out of the + terminal of the voltage source, and we put the + and − polarity symbols on the resistance voltages according to load-set conventions. Because i will be numerically positive, the v's across the resistances will also be positive. We can write KVL around the loop, going clockwise.

$$-V_s + v_1 + v_2 + v_3 = 0 \ \Rightarrow \ V_s = v_1 + v_2 + v_3 \tag{1.24}$$

Because the v's are positive and their sum is V_s, the battery voltage divides between v_1, v_2, and v_3. To show how this division depends on the values of the resistances, we introduce Ohm's law:

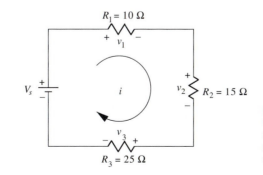

Figure 1.15 The three resistances have the same current and thus are in series. We may replace two or more resistances connected in series with a single equivalent resistance.

$$V_s = R_1 i + R_2 i + R_3 i$$
$$= (R_1 + R_2 + R_3)i$$
$$= R_{eq} i \tag{1.25}$$

where R_{eq} is an equivalent resistance. Thus resistances connected in series add to an equivalent resistance as

$$R_{eq} = R_1 + R_2 + \cdots + R_n \quad \text{(all series resistances)} \tag{1.26}$$

Figure 1.16 shows a circuit that is equivalent to that in Fig. 1.15, except that the three resistances have been replaced by R_{eq}.

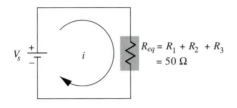

Figure 1.16 The 10-Ω, 15-Ω, and 25-Ω resistances connected in series are represented by an equivalent resistance of 50 Ω.

Voltage dividers. We set out to learn how the battery voltage, V_s, divides between the three resistances. We have determined the current

$$i = \frac{V_s}{R_{eq}} = \frac{V_s}{50} \tag{1.27}$$

We can now obtain the voltage across the individual resistances from Ohm's law applied to the original circuit:

$$v_1 = R_1 i = R_1 \frac{V_s}{R_{eq}} = \frac{R_1}{R_{eq}} \times V_s = \frac{10}{50} V_s \tag{1.28}$$

Formulas for v_2 and v_3 can be written in a similar fashion. For example, if V_s were 60 V, v_1 would be 12 V, v_2 would be 18 V, and v_3 would be 30 V. Our results are easily generalized to include an arbitrary number of series resistances; indeed, the voltage across the ith resistance would be

$$v_i = \frac{R_i}{R_{eq}} \times V_s, \tag{1.29}$$

where $R_{eq} = R_1 + R_2 + \cdots + R_i + \cdots$ (all series resistors)

As an intermediate result we learned that series resistances can be replaced by a single resistance whose value is the sum of the values of the resistances in series.

EXAMPLE 1.2 **Voltage dividers**

The circuit in Fig. 1.17 models a flashlight. The shaded boxes represent the batteries, but in this case we have included the internal resistance of the battery in addition to its internal voltage source. Calculate the voltage across the 2.5-Ω resistance representing the flashlight bulb when the switch is closed.

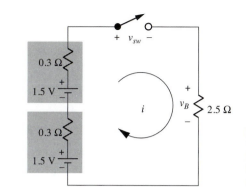

Figure 1.17 Model of flashlight circuit. The shaded boxes represent the two dry-cell batteries, which not only produce voltage but also have internal resistance. The 2.5-Ω resistance represents the bulb.

SOLUTION:

In the circuit in Fig. 1.17, all elements are in series because they have the same current. Kirchhoff's voltage law for the circuit, starting at the bottom and going clockwise, is

$$-(1.5) + 0.3i - (1.5) + 0.3i + v_{sw} + 2.5i = 0 \qquad (1.30)$$

We wrote the equation to show that the voltage from the batteries add to 3.0 V. With the switch closed, Eq. (1.29) gives the voltage across the bulb as

$$v_B = 3.0 \times \frac{2.5}{0.3 + 0.3 + 2.5} = 2.42 \text{ V} \qquad (1.31)$$

That the series resistances are separated from each other by voltage sources is irrelevant; the resistances share the same current and hence are connected in series.

WHAT IF? What if the switch was corroded and had a resistance of 1.2 Ω when ON. What then would be the bulb voltage?[6]

Parallel Resistances and Current Dividers

Parallel resistances. Circuit elements are connected *in parallel* when they have the same voltage across them. Figure 1.18 shows a parallel connection of three resistances and a current source. We have defined the currents through the resistances with reference directions such as the currents would be positive, assuming that I_s is positive. We write KCL for node a

$$-I_s + i_1 + i_2 + i_3 = 0 \implies I_s = i_1 + i_2 + i_3 \qquad (1.32)$$

Because the current reference symbols are directed into the + end of v on the resistances, we have load sets; hence we introduce Ohm's law, Eq. (1.21), into Eq. (1.32):

---

[6] 1.74 V.

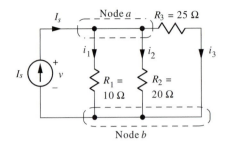

Figure 1.18 Three resistors connected in parallel.

$$I_s = \frac{v}{R_1} + \frac{v}{R_2} + \frac{v}{R_3}$$

$$= v\left(\frac{1}{R_1} + \frac{1}{R_2} + \frac{1}{R_3}\right)$$

$$= v\frac{1}{R_{eq}} \tag{1.33}$$

Equivalent Circuits

where R_{eq} is the equivalent resistance of R_1, R_2, and R_3 connected in parallel. The third form of Eq. (1.33) introduces a resistance, R_{eq}, that is equivalent to the three parallel resistances

$$\frac{1}{R_{eq}} = \frac{1}{R_1} + \frac{1}{R_2} + \frac{1}{R_3} \Rightarrow R_{eq} = \frac{1}{\dfrac{1}{R_1} + \dfrac{1}{R_2} + \dfrac{1}{R_3}} \tag{1.34}$$

Many students have memorized the special case of Eq. (1.34) that applies to two resistances in parallel—the product over the sum. We would encourage exclusive use of Eq. (1.34) for two reasons:

- Eq. (1.34) is the correct form regardless of the number of resistances.
- Eq. (1.34) is easier to implement on a calculator than the "product over the sum" for two parallel resistances.

We can express the combination of parallel resistors with the notation: $R_{eq} = R_1 \| R_2 \| R_3$. In this case

$$R_{eq} = 10\|20\|25 = \frac{1}{\dfrac{1}{10} + \dfrac{1}{20} + \dfrac{1}{25}} = 5.26\ \Omega \tag{1.35}$$

Thus we may replace the three resistances in Fig. 1.18 with an equivalent resistance of 5.26 Ω. We now determine how the current, I_s, divides between the three resistances.

Current dividers. We wish to use Eq. (1.33) in determining how the current I_s divides between the three resistances. We can find the voltage from the last form of Eq. (1.33) and solve for the current in, for instance, R_1 from Ohm's law:

$$v = R_{eq}I_s \Rightarrow i_1 = \frac{v}{R_1} = \left(\frac{R_{eq}}{R_1}\right) \times I_s = \left(\frac{\frac{1}{R_1}}{\frac{1}{R_1} + \frac{1}{R_2} + \frac{1}{R_3}}\right) \times I_s \qquad (1.36)$$

The last form of Eq. (1.36) looks awkward but is the easiest form to implement on a calculator. Substituting the numbers, we learn that the current through R_1 is $0.526I_s$, or 52.6% of the total current through the three parallel resistances. You can confirm that 26.3% passes through R_2 and 21.1% through R_3. These results are easy to generalize: For any number of resistances that are connected in parallel, the current through the ith resistance, R_i, is

$$i_i = \frac{R_{eq}}{R_i} \times I_T = \frac{1/R_i}{1/R_1 + 1/R_2 + \cdots \text{ (all parallel resistances)}} \times I_T \qquad (1.37)$$

where I_T is the total current entering the parallel combination.

Principle of Superposition

Principle stated. The principle of superposition may be stated as follows: *The response of a circuit due to multiple sources can be calculated by summing the effects of each source considered separately, all others being turned OFF.* By OFF we mean that current sources are replaced by open circuits and voltage sources are replaced by short circuits.

turned-OFF sources

Turned-OFF sources. The concept of a *turned-OFF source* is important and bears elaboration. We will start with the graphical presentation of the i/v characteristics of a resistance in Fig. 1.11. Because the slope of the resistance characteristic is $1/R$, a zero-resistance line would have an infinite slope. Thus any amount of current can pass without producing a voltage. On the other hand, the line for an infinite resistance would have zero slope, implying zero current no matter how large the voltage. In Fig. 1.11 we have called zero resistance a short circuit and infinite resistance an open circuit. The circuit symbol for a short circuit is merely a line, and the circuit symbol for an open circuit is a break in the line, indicating that there is no path for current.

Turned OFF voltage sources. From our earlier definition of a voltage source, we know that a voltage source establishes a certain voltage at its terminals, independent of the current. In Fig. 1.19 this is represented by a vertical line at V_s. It follows that a voltage source of zero volts would be represented by a vertical line of infinite slope

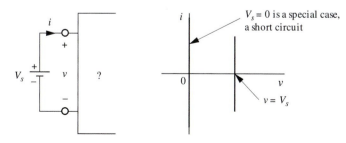

Figure 1.19 The voltage source has a vertical characteristic. Zero voltage ($V_s = 0$) is equivalent to a short circuit.

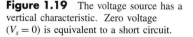

through the origin, which is the same as the characteristic of a short circuit. Thus a turned-OFF voltage source is a short circuit.

Turned OFF current courses. Figure 1.20 shows the graphical characteristics of a current source, a horizontal line of constant current. A current source of zero value produces a horizontal line through the origin, the same characteristic as an open circuit. Thus a turned OFF current source is identical to an open circuit.

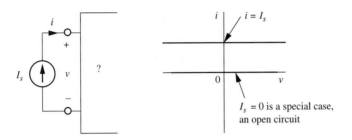

$I_s = 0$ is a special case, an open circuit

Figure 1.20 The current source has a horizontal characteristic. Zero source current ($I_s = 0$) is equivalent to an open circuit.

EXAMPLE 1.3	**Superposition**

Determine the voltage across the 4-Ω resistor in Fig. 1.21 using the principle of superposition.

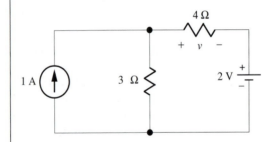

Figure 1.21 We will determine the voltage across the 4-Ω resistor, with the polarity shown, using superposition.

SOLUTION:

First we determine the voltage due to the current source, v_1, with the voltage source turned OFF (replaced by a short circuit).

The resulting circuit is shown in Fig. 1.22 and the two resistors are evidently in parallel

$$v_1 = +1 \text{ A} \times 3\,\Omega \| 4\,\Omega = +1.714 \text{ V} \tag{1.38}$$

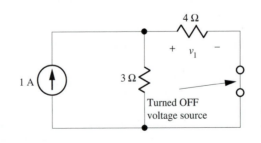

Turned OFF voltage source

Figure 1.22 The voltage source is OFF, replaced by a short circuit.

Next we turn OFF the current source (replace with an open circuit) and calculate the voltage in Fig. 1.23.

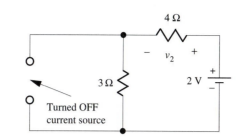

Turned OFF
current source

Figure 1.23 The turned OFF current source is an open circuit.

We see that v_2 can be determined by voltage division

$$v_2 = 2 \text{ V} \times \frac{4}{4+3} = +1.143 \text{ V} \tag{1.39}$$

Finally we add the results in Eq. (1.38) and (1.39), noting that v_2 has polarity opposite to the required voltage in Fig. 1.21.

$$v = v_1 - v_2 = 1.714 - 1.143 = +0.571 \text{ V} \tag{1.40}$$

| **WHAT IF?** | What if both resistors were $4 \, \Omega$?[7] |

Limitations of superposition. Superposition works because Kirchhoff's laws and Ohm's law are linear equations. In all KVL and KCL equations, voltage and current appear in the first power; there are no square roots or functions as e^v.

Superposition does not work for power calculations because power calculations involve multiplication of voltage by current, or squaring a variable like i^2R. Thus you will not get the correct total power by adding the power due to each source considered separately. Nor will superposition yield correct answers for circuits containing *nonlinear* electronic devices such as diodes or transistors.

Check Your Understanding

1. Is an infinite resistance equivalent to a turned-OFF voltage or current source?

2. A turned-OFF voltage source is equivalent to a resistance of what value?

3. Does superposition of powers give the correct value for computing the power out of a dc voltage source?

Answers. (1) Current source; (2) zero ohms; (3) yes. Generally, you *cannot* superpose powers, but it works in this case because the voltage is the same in each calculation.

---

[7] $v = 1$ V.

Thévenin's and Norton's Equivalent Circuits

load

LEARNING OBJECTIVE 4.

To understand basic methods for analyzing electric circuits

We will analyze the circuit shown in Fig. 1.24 using Thévenin's equivalent circuit. We are to determine the current in the variable load resistor R_L as a function of that resistance. A *load* is an element in a circuit of particular importance; in one sense the circuit exists to supply the load.[8]

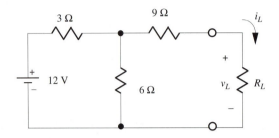

Figure 1.24 Solve for the current in R_L as a function of R_L, $i(R_L)$.

Our method uses an equivalent circuit first proposed in the 1880s by a French telegrapher named M. L. Thévenin. Thévenin's equivalent circuit leads to one of the most useful ideas of electrical engineering, namely, the idea of the "output impedance"[9] of a circuit. This concept influences the thinking of electrical engineers in much of the work they do.

The Thévenin equivalent circuit, shown in Fig 1.25, consists of a voltage source, V_T, in series with an equivalent resistor, R_{eq}, the output impedance of the circuit.

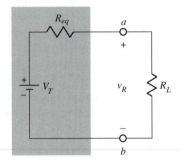

Figure 1.25 The Thévenin equivalent circuit consists of a voltage source, which is the open-circuit voltage, in series with the output impedance of the circuit.

Calculation of the Thévenin voltage. The Thévenin voltage is the open-circuit voltage calculated or measured at the circuit output with the load removed. When we remove the load in Fig. 1.24 the 9-Ω resistor has no current and hence no voltage across it; hence the Thévenin voltage is determined by voltage division

$$V_T = 12 \text{ V} \times \frac{6}{6+3} = 8.00 \text{ V} \tag{1.41}$$

---

[8] If you are making toast, the power system exists for the sake of the toaster.

[9] *Impedance* is a generalized form of resistance. The idea will be defined precisely in Sec. 1.5. For here, we introduce the word as a synonym for resistance.

Calculation of the output impedance. The output impedance is calculated looking back into the circuit from the load, with all sources turned OFF. In Fig. 1.24, turning OFF the voltage source, replacing it with a short circuit, leads to the circuit in Fig. 1.26.

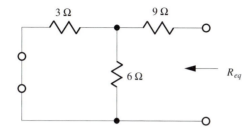

Figure 1.26 The output impedance is calculated looking back into the circuit, with all internal sources OFF.

The output impedance is thus

$$R_{eq} = 9 + 3\|6 = 11.0 \ \Omega \tag{1.42}$$

We may thus replace the circuit in Fig. 1.24 by a Thévenin equivalent circuit as shown in Fig. 1.27.

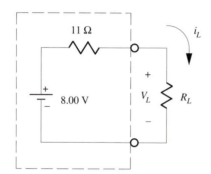

Figure 1.27 Thévenin equivalent to the circuit in Fig. 1.24.

Finally we solve for i_L by inspection

$$i_L = \frac{8}{11 + R_L} \tag{1.43}$$

Benefits of Thévenin equivalent circuit. Of course, with modest effort we could have computed the result in Eq. (1.43) directly from the original circuit in Fig. 1.24. With a Thévenin equivalent circuit, however, we gain the freedom to ask many additional questions such as: What value of R_L makes the voltage 10 V? or What value of R_L withdraws the most power from the circuit? These questions lead to mathematical complexities with the original circuit but can be answered simply with the Thévenin equivalent circuit. The answer to the first question is that no value of R_L will give 10 V. The investigation of the second question leads to an interesting and important result to which we soon turn.

Equivalent Circuits. First, let us consider further what we mean by a Thévenin *equivalent* circuit. The equivalent circuit replaces the circuit within the box only for effects *external* to the box. We can no longer ask questions about the circuit in the box after we have replaced it by an equivalent circuit. For example, if we are interested in the current in the 3-Ω resistor or the total power consumed by the resistors in the box, the equivalent circuit is useless.

Maximum power transfer.

Let us now investigate the question of maximizing the power in R_L in Fig. 1.24. We let the load resistance, R_L, be the independent variable, the power in the load be the dependent variable, P, and find $P(R_L)$ using basic circuit techniques.

$$P(R_L) = i_L^2 R_L = \frac{V_T^2 R_L}{(R_L + R_{eq})} = \frac{(8.00)^2 R_R}{(R_L + 11)^2} \tag{1.44}$$

To maximize $P(R_L)$ as given in Eq. (1.44), we set the derivative with respect to R_L to zero. You can confirm that the maximum (or minimum) occurs at R_L equals R_{eq} (11 Ω). **impedance match** This is called an *impedance match*, when the load is equal to the output impedance of the circuit. To confirm that we have a maximum and not a minimum, a second derivative can be taken, but an easier way in this case is to make a simple sketch of $P(R_L)$ to show that we have found a maximum. Figure 1.28 shows a plot of Eq. (1.44).

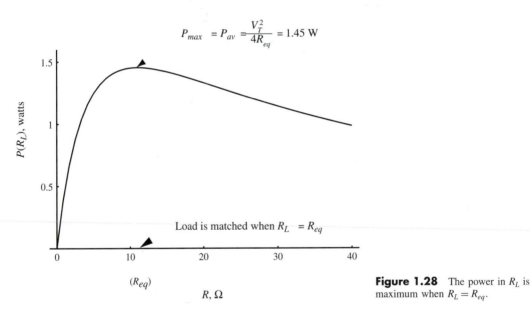

Figure 1.28 The power in R_L is maximum when $R_L = R_{eq}$.

Importance of maximum power transfer.

Obtaining the maximum power out of a circuit is important because often we deal in electronics with small amounts of power and wish to make full use of the power that is available. On a TV set, for example, we pull out the "rabbit ears" antenna to receive power from radio waves originating at a transmitter many miles away. The antenna does not collect much power, so the TV receiver is designed to make maximum use of the power provided by the antenna. Although our results were derived for a simple battery and resistor, they can be applied to a TV antenna. Figure 1.28 shows that we should design the receiver input circuit, represented by a load resistor, to match the output impedance of the antenna for withdrawing the maximum power from the antenna.

Optimum, Optimization

Optimization. When a design is the best possible under the constraints, the design is said to be *optimum*. Thus, the TV receiver input is *optimized* when its input impedance is matched to the output impedance of the antenna because this gives maximum power to the receiver.

Norton's equivalent circuit. An American engineer named E. L. Norton (1898–1983) came up with an alternative equivalent circuit. Norton's equivalent circuit consists of a current source connected in parallel with an output impedance, R_{eq}, as shown in Fig. 1.29. The output impedance, R_{eq}, is the same as before—namely, the resistance of the circuit presented to the load with all internal sources turned OFF.

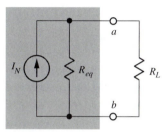

Figure 1.29 The Norton equivalent circuit appears in the box.

Norton current. The Norton current source has a magnitude identical to the current that would flow in a short circuit connected to the output terminals. We will calculate the Norton current, or short-circuit current, for the circuit and load in Fig. 1.24, with the load resistance replaced by a short circuit, Fig. 1.30.

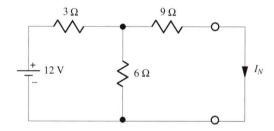

Figure 1.30 The circuit in Fig. 1.24 with the load replaced by a short circuit for the calculation of the Norton current.

We may calculate the Norton current by determining the current in the 12-V source and using a current divider, with the results

$$I_N = \frac{12}{3 + 6\|9} \times \frac{\dfrac{1}{9}}{\dfrac{1}{9} + \dfrac{1}{6}} = 1.818 \text{ A} \times 0.400 = 0.7273 \text{ A} \tag{1.45}$$

Relationship between Thévenin's and Norton's equivalent circuits. If two circuits are equivalent to the same circuit, they must be equivalent to each other. Thus if Norton's circuit in Fig. 1.29 is open-circuited, the voltage must be V_T, the same as the voltage source in the Thévenin circuit of Fig. 1.27. You can see that this proves true in the example since $0.7232\,\text{A} \times 11\,\Omega = 8.00\,\text{V}$. In general, it must be true that

$$V_T = I_N R_{eq} \tag{1.46}$$

Equation (1.46) is useful in theoretical work, but it also can be applied in the laboratory to find the output impedance of a source. In practice, we are unable to get inside and turn OFF internal sources in a battery or an electronic circuit, but we can measure the open-circuit output voltage and the current that flows when the output terminals are shorted. From the measured voltage and current, we can compute the output impedance from Eq. (1.47)

$$R_{eq} = \frac{V_T}{I_N} \tag{1.47}$$

Graphical interpretation. Figure 1.31(b) shows the output characteristic of any linear circuit, Fig. 1.31(a). The intercept on the voltage axis ($i_{\text{out}} = 0$) is the Thévenin, or open-circuit voltage, and the intercept on the current axis ($v_{\text{out}} = 0$) is the Norton, or short-circuit current. The slope is the negative of the reciprocal of the output impedance as required by Eq. (1.47).

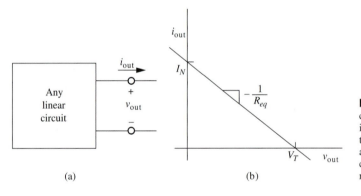

(a) (b)

Figure 1.31 (a) An arbitrary linear circuit with the output voltage and current indicated. (b) The output characteristic of the circuit in graphical form. The intercepts are the Thévenin voltage and the Norton current, and the slope is the negative of the reciprocal of the output impedance.

available power

Available power. The maximum power that can be extracted from a circuit is called the *available power*, P_{av}, and is given by

$$P_{av} = P(R_{eq}) = \frac{V_T^2}{4R_{eq}} \tag{1.48}$$

as shown in Fig. 1.28.

loading of a circuit, voltmeter

Loading of a circuit. We also are concerned about the *loading* of circuits, which occurs when the output voltage is changed by a load. Figure 1.32(a) shows a voltage divider creating a voltage v and a *voltmeter* to measure the voltage. The voltmeter has an input resistance of $10\,\text{M}\Omega$, and hence connecting it will change the voltage to be measured. We may determine the loaded voltage, v', from the Thévenin equivalent circuit in Fig. 1.32(b).

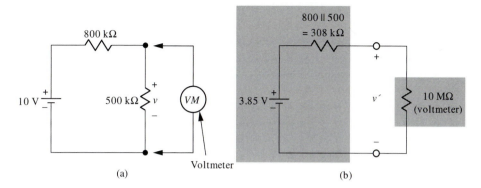

Voltmeter

(a)   (b)

Figure 1.32 (a) The amount of loading from the meter depends on the output impedance of the circuit at the point of measurement and the resistance of the meter; (b) Thévenin equivalent circuit showing the effect of meter loading.

$$v' = 3.85 \times \underbrace{\frac{10 \text{ M}\Omega}{10 \text{ M}\Omega + 308 \text{ k}\Omega}}_{0.970} = 3.73 \text{ V} \tag{1.49}$$

The loaded voltage, which is what the voltmeter will indicate, proves to be 3% lower than the unloaded voltage of 3.85 V, which is what we are seeking to measure with the voltmeter. In practice, we can correct for loading error if we know the output impedance of the circuit and the input impedance of the meter.

Impedance Level

impedance level

Definition of impedance level. Impedance level is a broad concept that is better illustrated than defined. The impedance level of a load is its resistance, and the impedance level of a circuit to a load is the output impedance of the circuit seen by that load. Thus the *impedance level* describes the approximate ratio of voltages to currents in a circuit or portion of a circuit. The interaction of a circuit with a load depends on the relative values of their respective impedance levels.

Circuit model. Consider the Thévenin equivalent circuit shown in Figure 1.33, bearing in mind that this represents the most general circuit/load interaction possible. We now investigate the effect of the impedance levels of circuit and load upon the transfer of power, voltage, and current.

Power. When the impedance level of the load is equal to the impedance level of the circuit, the power in the load is maximum. Hence it would be fitting to say that this condition yields a transfer of power to the load. Figure 1.28 shows that the amount of power depends reciprocally on the output impedance level of the source and will be relatively insensitive to the load impedance, especially if the load impedance is greater than the source impedance.

Voltage. The voltage across the load in Fig. 1.33 is determined by voltage division to be

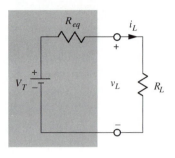

Figure 1.33 This circuit models all circuit/load interactions.

$$v_L = V_T \times \frac{R_L}{R_L + R_{eq}} = V_T \times \frac{1}{1 + \dfrac{R_{eq}}{R_L}} \tag{1.50}$$

When $R_L \gg R_{eq}$, we have $v_L \approx V_T$. Hence when the impedance level of the load is large relative to that of the source, the voltage transferred to the load will be approximately equal to the open-circuit voltage and will be relatively insensitive to the load impedance level. Hence this condition transfers voltage to the load. This is the desired condition when we measure voltage, as shown above.

Current. When $R_L \ll R_{eq}$, the voltage to the load will be small but the current will be

$$i_L = \frac{V_T}{R_L + R_{eq}} = \frac{V_T}{R_{eq}} \times \frac{1}{1 + \dfrac{R_L}{R_{eq}}} = I_N \times \frac{1}{1 + \dfrac{R_L}{R_{eq}}} \tag{1.51}$$

Hence when the load impedance is small compared with that of the source, the current will be approximately the Norton current, independent of R_L. Current is transferred to the load in this condition. Thus, a meter to measure current, an *ammeter*, should have an impedance level that is low compared with the impedance level of the circuit to be measured.

ammeter

Importance of impedance level. Consideration of the impedance level has applications for many circumstances. Consider the power system supplying energy for lighting, electric motors, etc. These appliances are designed to operate at a prescribed voltage; hence, the impedance level of the power system should be small compared with all the loads placed on it. Indeed, since all loads are placed in parallel, the output impedance of the power system should be small compared with the parallel combination of all loads on the system at any given time. If this were not the case, the voltage of the power system would fluctuate seriously.

We showed above that in electronics the impedance levels are controlled to optimize transfer of power between parts of the electronic circuit. Hence in power, electronics, metering, and many other contexts, the impedance levels of the various parts of the circuits are critical to performance.

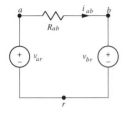
Figure 1.34 The reference node is marked r.

reference node, node voltage

Node-Voltage Analysis

Node voltages. Node-voltage analysis, or nodal analysis, is based on Kirchhoff's current law. We write KCL equations at all nodes except one, but we write these equations in such a way that current variables are never formally defined. We avoid defining current variables by expressing the currents in terms of the "node voltages" in a special way. Figure 1.34 shows a simple circuit with two voltage sources and a resistor. The point at the bottom we have marked r and the points at the ends of the resistor we have marked a and b. Using KVL and Ohm's law, we may express i_{ab} thus:

$$-v_{ar} + v_{ab} + v_{br} = 0 \Rightarrow v_{ab} = v_{ar} - v_{br} \tag{1.52}$$

$$i_{ab} = \frac{v_{ab}}{R_{ab}} = \frac{v_{ar} - v_{br}}{R_{ab}} \tag{1.53}$$

We will change the appearance of Eq. (1.53) by simplifying our notation: We drop the r in the voltage subscript. The *reference node*, labeled r, is like an elevation datum in surveying; we measure all elevations (voltages) relative to it. When we talk about the *node voltage* at a, v_a, we are referring to v_{ar}, the voltage between node a and the reference node, which is thus assigned a voltage of zero. With this change in notation, the form of Eq. (1.53) becomes

$$i_{ab} = \frac{v_a - v_b}{R_{ab}} \tag{1.54}$$

Pattern of subscripts. Equation (1.54) can be stated in words as follows: The current from a to b is the voltage at a, minus the voltage at b, divided by the resistance between a and b.[10] The pattern established in Eq. (1.54) is so simple and intuitive that with it we can express currents without defining current variables. That is, we can keep the "current from a to b" in our heads and write on the paper the voltage at a, minus the voltage at b, divided by the resistance between a and b, the right-hand side of Eq. (1.54).

We will illustrate nodal analysis to determine v_a in the circuit in Fig. 1.35.

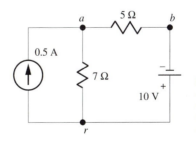

Figure 1.35 The circuit has marked a reference node and two other nodes, a and b. But node b is not an independent node because it is connected to the reference node with a known voltage.

---

[10]This applies only when there is a resistor between a and b.

Kirchhoff's current note at node a is written in the form

$$\Sigma \text{ currents leaving in resistors} = \Sigma \text{ currents entering from current sources} \qquad (1.55)$$

When we apply this form of KCL to node a in Fig. 1.35, using the patterns of subscripts in Eq. (1.54), we obtain

$$\frac{v_a - (0)}{7\,\Omega} + \frac{v_a - v_b}{5\,\Omega} = 0.5 \text{ A} \qquad (1.56)$$

where we have written (0) for the voltage of the reference node. We may determine v_b from writing KVL from b to r and back through the source

$$v_{br} + (+10) = 0 \ \Rightarrow \ v_{br} = v_b = -10 \text{ V} \qquad (1.57)$$

Thus from Eq. (1.56)

$$v_a\left(\frac{1}{7} + \frac{1}{5}\right) = 0.5 + \left(\frac{-10}{5}\right) \ \Rightarrow \ v_a = -4.375 \text{ V} \qquad (1.58)$$

independent node

Independent nodes. In general, a nodal equation must be written at each independent node in the circuit. An *independent node* is a node whose voltage cannot be derived from the voltage of another node. When we analyze a circuit using the method of node voltages, we will have as many unknowns, and equations to solve, as we have independent nodes.[11]

Here is a rule for counting independent nodes: Turn OFF all sources and count the nodes that remain separated by resistors. Turning OFF all sources means that voltage sources are replaced by short circuits and current sources are replaced by open circuits. The number of independent nodes is one less than the number of remaining nodes. For example, in Fig. 1.34 with both sources OFF, we are left with two nodes, the reference node and one independent node. Thus only one equation need be written and solved.

Critique of nodal analysis. Nodal analysis can be implemented somewhat routinely and always works. It is probably the favorite method of electrical engineers for analyzing electronic circuits because most circuit components are connected to the electronic ground, which is used for a reference node.

The physical meaning of the node voltages is clarified by considering a voltmeter. The black voltmeter lead, often marked "common", should be attached to the reference node. If the red voltmeter lead is then touched to the nodes in the circuit, the voltmeter will indicate the node voltages.[12] Thus node voltages are easily measured. Loop currents, the next method we present, also qualifies as a popular and powerful method, but measuring currents involves disconnecting the circuit and hence the loop currents are not easily measured. As you will see, the loop currents may not actually exist in the circuit.

---

[11] The number of independent nodes may sometimes be reduced by combining resistances in series.

[12] This assumes a modern electronic voltmeter that indicates the sign of the voltage. An old-fashioned analog meter might have to be reversed to measure node voltages that are negative.

ground node

The reference node versus "ground."

When the node voltage method is presented in books and used in practice, the reference node is sometimes called the *ground node* and given the symbol ⏚. Strictly speaking, the ground in an electrical circuit identifies the point that is physically connected via a thick wire to the moist earth, usually for safety purposes.

The grounded portion of an electrical circuit usually has many wires connected to it, and hence the electrical ground is often designated the reference node in a nodal analysis. But this is a mere coincidence: The reference node and the electrical ground are totally different concepts. We have avoided calling the reference node the "ground" to establish the concept of the reference node independently of the concept of electrical grounding.

potential difference, potential rise, potential drop

Node voltages and electrical potential.

Node voltages are a measure of the electrical potential of the various nodes in an electrical circuit, relative to the reference node. What we have defined as the voltage between two points can also be called the *potential difference* between the points. Likewise, we can define *potential rises* and *potential drops* in a circuit; for example, the potential rises across a battery (going from − to +) and drops across a resistor (going from + to −). However, we will not speak of potential rises and drops in our development of circuit theory. Our definition of voltage is that of a potential drop.

Check Your Understanding

1. What is the name of the point in a circuit defined to have zero volts?

2. If $v_{rb} = +5$ V, where r is the reference node, what is the node voltage at node b?

3. If node a is connected to the reference node by a 40-V voltage source with the − at node a, what is v_a?

Answers. (1) The reference node, not "ground"; (2) -5 V; (3) -40 V.

LEARNING OBJECTIVE 4.

To understand basic methods for analyzing electric circuits

Loop-Current Analysis

Introduction.

The method of loop currents is similar to nodal analysis, except that the variables are currents, not voltages, and the equations are based on KVL. The goal is to avoid defining unnecessary variables and to systematically write simultaneous equations for the unknown loop currents.

We will present the method by analyzing the circuit shown in Fig. 1.36(a); we are to solve for the voltage across the 2-Ω resistance using loop currents.

independent loop, loop current

Define and label loop currents.

An *independent loop* is a loop that does not pass through a current source. A *loop current* flows in a closed path following a loop. We define loop currents i_1 and i_2 going clockwise around the loops, shown in Fig. 1.36(b).

Conservation of Energy

Write KVL.

We next write KVL following the currents around each loop in a special form:

$$\sum \text{voltages across resistors in loop} = \sum (+ \text{ or } -) \text{ voltage sources in loop} \qquad (1.59)$$

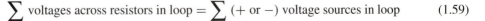

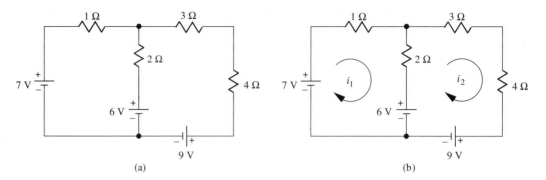

Figure 1.36 (a) The voltage across the 2-Ω resistor is to be determined. (b) Same circuit with loop currents i_1 and i_2 defined.

On the right side of Eq. (1.59) we use + if the voltage aids the loop current for that loop and − if the voltage source opposes the loop current for that loop. The KVL equation for loop 1 is

$$i_1(1) + (i_1 - i_2)(2) = +(+7) - (+6) \qquad (1.60)$$

The first term in Eq. (1.60) is the voltage across the 1-Ω resistance. We are going with the current and we are using a load set for the voltage across the resistance; thus we automatically get a + sign for that voltage. The second term is the voltage across the 2-Ω resistance. The downward current in that resistance is $i_1 - i_2$, the difference between the two loop currents. We are going with i_1 so we write the voltage across that resistance as $+(i_1 - i_2) \times R$. Thus the + sign in the second term is automatic, just like the + sign on the first term. The $+(+7)$ on the right side is due to the 7-V source. It has a + sign because that source tends to force the loop current in that loop (i_1) in its positive direction. The $-(+6)$ is due to the 6-V source. It has the minus sign outside the parentheses because it opposes i_1.

Of course, there are not two physical currents in the 2-Ω resistance, one going up and the other going down. Loop "currents" are mathematical variables that may or may not be identical to a current somewhere in the circuit. In this case i_1 is the current in the 1-Ω resistance, i_2 is the current in the 3-Ω and 4-Ω resistances, and $i_1 - i_2$ is the current referenced downward in the 2-Ω resistance.

The KVL equation for the second loop is written similarly

$$i_2(3) + i_2(4) + (i_2 - i_1)(2) = +(+6) - (+9) \qquad (1.61)$$

We now have two equations in two unknowns:

$$(1 + 2)i_1 - (2)i_2 = +1 \qquad (1.62)$$

$$-(2)i_1 + (3 + 4 + 2)i_2 = -3 \qquad (1.63)$$

Therefore, i_1 is 0.130 A and i_2 is −0.304 A.

This result is not the end of the problem, however, for we set out to calculate the voltage across the 2-Ω resistance. If we put the + polarity symbol at the top of the 2-Ω resistance, the voltage would be $+(i_1 - i_2)(2\ \Omega)$ or $[0.130 - (-0.304)](2) = +0.870$ V.

Check Your Understanding

1. In a current divider consisting of three resistances, the resistances are in the ratio 1:2:3. What percent of the total current goes through the largest resistance?

2. Three resistances having the same value, R, are connected together to have an equivalent resistance of $1.5R$. How are they connected?

3. Is an infinite resistance equivalent to a turned-OFF voltage or current source?

4. A 1.5-V dry cell has a maximum (short-circuit) current of 300 mA. What is the internal resistance of the battery?

5. If $v_{rb} = +5$ V, where r is the reference node, what is the node voltage at node b?

Answers. **(1)** 18.2%; **(2)** one in series with the other two in parallel; **(3)** turned-OFF current source; **(4)** 5 Ω; **(5)** −5 V.

1.3 INDUCTORS AND CAPACITORS

Statics and dynamics. In mechanics, dynamics usually follows statics. Statics deals with the distribution of forces in a structure; time is not a factor. Dynamics deals with the exchange of energy between components in a system, and time is an important factor because energy must be exchanged as a time process.

With this section we begin the study of electrical circuits in which rates of energy exchanged between circuit components are important. We begin by introducing the two circuit components that store energy in electric circuits. The presence of inductors or capacitors in an electric circuit suggests a true dynamics problem. We first will identify the two types of energy that may be stored in a circuit.

electric energy, magnetic energy

Electric energy and magnetic energy storage. To understand what we mean by magnetic energy and electric energy, let us recall from electrical physics that there are two types of forces between electric charges: electrostatic and magnetic. If we move a charge in the presence of a magnetic force, magnetic energy is exchanged. To store much magnetic energy, we must bring many moving charges close together, which is what an *inductor* does. On the other hand, when we move charges in the presence of electrostatic forces, electric energy is exchanged. To store electric energy, we must separate charges, yet keep them close together, which is what a *capacitor* does.

inductor, capacitor

Inductor Basics

LEARNING OBJECTIVE 3.

To understand the properties of resistors, inductors, and capacitors

Physical inductors. Figure 1.37(a) shows a coil of wire, which acts as an inductor. When current flows in the wire, moving charges are close together, magnetic forces are large, and magnetic energy is stored.

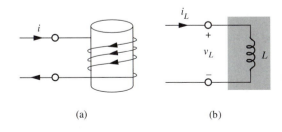

(a)

(b)

Figure 1.37 (a) A coil of wire gets moving charges close together and acts as an inductor. (b) Circuit model for an ideal inductor.

Circuit-theory definition of an inductor. Figure 1.37(b) shows the circuit symbol for an inductor. The equation describing the voltage-current characteristics of an inductor is based on Faraday's law of induction and may be stated in differential or integral form:

$$v_L(t) = +\frac{d}{dt}[Li_L(t)] \tag{1.64}$$

or

$$i_L(t) = i_L(0) + \frac{1}{L}\int_0^t v_L(t') \, dt' \tag{1.65}$$

The circuit symbol and the accompanying equations together define a circuit-theory model for an ideal inductor. Normally, the inductance, L, is considered a constant and brought outside the derivative in Eq. (1.64). Note in Fig. 1.37(b) that v_L and i_L form a load set.

Stored magnetic energy in an inductor. In Eq. (1.20) we compute the energy transferred by integrating the power. We thus can calculate the energy stored in the inductor by integrating its input power, $+v_L i_L$.

$$p_L = \frac{dW_m}{dt} = +v_L i_L = \left(L\frac{di_L}{dt}\right)i_L = \frac{d}{dt}\left(\tfrac{1}{2}Li_L^2\right) \tag{1.66}$$

$$W_m = \int p_L \, dt = \int \frac{d}{dt}\left(\tfrac{1}{2}Li_L^2\right)dt = \tfrac{1}{2}Li_L^2 \tag{1.67}$$

where p_L is the power into the inductor and W_m is the stored magnetic energy in the inductor. The constant of integration is set to zero because there is no stored energy if the current is zero.

Units of inductance. The inductance, L, depends on coil dimensions and the number of turns of wire in the coil. The inductance also depends on the material (if any) located near the coil. In particular, the magnetic properties of iron greatly increase the inductance of a coil wound on an iron core. The units of inductance are volt-seconds per ampere, but we use the name *henry* (H) for this unit; *millihenries* (10^{-3} H or mH) and *microhenries* (10^{-6} H or μH) are also in common use.

henry, millihenry, microhenry

Capacitor Basics

Physical capacitors. Electric (electrostatic) forces arise from interactions between separated charges, and the associated energy is called electric energy. To store electric energy, we must separate charges as in Fig. 1.38.

Circuit-theory definition of a capacitor. Figure 1.39 shows the circuit symbol for a capacitance. Capacitance is defined as the constant relating charge and voltage in a structure that supports a charge separation. If q is the charge on the $+$ side of the capacitor and the voltage is v_C, as shown in Fig. 1.39, the capacitance, C, is defined as

$$q = Cv_C \tag{1.68}$$

Figure 1.38 Structure having capacitance.

Figure 1.39 Circuit symbol for capacitance.

For every charge arriving at the + side of the capacitor, a charge of like sign will depart from the − side and the structure as a whole will remain charge-neutral. Thus KCL will be obeyed because charge flowing into the + terminal side is matched by charge flowing out of the − terminal, as for a resistor. If current is positive into the + terminal, positive charge will accumulate there and will be increasing in proportion to the current. From Eq. (1.3) we can relate the charge in Eq. (1.68) and current as follows:

$$i_C = \frac{dq(t)}{dt} \tag{1.69}$$

where $q(t)$ is the charge in the + side of the capacitor. We may thus define the relationship between current and voltage for a capacitor as

$$i_C = \frac{d}{dt}(Cv_C) = C\frac{dv_C}{dt} \tag{1.70}$$

where the last form of Eq. (1.70) is valid if the capacitance is constant. Note that we have used a load set in Fig. 1.39 for the voltage and current variables.

Units of capacitance. The unit of capacitance is coulomb/volt, but we use the name *farad* (F). Realistic capacitor values come small and usually are specified in microfarads (10^{-6} F or μF), nanofarads (10^{-9} F or nF), or picofarads (10^{-12} F or pF). When a capacitor is constructed from parallel plates, as in Fig. 1.38, the capacitance depends on the area, separation, and material (if any) between the plates.

farad, microfarad, nanofarad, picofarad

EXAMPLE 1.4   **Equation of *RC* circuit**

Derive the equation for the current in Fig. 1.40(a). Assume $v_C(0) = 0$.

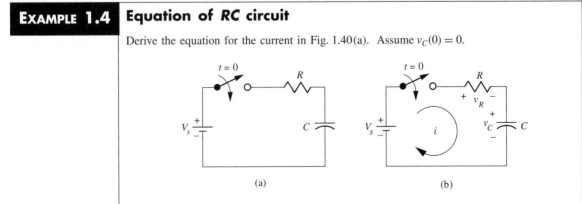

(a)   (b)

Figure 1.40   (a) Closing the switch will cause a momentary current; (b) same circuit with variables defined.

SOLUTION:

Figure 1.40(b) defines a loop current and the capacitor and resistor voltages in a load-set convention. With the switch closed, KVL is

$$-V_s + v_R(t) + v_C(t) = 0 \tag{1.71}$$

We may express both voltages in terms of the current if we differentiate Eq. (1.71):

$$0 + \frac{d}{dt}[Ri(t)] + \frac{dv_C(t)}{dt} = R\frac{di(t)}{dt} + \frac{i(t)}{C} = 0 \tag{1.72}$$

The last form in Eq. (1.72) is readily manipulated to the form

$$\frac{di}{i} = -\frac{1}{RC}\,dt \;\Rightarrow\; \ln i = -\frac{t}{RC} + \ln A \;\Rightarrow\; i(t) = Ae^{-t/RC} \tag{1.73}$$

where A is a constant. The current will charge the capacitor until the capacitor voltage is equal to the battery voltage. Equation (1.74) equates the final charge on C to the integral of the current.

$$q(\infty) - q(0) = CV_s = \int_0^\infty i(t)\,dt = \left.\frac{Ae^{-t/RC}}{-1/RC}\right|_0^\infty = A(RC) \tag{1.74}$$

Since $v_C(0) = 0$, it follows that $q(0) = 0$. Thus Eq. (1.74) yields

$$A = \frac{V_s}{R} \tag{1.75}$$

Hence

$$i(t) = \frac{V_s}{R}e^{-t/RC} \tag{1.76}$$

WHAT IF?

What if the switch is opened after being closed a long time? What would happen to the capacitor voltage?[13]

Integral *i–v* equation. Equation (1.70) is useful in determining the current through a capacitor, given the voltage as a function of time. If we know the current and wish to determine the voltage, we can integrate Eq. (1.70) from 0 to t, with the result

$$v_C(t) = v_C(0) + \frac{1}{C}\int_0^t i_C(t')\,dt' \tag{1.77}$$

---

[13] No current could flow after the switch opened, and according to Eq. (1.70) the voltage would remain constant. A real capacitor would discharge over time due to leakage current.

Stored energy in a capacitor. The charge separation in a capacitor stores electric energy. We may derive the stored electric energy in a capacitor by integrating the power into the capacitor,

$$W_e = \int p_C \, dt = \int v_C \times C \frac{dv_C}{dt} \, dt = \int \frac{d}{dt}\left(\tfrac{1}{2} C v_C^2\right) dt = \tfrac{1}{2} C v_C^2 = \tfrac{1}{2}\frac{q^2}{C} \qquad (1.78)$$

where p_C is the power into the capacitor and W_e is the stored electric energy in the capacitor. The constant of integration must be zero because the uncharged capacitor stores no energy.

1.4 DC TRANSIENT ANALYSIS

In this section we discuss the response of first-order[14] RL or RC circuits to sudden changes in the circuit, usually as the result of some switch action. Such responses are called "transient" because the transition of the circuit from its state before the change to its state after the change takes place during a limited period of time.

We will present two methods for solving first-order transient problems. The first of these is essentially that of the previous example: Write and solve the differential equation. The second avoids differential equations and derives solutions based on initial values, final values, and the time constant of the circuit.

Deriving the differential equation (DE). We will analyze the circuit in Fig. 1.41. With the switch open, the full 10 V appears across the switch because there is no current to cause a voltage across the resistor or the inductor. After the instant at which the switch is closed, KVL for the circuit is

$$-V_s + v_R(t) + v_L(t) = 0 \qquad (1.79)$$

Equation (1.79) becomes a DE when we express v_R and v_L in terms of the unknown current, $i(t)$.

$$L\frac{di(t)}{dt} + Ri(t) = V_s, \quad t > 0 \qquad (1.80)$$

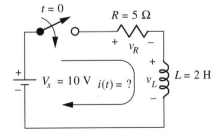

Figure 1.41 An *RL* transient problem. The switch is closed and a current builds up in the circuit. We will solve for the current after the switch is closed.

---

[14] First-order systems have one mode of energy storage. In this context, we refer to circuits with one capacitor or one inductor.

Form of the solution. Equation (1.80) is a linear DE with constant coefficients and a constant forcing term on the right side. The solution of a DE of this type usually proceeds by separating the unknown solution into two parts, the *forced* or *steady-state response* and the *natural* or *transient response*, which is always of the form $e^{\alpha t}$, where α is an unknown constant with the dimensions of time^{-1}. The general solution to Eq. (1.80) must therefore be of the form

steady-state or forced response

homogeneous solution, natural or transient response

$$i(t) = \underbrace{A}_{\substack{\text{forced} \\ \text{response}}} + \underbrace{Be^{\alpha t}}_{\substack{\text{natural} \\ \text{response}}} \tag{1.81}$$

where A, B, and α are unknown constants to be determined from Eq. (1.80) and the initial conditions of the circuit.

Determining the unknown constants. We can determine A and α by substituting Eq. (1.81) back into Eq. (1.80), with the result

$$LB\alpha e^{\alpha t} + R(A + Be^{\alpha t}) = V_s \tag{1.82}$$

The coefficient of the exponential term must vanish if the equation is valid for all times.

$$B(L\alpha + R)e^{\alpha t} + AR = V_s \Rightarrow \alpha = -\frac{R}{L} = -2.5 \text{ s}^{-1} \tag{1.83}$$

In engineering contexts, it is customary to express the exponential term in Eq. (1.83) in the form $e^{-t/\tau}$, where

$$\tau = -\frac{1}{\alpha} = \frac{L}{R} \quad \text{seconds} \tag{1.84}$$

time constant

for the circuit in Fig. 1.41. The *time constant*, τ, is a characteristic time for the transient; we will explore its significance shortly. Setting the coefficient of the exponential term to zero in Eq. (1.82) leads also to the value of A, the steady-state response.

$$AR = V_s \Rightarrow A = \frac{V_s}{R} = \frac{10}{5} = 2 \text{ A} \tag{1.85}$$

Initial condition. To determine B, we must consider the initial condition. The initial condition for this and for all such systems arises from consideration of energy. Time is required for energy to be exchanged and hence processes involving energy carry the system from one state to another, particularly when, as here, a sudden change occurs.

With the switch open, there is no current and no stored magnetic energy in the inductor. The closing of the switch will allow the inductor to store energy, but at the instant after the switch is closed, the stored energy must still be zero. Zero energy implies zero current because the stored energy in an inductor, Eq. (1.67), is $\frac{1}{2}Li^2$; hence, $i(0^+)$ is zero, where 0^+ denotes the instant *after* the switch is closed. This condition leads to the value for B:

$$0 = A + Be^{-0/\tau} \Rightarrow B = -A = -2 \text{ amperes} \tag{1.86}$$

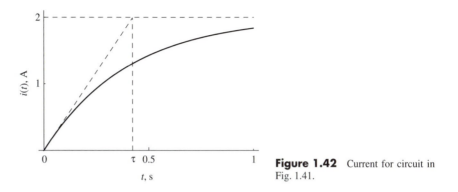

Figure 1.42 Current for circuit in Fig. 1.41.

The final solution is given in Eq. (1.87) and plotted in Fig. 1.42.

$$i(t) = \frac{V_s}{R} - \frac{V_s}{R}e^{-t/\tau} = 2 - 2e^{-t/0.4\text{ s}}\text{ A} \tag{1.87}$$

Physical interpretation of the solution. The mathematical solution is complete, but we wish to examine carefully the physical interpretation of the response, for we will develop our direct method from a physical understanding of this type of problem.

The current is zero before the switch is closed and approaches V_s/R as a final value. This final value does not depend on the value of the inductor, but would be the current if no inductor were in the circuit. The inductor cannot matter in the end because we have a dc excitation in the circuit and eventually the current must reach a constant value. The constant current renders the inductor to act as a short circuit since an inductor will exert itself only when its current is changing.

Without the inductor, the current would change instantaneously from zero to V_s/R when the switch is closed. The inductor effects a smooth transition between the initial and final values of the current. The smooth change moderated by the inductor takes place over a *characteristic time* of τ, the *time constant*.

Note from Fig. 1.42 that the current initially increases with a rate as if to arrive at the final value in one time constant. This time constant for the *RL* circuit, $\tau = L/R$, can be understood from energy considerations. The inductor will require little energy if L is small or R is large; hence the transition will be rapid for small L/R. But if the inductor is large or R is small, much energy will eventually be stored in the inductor; hence the transition period will be much longer for large L/R.

Summary. The response consists of a transition between two constant states, an initial state and a final state. The energy storage element, the inductor, effects a smooth transition between these states. Because of the form of the DE for this class of problems, the transition between states is exponential and is characterized by a time constant. The stored energy in the inductor carries the state of the circuit across the instant of sudden change and leads directly or indirectly to the initial condition.

A Direct Solution

We will now rework this problem using a more direct method. The goal is to write the circuit response directly based on physical understanding. There are four steps.

1. **Find the time constant.** We already know the time constant for this problem, $\tau = L/R$, but in general we can determine the time constant for a first-order system by putting the DE into the special form of Eq. (1.88). We divide by the coefficient of the linear term to obtain the form

$$\tau \frac{dx}{dt} + x = \text{constant} = x(\infty) \qquad (1.88)$$

In Eq. (1.88), x refers to the physical quantity being determined. It would represent a voltage or current in a circuit problem, but in other physical systems x might represent a temperature, a velocity, or something else. When we put the DE in this form, τ will always be the characteristic time, the time constant. To put Eq. (1.80) into this form, we have to divide by R to see that $\tau = L/R$.

Since the purpose of this method is to avoid DEs, we hesitate to suggest that you must write the DE to get started. Usually, you will know the equation of the time constant for the circuit or system from previous experience. For an *RL* circuit, for example, the time constant is always L/R. For the *RC* circuit in Fig. 1.40, described by the DE in Eq. (1.72), we multiply by C and compare with Eq. (1.88) to see

$$\tau = RC \qquad (1.89)$$

for an *RC* circuit. Because we solve circuits having only one energy storage element with this method, we have only one L or C. Below we show how to handle circuits with more than one resistor. Returning to the problem we are solving, we now know our time constant: $L/R = 0.4$ s for the circuit in Fig. 1.41.

2. **Find the initial condition.** The initial condition, $x(0)$ in general, $i(0)$ in this case, always follows from consideration of the energy condition of the system at the beginning of the transient. In this case, we argued earlier that because the inductor had no energy before the switch was closed, it must have no energy the instant after the switch is closed. Hence the initial current is zero, $i(0) = 0$. Later we will discuss generally how to determine initial values in circuits.

3. **Find the final value.** The final value, $x(\infty)$ in general, $i(\infty)$ in this case, follows from the steady-state solution of the system. Earlier we argued that the inductor approaches a short circuit because the final state of the circuit is a dc state, and the inductor has no voltage across it for a dc current. Thus application of Ohm's law to the circuit in Fig. 1.41 shows that the final current must be $i = 2$ A.

If the DE has been written and put into the form of Eq. (1.88), then the final value is the constant on the right-hand side of the DE. In the final state, time derivatives must vanish and Eq. (1.88) reduces to $x = x(\infty)$.

4. **Substitute the time constant and the initial and final values into a standard formula.**

$$x(t) = x(\infty) + [x(0) - x(\infty)]e^{-t/\tau} \qquad (1.90)$$

Equation (1.90) is the solution for all first-order systems having dc (constant) excitation. In our case, the $x(0)$ and $x(\infty)$ are known currents, so our solution is

$$i(t) = \frac{V_s}{R} + \left(0 - \frac{V_s}{R}\right)e^{-t/\tau} \tag{1.91}$$

$$= 2 + (0 - 2)e^{-t/0.4} \text{ A}$$

As before, we see that as $t \to \infty$, $i(t) \to V/R = 2$ A, and initially $i(0) = 0$ since $e^0 = 1$.

Figure 1.43 shows a generalized plot of Eq. (1.88). The curve starts at $x(0)$ and asymptotes to $x(\infty)$. Its initial slope is such as to move from $x(0)$ to $x(\infty)$ in one time constant, but it reaches only $(1 - e^{-1})$ or 63% of the way during the first time constant. We have shown a discontinuity at the origin because this can happen, although it does not happen in our present example.

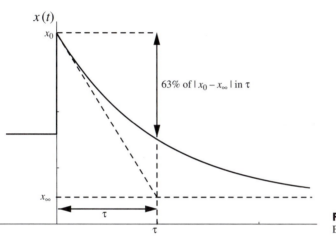

Figure 1.43 Generalized response in Eq. (1.90).

Summary. Our direct method consists of determining three constants and substituting these into a standard formula. The time constant can be determined from the DE, but usually the formula for τ is known from prior experience. The other two constants are the initial value of the unknown, which is determined from energy considerations, and the final value, which is determined from the dc solution. The energy storage element effects a smooth transition between the initial and final values and influences the initial value through energy considerations.

Equivalent Circuits

Circuits with Multiple Resistors

Thévenin equivalent circuit. Circuits with one inductor or capacitor but more than one resistor can be analyzed with the concept of equivalent circuits. The circuit in Fig. 1.44(a) with the switch closed, for example, can be reduced to the equivalent circuit in Fig. 1.44(b). From this Thévenin equivalent circuit, clearly the time constant is $R_{eq}C$, where R_{eq} is the output impedance of the circuit and the switch closed. Thus R_{eq} is the output impedance of the circuit to the capacitor as a load.

We may or may not be interested in the Thévenin equivalent circuit as a means for calculating the initial or final values of the unknown. That is a separate problem to be

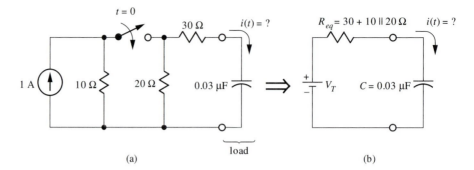

Figure 1.44 (a) The three resistors can be combined using a Thévenin equivalent circuit; (b) the time constant is easily identified in this equivalent circuit.

approached by the most efficient method. It is the *concept* of the Thévenin circuit, specifically the concept of the output impedance, or impedance level, that leads to the time constant for a transient circuit with multiple resistors. In general, the relevant resistance for the time constant is the output resistance presented to the energy storage element as a load.

Initial and Final Values

initial and final values

What we mean by "initial" and "final." We now look generally at how to calculate the initial and final values. By *initial* we mean the instant after a change occurs in the circuit, usually a switch closing or opening. By *final* we mean the steady-state condition of the circuit, its state after a large period of time. The final state may be hypothetical because the circuit may never reach that state due to subsequent switch action. A switch may close and then open before the circuit reaches the final state. The circuit cannot anticipate the second switch action and hence reacts to the first switch action as if it *would* reach steady state. In this section we develop guidelines for determining the final and initial states of the circuit.

Determining final values. Final values arise out of the eventual steady state of the circuit. All time derivatives must eventually vanish. Consequently, the current through capacitors and the voltage across inductors must approach zero, as suggested in Eq. (1.92)

$$v_C \to \text{constant} \Rightarrow i_C = C\frac{dv_C}{dt} \to 0 \text{ as } t \to \infty$$

$$i_L \to \text{constant} \Rightarrow v_L = L\frac{di_L}{dt} \to 0 \text{ as } t \to \infty$$

(1.92)

Thus capacitors act as open circuits and inductors act as short circuits in establishing final values as shown in Figs. 1.45 and 1.46, respectively.

Determining initial values. Initial values always follow from energy considerations. The fundamental principle is that the stored electric energy in a capacitor and the stored magnetic energy in an inductor must be continuous. From continuity of energy we conclude that the voltage across a capacitor and the current through an inductor

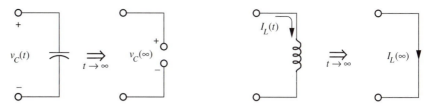

Figure 1.45 The capacitor behaves as an open circuit as time becomes large.

Figure 1.46 The inductor behaves as a short circuit as time becomes large.

must be continuous functions of time. Thus we always calculate capacitor voltage or inductor current before the switch is thrown and then carry this value over to the moment after the switch is thrown. From these known values of the capacitor voltage or inductor current, required circuit unknowns can be calculated.

A model for an energized capacitor at the instant after the switch action is shown in Fig. 1.47. For that first instant, a charged capacitor acts like a voltage source because the voltage across the capacitor cannot change instantaneously. As current flows through the capacitor, its voltage will change, and hence the capacitor acts as a battery *only* at that first instant. Similarly, a current source models an energized inductor at the instant after switch action, as shown in Fig. 1.48. Although these models are valid only for the first instant of the new regime, they suffice for the calculation of the initial values of the circuit unknowns of interest.

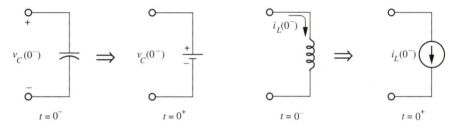

Figure 1.47 A battery models the initial voltage of a capacitor.

Figure 1.48 A current source models the initial current of an inductor.

Summary. We have shown a formal and a direct method for solving dc transients. The direct method builds the solution out of the initial and final values and the time constant. No DEs must be solved.

1.5 AC CIRCUIT ANALYSIS

The vast majority of all electrical power is generated, distributed, and consumed in the form of ac power. Before World War II, "electrical engineering" meant ac generators, motors, transformers, transmission lines, and the like. The ac power industry currently employs a mature technology and is a vital part of modern civilization.

In this section you will learn how to analyze ac circuits and, more broadly, how electrical engineers think about ac waveforms. The techniques of ac analysis, once mastered, are powerful for solving problems and stimulating insight. We begin by acquainting you with the sinusoidal waveform.

Sinusoids

Mathematical form for a sinusoid. We may describe the ac (alternating current) waveform mathematically as a sine function or a cosine function, but we will simply call it a *sinusoid*, or sinusoidal waveform. Electrical engineers have adopted the cosine function as the standard mathematical form for sinusoidal waveforms. Figure 1.49 shows the peak value, period, and phase of a sinusoidal waveform. The corresponding mathematical form is

$$v(t) = V_p \cos(\omega t + \theta) \tag{1.93}$$

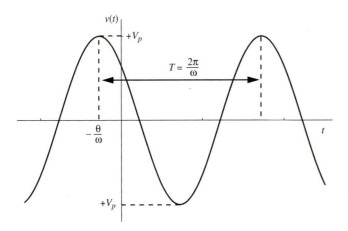

Figure 1.49 The sinusoidal function is defined by its peak value, its phase, and its period (or frequency).

period, event frequency

The peak value of the voltage is V_p. The sinusoid repeats with a *period*, T, which determines the frequency of the sinusoid. The *event frequency*, namely the number of cycles during a period of time, is the reciprocal of the period

$$f = \frac{1}{T} \text{ hertz} \tag{1.94}$$

The unit hertz, abbreviated Hz, has replaced cycles per second as the common unit for event frequency.

When most people talk about a frequency, they mean the event frequency. When we write mathematical expressions for sinusoidal waveforms, however, we more often deal with the proper mathematical measure, the angular frequency, ω (the Greek lowercase omega), in radians per second. Because there are 2π radians in a full circle, a cycle, the relationship between *angular frequency*, event frequency, f, and period, T, is

angular frequency

$$\omega = 2\pi f = \frac{2\pi}{T} \quad \frac{\text{radians}}{\text{second}} \tag{1.95}$$

The scientific dimensions of frequency are reciprocal seconds. The numerator of Eq. (1.95) represents an angle and is therefore dimensionless. Just as we retain radians or degrees to remind us which measure of an angle we are using, so we need as well to state the units for frequency to make explicit which frequency we mean, f or ω.

Units of phase. The phase of the sinusoid, θ, is what permits the waveform in Fig. 1.49 to represent a general sinusoid. We have drawn the curve for a phase of

$\theta = +50°$, but we can shift the position of the sinusoid by varying the phase. The phase is related to the time origin when we use a mathematical description, but in an ac problem what matters are the relative phases of the various sinusoidal voltages and currents.

Here you must tolerate one of the traditional inconsistencies of electrical engineers. The mathematical unit for ωt is radians and hence the correct unit for phase should also be radians. But electrical engineers usually speak of phase in degrees, as we did above ($+50°$) because they, like most people, still are more comfortable thinking about and sketching angles in degree measure. Probably you also think best in degree measure, so we will continue to express phase in degrees unless radians are required by the mathematics.

AC Time-Domain Analysis

The differential equation (DE). Figure 1.50 shows a typical ac circuit problem. The ac source has a frequency of 60 Hz or 120π rad/s and is connected by a switch that closes at $t = 0$. We wish to solve for the steady-state current, $i(t)$. After the switch is closed, we write KVL as

$$-v_s(t) + v_R(t) + v_L(t) = 0 \tag{1.96}$$

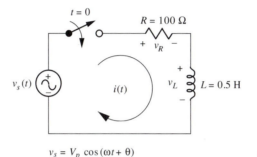

$v_s = V_p \cos(\omega t + \theta)$

$V_p = 100$ V, $f = 60$ Hz, $\theta = 30°$

Figure 1.50 Solve for the steady-state $i(t)$ after the switch closes.

We may introduce $i(t)$ through Ohm's law and the definition of inductance, with the result

$$L\frac{di}{dt} + Ri = V_p \cos(\omega t + \theta) \tag{1.97}$$

Equation (1.97) is a linear first-order DE that can be integrated directly with the aid of an integrating factor. This approach fails, however, when we try it on more complicated circuits and, besides, our goal is to learn how electrical engineers analyze ac problems. No electrical engineer would integrate this equation directly to find the steady-state solution. We will lead you down the traditional path that is based on frequency-domain techniques such as phasors and impedance.

LEARNING OBJECTIVE 7.

To be able to analyze AC circuits by frequency domain techniques

Representing Sinusoids with Phasors

Sinusoid in, sinusoid out. An important idea is suggested in Fig. 1.51. Here we have represented a circuit as a linear system—linear because the equations of R, L, and C are linear equations and a "system" because a circuit is an interconnection of such elements. Think of the voltage source as an input to this system (the circuit in Fig. 1.50) and the current $i(t)$ as an output of the system.

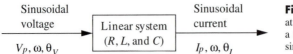

Sinusoidal voltage

V_P, ω, θ_V

Linear system
$(R, L, \text{ and } C)$

Sinusoidal current

I_P, ω, θ_I

Figure 1.51 A linear system responds at the frequency of excitation. The input is a sinusoidal voltage and the output a sinusoidal current of the same frequency.

The important idea is the following: If the input is a sinusoid, the output is also a sinusoid at the same frequency. This assertion can be justified through examination of Eq. (1.97) and reflection on the properties of sinusoids. The sinusoidal steady-state solution of Eq. (1.97) must be a function which, when differentiated and added to itself, will result in a sinusoid of frequency ω. The only mathematical function that qualifies is a sinusoid of the same frequency because the *shape* of the sinusoidal function is invariant to linear operations such as addition, differentiation, and integration.

New unknowns. The input voltage, being a sinusoid, is completely described by three numbers: the amplitude ($V_p = 100$ V), the frequency ($\omega = 120\pi$), and the phase ($\theta_V = 30°$). The output current must also be a sinusoid, and hence can also be described by an amplitude ($I_p = ?$), frequency ($\omega = 120\pi$), and phase ($\theta_I = ?$). Because the output frequency is known, only the amplitude and phase need to be derived to determine the steady-state current. Our object, therefore, is to develop an efficient method for finding the amplitude and phase of the output; I_p and θ_I become our new unknowns. We will now develop a mathematical model suited to finding the unknown amplitude and phase. The first step is to represent the amplitude and phase of a sinusoid by a complex number.

Phasors. Equation (1.98) relates a general sinusoid to its phasor representation

$$V_p \cos(\omega t + \theta) = \text{Re}\{\underbrace{\underline{\mathbf{V}}}_{\text{Phasor}} e^{j\omega t}\} \qquad (1.98)$$

where $j = \sqrt{-1}$ indicates an imaginary number. The right-hand side of Eq. (1.98) consists of three parts. The exponential term, $e^{j\omega t}$, may be considered a rotation operator in the complex plane. The term $\underline{\mathbf{V}}$ is a complex number with a magnitude of V_p, the peak value of the sinusoid and an angle in the complex plane of θ, the phase of the sinusoid. Thus the phasor $\underline{\mathbf{V}} = V_p \underline{/\theta}$, is a complex number that expresses the amplitude and phase of a sinusoid in a concise form. The product $\underline{\mathbf{V}} e^{j\omega t}$ is thus a rotating point in the complex plane, and the real part operation, $\text{Re}\{\}$, gives the projection of this rotating point on the real axis—simple harmonic motion tracing out the sinusoid as a function of time.

A rule of transformation. Equation (1.98) is mathematically rigorous but is also abstract and mysterious. We will treat it as a transforming relationship, expressed in Eq. (1.99)

$$v(t) = V_p \cos(\omega t + \theta) \Leftrightarrow \underline{\mathbf{V}} = V_p \angle \theta \qquad (1.99)$$

We thus may express a time-domain sinusoid by a frequency-domain phasor, which is a complex number containing the amplitude and phase of the sinusoid.[15]

---

[15] Although frequency does not appear explicitly on the right-hand side of Eq. (1.99), a frequency of ω is assumed. In the present approach there is only one frequency associated with the circuit. We will explore the frequency domain more fully in Chaps. 4 and 7.

Time derivatives. Since

$$\frac{d}{dt} e^{\alpha t} = \alpha e^{\alpha t} \tag{1.100}$$

we can differentiate Eq. (1.98) to obtain

$$\frac{d}{dt} V_p \cos(\omega t + \theta) = \text{Re}\{j\omega \underline{V} e^{j\omega t}\} \tag{1.101}$$

This result can be expressed as a transformation

$$\frac{d}{dt} \Leftrightarrow j\omega \tag{1.102}$$

and hence differentiation in the time domain corresponds to multiplication by $j\omega$ in the frequency domain.

Solving the DE. We now have the tools to solve Eq. (1.97) by transforming it into the frequency domain as shown in Eq. (1.103)

$$\underbrace{L\frac{d}{dt} i(t)}_{\Downarrow} + \underbrace{Ri(t)}_{\Downarrow} = \underbrace{V_p \cos(\omega t + \theta)}_{\Downarrow}$$

$$L(j\omega)\underline{I} + R\underline{I} = V_p \angle \theta \tag{1.103}$$

where $\underline{I}$ is a phasor representing the unknown steady-state current in Fig. 1.50. The transformed equation in Eq. (1.103) is an algebraic equation involving complex numbers, and the solution is

$$\underline{I} = \frac{V_p \angle \theta}{R + j\omega L} \tag{1.104}$$

For the voltage, frequency, resistance, and inductance in Fig. 1.50, the phasor current has the numerical value

$$\underline{I} = \frac{100 \angle 30°}{100 + j(120\pi \times 0.5)} = \frac{100 \angle 30°}{213 \angle 62.1°} = 0.469 \angle -32.1° \tag{1.105}$$

Interpretation of Eq. 1.105. The phasor current proves to be the phasor voltage divided by a quantity that combines the effects of the resistance and the inductance: $100 + j60\pi = 213 < 62.1°$. This combination is called the impedance of the circuit; we shall explore the impedance concept shortly. The impedance affects both the amplitude and the phase of the current. Numerically, Eq. (1.105) shows the amplitude of the current to 0.469 A and its phase to be $-32.1°$. Since we know the frequency of the current, the amplitude and phase give us all the information we need. If we wish to write the time-domain current, we may transform back as shown in Eq. (1.106)

$$\underset{\Downarrow}{\underline{\mathbf{I}}} = \underset{\Downarrow}{\underbrace{0.469}} \qquad \underset{\Downarrow}{\underbrace{\angle -32.1^\circ}}$$

$$i(t) = 0.469 \, \cos\,(120\pi t - 32.1^\circ) \tag{1.106}$$

Summary. We have demonstrated transformations based on Eq. (1.98) that lead directly to the steady-state solution of Eq. (1.97). The DE is transformed into an algebraic equation involving phasors and impedances, which are complex numbers. The solution yields the amplitude and phase of the unknown sinusoid, and the resulting phasor may be transformed back into the time domain in the manner of Eq. (1.106).

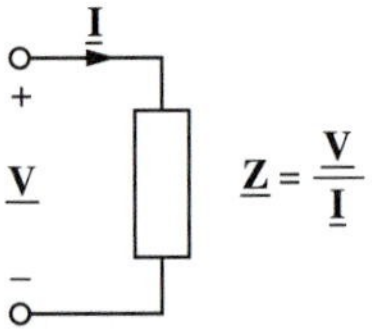

Impedance. The impedance of a circuit may be introduced directly by transforming R, L, and C directly into the frequency domain. We define the impedance of a circuit or circuit element as the phasor voltage divided by the phasor current, as shown in Fig. 1.52.

We may determine the impedance of resistors, inductors, and capacitors by transforming their defining equations into the frequency domain. The resistor is simplest: In Eq. (1.107) we transform the top line, the time-domain equation relating voltage and current in a resistor, into the second line, the frequency-domain description of a resistor:

Figure 1.52 The impedance is the phasor voltage divided by the phasor current. The impedance is a complex number describing the effect of the circuit or circuit element on the amplitude and phase of the current, given the voltage, or vice versa.

$$\underset{\Downarrow}{\underbrace{v_R(t)}} = \underset{\Downarrow}{\underbrace{R}} \; \underset{\Downarrow}{\underbrace{i_R(t)}}$$

$$\underline{\mathbf{V}}_R = R \; \underline{\mathbf{I}}_R \quad \Rightarrow \quad \underline{\mathbf{Z}}_R = \frac{\underline{\mathbf{V}}_R}{\underline{\mathbf{I}}_R} = R \tag{1.107}$$

Thus the impedance of a resistor is its resistance. For an inductor

$$\underset{\Downarrow}{\underbrace{v_L(t)}} = \underset{\Downarrow}{\underbrace{L}} \; \underset{\Downarrow}{\underbrace{\frac{d}{dt}}} \; \underset{\Downarrow}{\underbrace{i_L(t)}}$$

$$\underline{\mathbf{V}}_L = L(j\omega)\underline{\mathbf{I}}_L \quad \Rightarrow \quad \underline{\mathbf{Z}}_L = \frac{\underline{\mathbf{V}}_L}{\underline{\mathbf{I}}_L} = j\omega L \tag{1.108}$$

Thus the impedance of an inductor is pure imaginary, indicative of a 90° phase shift between voltage and current as a result of the differentiation. For a capacitor,

$$\underset{\Downarrow}{\underbrace{i_C(t)}} = \underset{\Downarrow}{\underbrace{C}} \; \underset{\Downarrow}{\underbrace{\frac{d}{dt}}} \; \underset{\Downarrow}{\underbrace{v_C(t)}}$$

$$\underline{\mathbf{I}}_C = C(j\omega)\underline{\mathbf{V}}_C \quad \Rightarrow \quad \underline{\mathbf{Z}}_C = \frac{\underline{\mathbf{V}}_C}{\underline{\mathbf{I}}_C} = \frac{1}{j\omega C} = \frac{1}{\omega C} \angle -90^\circ \tag{1.109}$$

Thus the impedance of a capacitor is also pure imaginary with a 90° phase shift between current and voltage because the capacitor integrates the current to give the voltage.

Using impedance. We may transform the time-domain circuit in Fig. 1.50 directly into the frequency domain by replacing sinusoids by phasors and circuit elements by their impedances, as shown in Fig. 1.53.

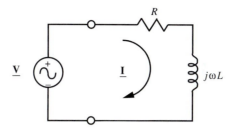

Figure 1.53 The circuit in Fig. 1.50 with sinusoids represented by phasors and circuit elements represented by impedances.

Impedances can be combined like complex resistors, so the solution is

$$\mathbf{I} = \frac{\mathbf{V}}{R + j\omega L} \tag{1.110}$$

Using impedance directly gives the same result as Eq. (1.104), which we derived by transforming the DE into the frequency domain.

Power in AC circuits. The instantaneous power in a circuit or circuit element is the product of the sinusoidal voltage and current

$$\begin{aligned} p(t) = v(t)i(t) &= V_p \cos(\omega t) \times I_p \cos(\omega t + \theta) \\ &= \frac{1}{2} V_p I_p [\cos\theta + \cos(2\omega t + \theta)] \end{aligned} \tag{1.111}$$

where we have let the voltage be the phase reference. Equation (1.111) reveals a steady part and a fluctuating part that averages zero. The time average of Eq. (1.111) is

$$P = \frac{1}{2} V_p I_p \cos\theta = V_e I_e \times PF \tag{1.112}$$

where $V_e = V_p/\sqrt{2}$ is the effective voltage,[16] $I_e = I_p/\sqrt{2}$ is the effective current, and *PF*, the power factor, is the cosine of the phase difference between the voltage and current. In Eq. (1.112) $V_e I_e$ is called the *apparent power* and carries units of volt-amperes.

apparent power

Transformers

Voltage transformation. A transformer is a highly efficient device for changing ac voltage from one value to another. Transformers come in all sizes, from the enormous transformers used in power substations to the small transformers used for doorbells. The transformer gives ac a feature lacking in dc power systems. Using a transformer, we can efficiently change ac voltage from small amplitudes to large amplitudes or vice versa. Such changes are not simply accomplished with dc voltage.

Current and impedance transformation. As we will soon prove, a transformer also transforms current, and as a consequence of transforming voltage and current, transforms impedance level as well. Indeed, the transformation of impedance level is perhaps the most important property of the transformer.

---

[16] Also called the rms voltage, for root mean square.

Transformer construction. Figure 1.54 shows a simple transformer. Two coils are coupled by time-varying magnetic flux, which is channeled by an iron core. The coil connected to the ac source is called the *primary* and the one connected to the load the *secondary*. There is nothing special about the two sides, for the transformer can convey power either way. In most applications the power flows in only one direction and hence these names are useful.

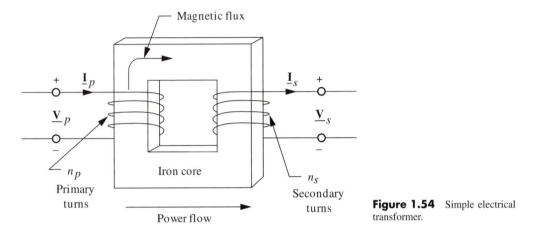

Magnetic flux

$\mathbf{I}_p$

$\mathbf{I}_s$

$\mathbf{V}_p$

$\mathbf{V}_s$

n_p
Primary
turns

Iron core

n_s
Secondary
turns

Power flow

Figure 1.54 Simple electrical transformer.

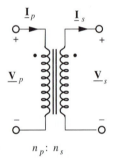

$\mathbf{I}_p$ $\mathbf{I}_s$

$\mathbf{V}_p$ $\mathbf{V}_s$

$n_p : n_s$

Figure 1.55 Circuit symbol for an ideal transformer. The primary voltage and current form a load set, and the secondary voltage and current form a source set.

ideal transformer

The ideal transformer. The ideal transformer is a circuit element that models a transformer. Primary and secondary voltage and current variables are defined in Fig. 1.55, which shows the circuit-theory symbol for an ideal transformer. The primary variables form a load set and the secondary variables form a source set. These definitions are customary and indicate that the primary acts as a *load* to the power system supplying power to the transformer, but the secondary acts as a *source* to the loads connected to it. We also define n_p and n_s, the turns on the primary and secondary.

Voltage and current transformation in ideal transformers. The primary and secondary voltages in the *ideal transformer* are related by

$$\frac{\mathbf{V}_p}{n_p} = \frac{\mathbf{V}_s}{n_s} \qquad (1.113)$$

Thus the side of the transformer with the larger number of turns has the larger voltage; indeed, the voltage per turn is constant for a given transformer. The primary and secondary currents in the *ideal transformer* are related by

$$n_p \mathbf{I}_p = n_s \mathbf{I}_s \qquad (1.114)$$

Thus the side of the transformer with the larger number of turns has the *smaller* current. For example, a transformer to increase the voltage would have a primary with few turns of large wire (small voltage, large current) and the secondary would have many turns of small wire (large voltage, small current).

Equations (1.113) and (1.114) define the ideal transformer as a circuit element, with one restriction—no dc. The primary of the transformer is, like an inductor, a short circuit to dc and provides no coupling to the secondary unless voltage and current are changing.[17]

Conservation of Energy (IDEA 2)

Conservation of power.
Multiplication of the left and right sides of Eqs. (1.113) and (1.114) shows that the apparent power into the primary is equal to the apparent power out of the secondary.

$$\frac{\mathbf{V}_p}{n_p} \times n_p \mathbf{I}_p = \frac{\mathbf{V}_s}{n_s} \times n_s \mathbf{I}_s \tag{1.115}$$

Thus apparent power is conserved and it follows that the *ideal* transformer has no losses and stores no energy.

Impedance Level (IDEA 5)

Impedance transformer.
Equations (1.113) and (1.114) reveal a very useful property of transformers—impedance transformation. Figure 1.56 shows an ideal transformer, a load impedance, $\underline{\mathbf{Z}}_L$, connected to the secondary, and an equivalent impedance, $\underline{\mathbf{Z}}_{eq}$, defined at the primary. The equations of the circuit are those of the ideal transformer plus the definition of impedance in Fig. 1.52. Dividing Eq. (1.113) by Eq. (1.114) yields

$$\frac{1}{n_p^2} \frac{\mathbf{V}_p}{\mathbf{I}_p} = \frac{1}{n_s^2} \frac{\mathbf{V}_s}{\mathbf{I}_s} \tag{1.116}$$

But $\underline{\mathbf{V}}_s/\underline{\mathbf{I}}_s = \underline{\mathbf{Z}}_L$, the load impedance, and $\underline{\mathbf{V}}_p/\underline{\mathbf{I}}_p = \underline{\mathbf{Z}}_{eq}$ defines an equivalent impedance looking into the primary, so Eq. (1.116) leads to the value of the equivalent impedance

$$\underline{\mathbf{Z}}_{eq} = \left(\frac{n_p}{n_s}\right)^2 \underline{\mathbf{Z}}_L \tag{1.117}$$

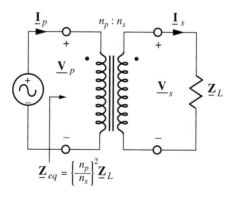

$$\underline{\mathbf{Z}}_{eq} = \left\{\frac{n_p}{n_s}\right\}^2 \underline{\mathbf{Z}}_L$$

Figure 1.56 The transformer will change $\underline{\mathbf{Z}}_L$ to an equivalent resistance $\underline{\mathbf{Z}}_{eq}$.

---

[17] Practical transformers have a frequency below which the coupling between primary and secondary begins to decrease, decreasing to zero at dc.

turns ratio

Thus the impedance level is transformed by the square of the *turns ratio*, n_p/n_s.[18] By using a transformer, we can make a large impedance appear small, or we can make a small impedance appear large. This property is useful in both power and electronics applications of transformers.

Summary of time and frequency domain. Table 1.1 summarizes the relationships between the time domain and the frequency domain that we have developed in this section.

TABLE 1.1 Time-Domain and Frequency-Domain Transforms	
Time Domain	*Frequency Domain*
Sinusoid	Phasor
Phase angle	Angle in complex plane
$V_p \cos(\omega t + \theta)$	$\mathbf{V} = V_p \angle \theta$
DEs	Arithmetic with complex numbers
d/dt	$j\omega$
R, L, and C	Impedances
R	$\mathbf{Z}_R = R$
L	$\mathbf{Z}_L = j\omega L = \omega L \angle +90°$
C	$\mathbf{Z}_C = \dfrac{1}{j\omega C} = \dfrac{1}{\omega C} \angle (-90°)$

Check Your Understanding

1. If eight cycles of a sinusoidal waveform take 2 ms, find the angular frequency, ω.

2. In transforming a DE into the frequency domain, what replaces d/dt?

3. What is the impedance, including units, of a 0.7-H inductor at 50 Hz?

4. A capacitor at 1 kHz has an impedance with a magnitude of 20 Ω. What is the magnitude of the impedance at 2 kHz?

5. What is the magnitude of the impedance of a 20-Ω resistor in series with a 20-mH inductor at 80 Hz?

Answers. (1) 25,100 rad/s; (2) $j\omega$; (3) $j220\ \Omega$; (4) 10 Ω; (5) 22.4 Ω.

CHAPTER SUMMARY

This chapter reviews the fundamentals of electric circuits. This survey includes the basic definitions and physical laws, such as Kirchhoff's voltage and current laws; the definitions of common circuit elements such as switches, resistors, inductors, and

---

[18] Note that the ratio of primary to secondary turns determines voltage, current, and impedance transformation properties. For this reason, the *turns ratio* is often stated as $1:n$ (or $n:1$), where n is not necessarily an integer and can be less than unity.

capacitors; the presentation of the common techniques used to analyze circuits such as voltage and current dividers, Thévenin's equivalent circuit, nodal analysis, and loop analysis; solution techniques for first-order transients; and ac circuit theory. All this information is needed to analyze electronic circuits.

Objective 1: To understand the physical variables used to describe electric circuits, particularly their relationship to energy. Voltage and current are defined physically in terms of fundamental quantities: charge, length, energy, and time. From the definitions it follows that power is voltage times current.

Objective 2: To understand how to apply Kirchhoff's Voltage Law and Kirchhoff's Current Law to electric circuits. Kirchhoff's laws describe conservation of energy and charge in electric circuits. The laws are applied through identifying loops or nodes and summing variables, paying special attention to reference directions and the resulting signs.

Objective 3: To understand the properties of resistors, inductors, and capacitors. The terminal voltage–current characteristics of resistors, inductors, and capacitors are defined. The role of these circuit elements in modeling energy processes is established: resistors model the loss of energy to the circuit; inductors model the storage of magnetic energy, and capacitors model the storage of electric energy.

Objective 4: To understand basic methods for analyzing electric circuits. All methods utilize Kirchhoff's laws but differ primarily in the way in which variables are defined. The method of voltage and current dividers uses the topology of the circuit to guide the solution. Thévenin's equivalent circuit allows the circuit to be represented to the load. Nodal and loop analysis are formal systems that represent the entire circuit at once.

Objective 5: To understand the significance of impedance level in interactions between circuits. The concept of impedance level, which arises primarily from the Thévenin equivalent circuit, guides electrical engineers in design and analysis of electrical systems. The interaction of two circuits, when connected together, depends on their respective impedance level.

Objective 6: To be able to formulate and solve first-order transients. The method emphasized uses the time constant, plus the initial and final values, to give a quick solution to transient problems. This method avoids writing a differential equation, but is limited to first order circuits.

Objective 7: To be able to analyze AC circuits by frequency domain techniques. The traditional methods represent sinusoids by phasors, circuit elements by impedances, and generally find and interpret solutions in the frequency domain.

Objective 8: To understand the voltage, current, and impedance transforming properties of transformers. Transformers are found in many electrical systems due to their remarkable ability to transform impedance level. The ideal transformer is defined, its properties are explored, and circuit analysis techniques for circuits containing transformers are developed.

GLOSSARY

Angular frequency, p. 44, 2π times the event frequency, radians/second.

Apparent power, p. 49, the product of the effective values of the voltage and current.

Available power, p. 26, the maximum power that can be extracted from a circuit.

Capacitance, capacitor, p. 34, the capacity of a circuit element to store electric energy.

Charge, p. 4, a property of matter. There are two types of charge, positive and negative.

Conductance, p. 4, the reciprocal of resistance.

Current, p. 4, the movement of charge, usually in a wire or electrical device. If in a period of time Δt, ΔQ coulombs move past a cross-section in a wire in an indicated direction, the **current** is $i = \dfrac{\Delta Q}{\Delta t}$ C/s or ampere, A.

Effective value of a time-varying waveform, p. 49, the equivalent dc value that would heat a resistor as hot as the time-varying waveform heats it. Also called the root-mean-square (rms) value.

Energy, p. 4, the medium of exchange in a physical system. In mechanics it takes force and movement to do work (exchange energy), and in electricity it takes electrical force (voltage) and movement of charges (current) to do work (exchange energy).

Equivalent circuit, pp. 16, 23, a circuit that represents in a specified characteristic the properties of another, more complicated circuit.

Equivalent resistance, p. 16, a single resistance that has the same resistance as a two-terminal network of resistors.

Event frequency, p. 44, the number of cycles during a period of time, the reciprocal of the period, cycles/second = hertz.

Final value, p. 42, the steady-state condition of the circuit, its state after many time constants.

Forced response, p. 38, the response of a system or electrical circuit to a periodic source, including a dc source. Also called steady-state response.

Frequency domain, p. 52, circuit analysis techniques and notations arising out of using frequency as the independent variable.

Ground node or ground, p. 31, the point in an electrical circuit that is physically connected via a thick wire to the moist earth, usually for safety purposes.

Ideal current source, p. 14, a current source that produces its prescribed current independent of its output voltage.

Ideal switch, p. 13, a path that can be changed from a short circuit to an open circuit to turn an electrical device ON or OFF.

Ideal transformer, p. 50, a lossless transformer that stores no energy. Voltage and current are transformed reciprocally.

Ideal voltage source, p. 13, a voltage source that maintains its prescribed voltage independent of its output current.

Impedance, p. 48, a complex number, the phasor voltage divided by the phasor current. The magnitude of the impedance is the ratio of the peak (or effective) values of the voltage and current. The angle of the impedance is the phase of the voltage minus the phase of the current.

Impedance level, p. 27, the approximate ratio of voltages to currents in a circuit or portion of a circuit. The interaction of a circuit with a load depends on the relative values of their respective impedance levels.

Impedance match, p. 24, where the load impedance is the complex conjugate to the output impedance of the circuit, required for maximum power transfer.

Independent node, p. 30, a node whose voltage cannot be derived from the voltage of another node.

Inductor, inductance, p. 33, the capacity of a circuit element to store magnetic energy, normally associated with a coil of wire.

Initial value, p. 42, the value of a circuit variable the instant after a change occurs in the circuit, usually a switch closing or opening.

Kirchhoff's current law (KCL), p. 6, the constraint imposed at a node by conservation of charge and charge neutrality: The sum of the currents leaving a node is zero at all times. An equivalent statement would be that the sum of the currents entering a node is equal to the sum of the currents leaving that node at any instant of time.

Load, p. 22, An element in a circuit of particular importance; in one sense the circuit exists to supply the load.

Load set, p. 10, where the current reference direction is into the + polarity marking (or the first subscript) of the voltage reference direction. This is the convention normally used with a resistance or some other element that receives power from a circuit.

Loading of a circuit, p. 26, the change in output voltage when a load or meter is attached. Also apply a change in current when an ammeter is inserted.

Loop, p. 31, any closed path in an electrical circuit.

Loop current, p. 31, a current that flows in a closed path.

Matching impedances, p. 24, when the load has the same resistance as the output impedance of the source but the opposite reactance, required for maximum power transfer.

Natural response, p. 38, the response of a system or electrical circuit stimulated by external force, but not dependent on the form of that force, that arises out of internal energy exchanges. Also called transient response.

Node, p. 6, the junction of two or more wires.

Open circuit, p. 13, a "path" that permits no current flow, regardless of the voltage across it, equivalent to infinite resistance.

Open-circuit voltage, p. 22, the voltage between two terminals with the load removed, that is, as an open circuit.

Parallel connection, p. 17, a connection of two or more circuit elements in which all share the same voltage.

Phasor, p. 46, a complex number representing the amplitude and phase of a sinusoid.

Physical current, p. 5, the magnitude of the current referenced in the direction in which conventional current is positive. Physical current goes opposite to the direction of electron motion.

Physical polarity of the voltage, p. 7, the reference direction markings that give a positive voltage. Thus the physical voltage is always positive and has polarity markings assigned accordingly.

Potential difference, p. 31, the voltage between two points in a circuit. Our definition of voltage is equivalent to a potential drop.

Power, p. 9, the rate of energy exchange: The electrical power is the product of voltage and current

$$v\left(\frac{\text{work}}{\text{charge}}\right) \times i\left(\frac{\text{charge}}{\text{time}}\right) = \frac{\text{work}}{\text{time}} = \text{power}$$

Power factor, p. 49, the real power divided by the apparent power. The power factor is the cosine of the phase angle between the voltage and current.

Primary, p. 50, the input transformer winding connected to the ac source.

Real power, p. 49, the time average of the instantaneous power for a load in sinusoidal steady state.

Reference node, p. 29, the point in an electrical circuit from which all node voltages are measured. By definition, the node voltage of the reference node is zero.

Resistance, resistor, p. 12, the circuit property or element that models conversion of electrical energy to nonelectrical energy.

Secondary, p. 50, the output transformer winding supplying electrical power to a load.

Series connection, p. 15, a connection of two or more circuit elements in which all have the same current.

Short circuit, p. 13, an ideal connection for current flow, equivalent to zero ohms resistance.

Sinusoid, or sinusoidal waveform, p. 44, a variable that varies as the sine or cosine of an angle that increases linearly with time, with any phase angle.

Source set, p. 10, where the current reference direction is out of the + polarity marking (or the first subscript) of the voltage. This combination is normally used for a source of energy to the circuit, such as a battery.

Time constant, p. 38, a characteristic time for a transient, associated with the natural response of a system or electrical circuit. For a dc transient, changes are 63% complete in one time constant.

Time domain, pp. 45, 52, circuit analysis techniques and notations arising out of using time as the independent variable.

Transient, p. 37, the transition of a circuit from one steady state to another, usually resulting from a switch action.

Turned-OFF current source, p. 20, an open circuit, equivalent to infinite resistance.

Turned-OFF voltage source, p. 19, a short circuit, equivalent to zero resistance.

Turns ratio, p. 52, ratio of primary to secondary turns in a transformer ratios of voltage and current transformation.

Voltage, p. 7, the work done by the electrical system in moving a charge from one point to another in a circuit, divided by the charge.

PROBLEMS

Section 1.1: Charge, Current, Voltage, Energy, and Power

1.1. The current in a wire is zero for $t < 0$ but begins increasing at a rate of 5 Amperes/second at $t = 0$, as shown in Fig. P1.1.
  (a) At what time is the current 7.2 A?
  (b) At what time has 12 C of charge passed a cross section of the wire?

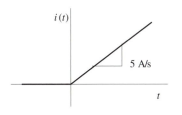

Figure P1.1

1.2. (a) Write KVL to determine v_{ab} in the circuit in Fig. P1.2.
  (b) If a battery (voltage source) were inserted between c and d to make $v_{ab} = 12.3$ V, draw the circuit showing the battery polarity and voltage magnitude.

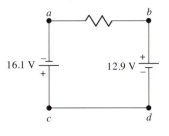

Figure P1.2

1.3. In Fig. P1.3, the power into the 5-V battery is $+10$ W. Determine the power into R and the power out of the 9-V battery.

1.4. A car radio designed to operate from a 6.3-V system uses 4.5 A of current, as shown in Fig. P1.4.
  (a) What resistance should be placed in series with this radio if it is to be used in a 12.6-V system?
  (b) What should be the power rating of this resistance?

1.5. Figure P1.5 shows a voltage source and a switch.
  (a) Is the switch single- or double-pole?
  (b) Is the switch single- or double-throw?

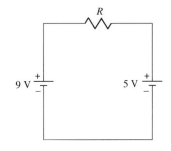

Figure P1.3

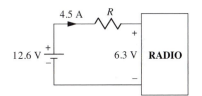

Figure P1.4

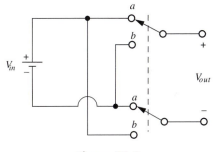

Figure P1.5

  (c) What is the output voltage for the switch in position a and in position b?

1.6. Figures 1.6(a) and 1.7(b) show a resistance and a voltage source or current source, respectively. Make

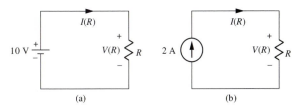

Figure P1.6

a plot of $V(R)$ and $I(R)$ versus R for the range $1 < R < 10\ \Omega$ in both cases. *Note:* Commonly we consider $V(R)$ to mean that V *depends* on R, so V cannot be constant as R is varied. But actually $V(R) = $ constant is a perfectly good mathematical function that happens to have a zero derivative everywhere.

1.7. Figure P1.7 shows the voltage and current for a semiconductor switch closure. Note the switch is not ideal in that voltage and current do not change instantaneously. This means that the switch will dissipate some energy during its operation.

(a) How much energy is used up in each switching operation? (Assumes a source set of reference directions.)

(b) If the same energy were used in opening the switch and the switch went through 5000 cycles each second, what is the average power lost in the switch? Each cycle involves opening and closing the switch.

1.8. Figures 1.8(a)–(k) show circuits containing voltage and current sources and resistances. Some circuits are valid (noncontradictory) for all values of the variables; some are valid for no values; and some are valid for specific values of the variables.

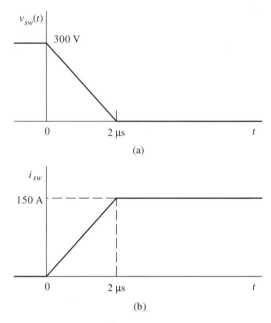

(a)

(b)

Figure P1.7

Classify each and tell what, if any, values of the Vs, Is, and Rs lead to circuits compatible with the definitions of the sources and resistance.

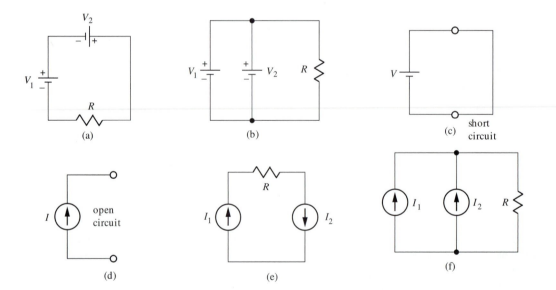

(a)

(b)

(c)   short circuit

(d)   open circuit

(e)

(f)

Figure P1.8

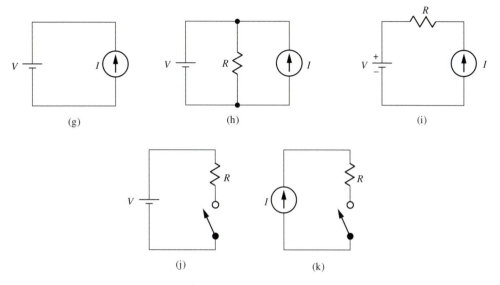

(g)   (h)   (i)

(j)   (k)

Figure P1.8 *(Continued)*

Section 1.2: Circuit Analysis Techniques

1.9. For the circuit shown in Fig. P1.9, the arrow indicates that the resistor R is variable; hence the voltage across R, $v(R)$ will depend on the value of R as the notation indicates. Determine the function and plot $v(R)$ for $0 < R < 40\ \Omega$ and find the value of R to make the voltage 5.5 V.

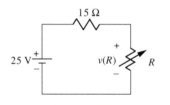

Figure P1.9

1.10. (a) Find the equivalent resistance of the parallel combination shown in Fig. P1.10.
(b) If 10 A enters the parallel combination, referenced in at a and out at b, what is the current referenced downward in the 6-Ω resistance?
(c) What is the current referenced upward in the 2-Ω resistance?
(d) What is the voltage, v_{ab} for this current?

1.11. Design a current divider that has an equivalent resistance of 100 Ω and divides the current in the ratio of 5:2. This problem is summarized in Fig. P1.11.

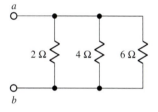

Figure P1.10

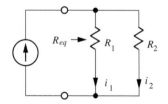

$$R_{eq} = R_1 \parallel R_2 = 100\ \Omega$$

$$2i_1 = 5i_2$$

Figure P1.11

1.12. Find the current in the 20-Ω resistor in Fig. P1.12 using superposition.

1.13. The starter motor on a car draws 75-A starting current, which lowers the battery voltage from 12.6

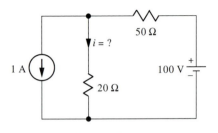

Figure P1.12

to 9.1 V. What would be the battery voltage if it
were being charged at 30 A?

1.14. Using a Thévenin equivalent circuit, find R in
Fig. P1.14 such that $i = 0.5$ A.

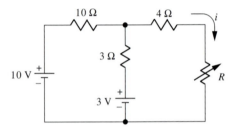

Figure P1.14

1.15. (a) For the circuit shown in Fig. P1.15, solve for v_a
and v_b using node-voltage techniques.
  (b) Find the voltage across the three resistors (you
supply reference directions) and show that KVL
is satisfied.
  (c) Now double v_a and repeat **(b)**. Note that KVL
is still satisfied, even if *incorrect* node voltages
are used. Of course, KCL would be violated.

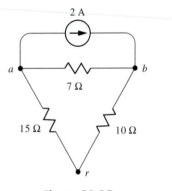

Figure P1.15

1.16. Using node-voltage analysis, solve for the indicated

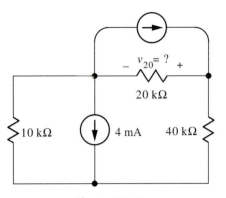

Figure P1.16

unknowns in Fig. P1.16. *Hint*: note that kV, mA,
and volts make a consistent set of units.

1.17. For the circuit shown in Fig. P1.17.
  (a) Using nodal analysis, find the current in the
10-Ω resistor for $R = 6\,\Omega$.
  (b) Derive a Thévenin equivalent circuit with R as
the load.
  (c) Find the value of R that gives 10% of the
current that would flow if the load were
replaced by a short circuit.

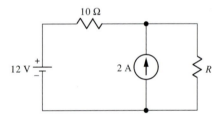

Figure P1.17

1.18. Solve for all branch currents in the circuit in
Fig. P1.18, using loop-current analysis.

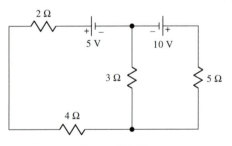

Figure P1.18

1.19. Solve for i in Fig. P1.19 using the loop-current
method.

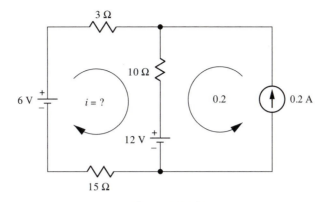

Figure P1.19

Section 1.3: Inductors and Capacitors

1.20. A 6.3-V battery is connected to an ideal 0.3-H inductor, as shown in Fig. P1.20.
  (a) Calculate the current as a function of time after the switch is closed.
  (b) At what time does the stored energy in the inductor reach 10 J? Verify the stored energy by integrating the input power (product of v_L and i_L) from $t = 0$ to the time you calculate.

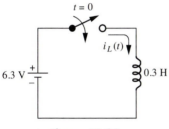

Figure P1.20

1.21. Figure P1.21 shows 20-μF capacitor that has a voltage

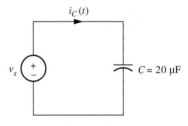

Figure P1.21

$$v_s(t) = 0, \; t < 0$$

$$v_s(t) = 10^4 t^2, \; 0 < t < 0.1 \text{ s}$$

$$v_s(t) = 100, \; t > 0.1 \text{ s}$$

  (a) Find the current, $i_C(t)$.
  (b) At what time is the energy in the capacitor 50 mJ?
  (c) At what time is the power into the capacitor 1 watt?

1.22. A 0.1-μF capacitor is charged with a 1-μs pulse of current, as shown in Fig. P1.22. Find the voltage

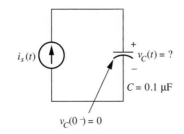

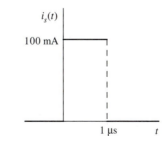

Figure P1.22

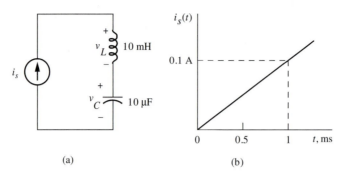

(a)

(b)

Figure P1.23

across the capacitor with the polarity shown as a function of time.

1.23. An inductor and a capacitor are placed in series with a current source whose current increases with time as shown in Fig. P1.23(b). The circuit is shown in Fig. P1.23(a).

(a) Find the voltage across the inductor, $v_L(t)$.

(b) Assuming no initial charge on the capacitor, find its voltage, $v_C(t)$.

(c) Calculate the instant when the stored energy in the capacitor first exceeds that in the inductor.

1.24. In the circuit shown in Fig. P1.24, the initial voltage on the capacitor is 10 V, with the + at the bottom, the initial current in the inductor is zero. The current into the capacitor is -5 A dc for $t > 0$, as shown.

Section 1.4: DC Transient Analysis

1.25. A system is described by the DE

$$\frac{1}{3}\frac{dx}{dt} + 5x = 10$$

(a) What is the time constant for this system?

(b) What would be the final value, x?

(c) If $x(0) = -2$, find and sketch $x(t)$ for $t > 0$.

1.26. A current source is placed in series with a resistor and inductor, as shown in Fig. P1.26. During this period the switch is open. Then the switch is closed, and the circuit is separated into two independent loops that share a common short circuit but do not interact.

(a) Calculate the current in the right-hand loop after the switch closes at $t = 0$.

(b) Plot the current in the switch, referenced downward.

(a) Find the source voltage.

(b) Determine the *two* times after $t = 0$ when the power out of the voltage source is zero.

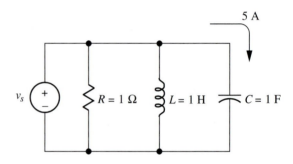

Figure P1.24

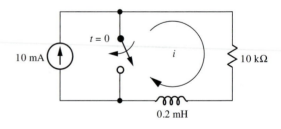

Figure P1.26

1.27. For the circuit shown in Fig. P1.27, assume that the switch has been in the position a for a long time and then is changed to position b.

(a) Find and sketch $i(t)$.

(b) Calculate by integration the energy lost in the 100-Ω resistor during positive time. Confirm that all the energy stored initially in the

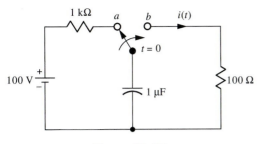

Figure P1.27

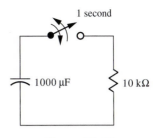

Figure P1.30

capacitor is accounted for by the loss in the resistor.

1.28. The switch in Fig. P1.28 is in position *a* for negative time, moved to *b* at $t = 0$, and to *c* at $t = 10$ ms. Sketch the voltage across the capacitor for $0 < t < 30$ ms. At what time should the switch be switched to *c* for no "transient"?

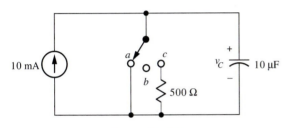

Figure P1.28

1.29. The switch in Fig. P1.29 is closed at $t = 0$. Find the time at which the power into the inductor is maximum.

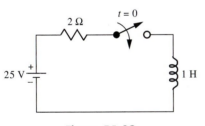

Figure P1.29

1.30. For the circuit in Fig. P1.30, the original energy stored in the capacitor is 500 J and the switch is open. The switch is then closed for 1 second and

opened again after 1 second. Find the final energy stored in the capacitor.

1.31. After being open a long time, the switch in Fig. P1.31 closes at $t = 0$. Find the power out of the 6-V source at $t = 10$ ms.

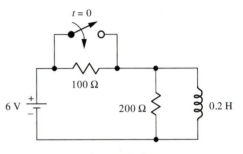

Figure P1.31

1.32. The circuit in Fig. P1.32 has had the switch closed a long time. After the switch is opened, find:
  (a) τ
  (b) $i(0)$
  (c) $i(\infty)$
  (d) $i(t)$

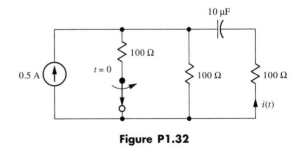

Figure P1.32

Section 1.5: AC Circuit Analysis

1.33. A sinusoidal function is shown in Fig. P1.33. Determine the frequency, phase, and amplitude for

expressing this sinusoid in the standard form $i(t) = I_p \cos(\omega t + \theta)$

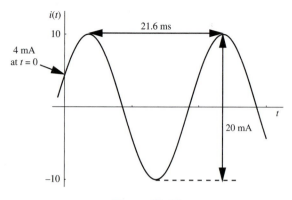

Figure P1.33

1.34. Find the steady-state value of the voltage across the resistance in the circuit of Fig. P1.34.
 (a) Write the DE for $v_R(t)$.
 (b) Transform into the frequency domain.
 (c) Solve for the unknown phasor representing $v_R(t)$.
 (d) Transform back into the time domain.

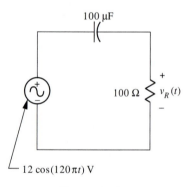

Figure P1.34

1.35. (a) What value of capacitance and what value of inductance have an impedance with a magnitude of $12\,\Omega$ at a frequency of 800 Hz?
 (b) What would be the magnitudes of the impedance of this C and this L at 1.6 kHz?
 (c) What would be the impedance of this inductance and capacitor connected in series at a frequency of 1.2 kHz?

1.36. For the circuit shown in Fig. P1.36, determine $v(t)$ using phasor techniques. Sketch $v(t)$ in the time domain.

1.37. Consider a 4-μF capacitor and a 10-Ω resistor.
 (a) They are connected in series. At what frequency

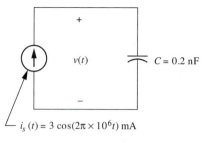

Figure P1.36

in hertz is their series impedance $20\,\Omega$ in magnitude?
 (b) If the resistor and capacitor are now placed in parallel, find the frequency at which their combined impedance is $7\,\Omega$ in magnitude.
 (c) Still connected in parallel, at what frequency is the angle of the impedance $-45°$?

1.38. Figure P1.38(a) shows a circuit in sinusoidal steady state, with the phasor diagram in Fig. P1.38(b). Assume the load consists of a resistor in series with a reactive component. The frequency is 60 Hz.
 (a) What is the voltage at $t = 0$?
 (b) What is the magnitude of the impedance?
 (c) What is the resistance of the circuit?
 (d) What is the reactive component (type and value)?

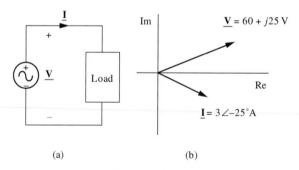

Figure P1.38

1.39. A resistor in series with a capacitor or inductor has a current of $i(t) = 1.2\cos(1000t + 75°)$ mA and a voltage across the series combination of $v(t) = 0.8\cos(1000t + 47°)$. What is the value of the resistance and the capacitor or inductor?

1.40. The circuit in Fig. P1.40 is to be represented by a Norton equivalent circuit. Determine $\mathbf{I}_N$ and $\mathbf{Z}_{eq}$.

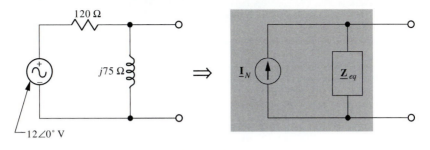

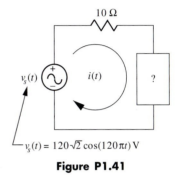

Figure P1.40

1.41. In Fig. P1.41, the design goal is to have $i(t)$ lead $v_s(t)$ by 55° of phase.
   (a) What is in the box: R, L, or C?
   (b) What is its numerical value?
   (c) What is the peak value of $i(t)$?
   (d) If the frequency were doubled, what would be the phase difference between $i(t)$ and $v_s(t)$? Consider the phase positive if $i(t)$ leads $v_s(t)$.

Figure P1.41

Answers to Odd-Numbered Problems

1.1. (a) 1.44 s; (b) 2.19 s.

1.3. +18 W from source, +8 W to resistor.

1.5. (a) double pole; (b) double throw; (c) $V_{out} = +V_{in}$ for a, $V_{out} = -V_{in}$ for b.

1.7. (a) 0.0150 J; (b) 150 W.

1.9. 4.23 Ω.

1.11. $R_1 = 140$ Ω, $R_2 = 350$ Ω.

1.13. 0.714 A

1.15. (a) $v_a\left(\frac{1}{7} + \frac{1}{15}\right) + v_b\left(-\frac{1}{7}\right) = -2$;

$v_a\left(-\frac{1}{7}\right) + v_b\left(\frac{1}{7} + \frac{1}{10}\right) = +2$

(b) $v_{ar} = -6.56$ V $v_{vbr} = +4.38$ V, $v_{ab} = -10.9$ V;
(c) $v'_{ar} = -13.1$ V, $v'_{br} = 4.38$ V, $v'_{ab} = -17.5$ V.

1.17. (a) 0 A; (b) 32 V, 10 Ω; (c) 90 Ω.

1.19. −0.286 A.

1.21. (a) $0.4t$ for $0 < t < 0.1$ s, 0 elsewhere; (b) 0.0841 s; (c) 0.0630 s.

1.23. (a) 1 V; (b) $10^7 t^2/2$; (c) 0.632 ms.

1.25. (a) 1/15 s; (b) +2; (c) $x(t) = 2 - 4e^{-15t}$.

1.27. (a) $i(t) = 1e^{-t/100\mu s}$A; (b) 5 mJ.

1.29. 0.347 s.

1.31. 3.24 W.

1.33. 10 mA peak, 291 rad/s, −113.6°.

1.35. (a) 16.6 μF; (b) $j10$ Ω.

1.37. (a) 2300 Hz; (b) 4060 Hz; (c) 3980 Hz.

1.39. 589 Ω, 3.20 μF capacitance.

1.41. (a) C; (b) 186 μF; (c) 9.73 A; (d) +35.5°.

Semiconductor Devices and Circuits

1. To understand how to analyze half- and full-wave rectifier circuits, with and without a filter capacitor
2. To understand the physical processes and the *i-v* characteristic of a *pn* junction
3. To understand the semiconductor processes in the *npn* bipolar junction transistor and the resulting input and output characteristics
4. To understand how to analyze a common-emitter amplifier-switch circuit.
5. To understand how to analyze a common-emitter small-signal amplifier for operating point and gain
6. To understand FET operations and applications

objectives

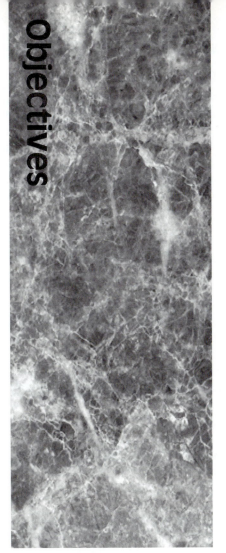

In addition to resistors, capacitors, and inductors, electronic circuits use diodes, transistors, and other nonlinear devices. We analyze rectifier circuits that convert ac power to dc power and circuits that use transistors as switches or amplifiers.

Introduction to Electronics

power,
electronics

Electronics and information. Electrical engineering may be divided into two branches, power and electronics, which differ in their use of electrical energy. In the power industry, electrical energy is generated from a primary source such as coal or water power and eventually converted into heat, illumination, mechanical work, or some other useful form. In electronics, electrical energy is used to symbolize, transport, and process information. Think of a telephone system, a radio, a computer, or a traffic control system—all use electrical energy to convey, process, and use information.

signal

In electronics, voltage and current become electrical *signals*; that is, they signify something else, as the root word "sign" suggests. The voltage generated when you speak into a telephone is a signal because it reproduces the acoustic vibrations in the air, these sounds have meaning, and information is exchanged in the conversation.

This book discusses the main ideas used in electronics at the present time. As you will see from the historical survey to follow, electronics is a young enterprise. The current rate of progress in the field makes it difficult to anticipate what the future holds. But there are some major themes that, once understood, will give you a good grasp of the nature of electronics. Some of the important factors we address are (1) analog and digital representation and processing of information, (2) expanded use of the frequency domain, (3) application of the electrical properties of semiconductor materials, (4) feedback, and (5) utilization of nonlinear effects in circuits.

History of electronics. Before World War II, electronics had commercial importance primarily in radio broadcasting and in the telephone and telegraph industries. Most people who were active in these fields were educated through experience, including many physicists and electrical engineers. In those days, to study electrical engineering in college meant to study about power: motors, generators, lighting, transformers, and transmission lines.

World War II profoundly changed electronics. In addition to the obvious need to improve radio communication, one of the Allies' top-secret war projects focused on electronics. Everybody knows about the Manhattan Project and the atomic bomb. Through the work of the Radiation Laboratory in developing microwave radar, scientists and engineers made an equally important contribution toward ending the conflict. The best technical minds in the country were employed in these two projects and spectacular success crowned both efforts.

The postwar fruit of the Manhattan Project was more and bigger bombs, and even the peaceful application of nuclear technology continues to excite controversy. But widespread and, for the most part, benevolent have been the postwar fruits of the work of the Radiation Laboratory. The cathode-ray tubes (CRTs) that were developed as radar displays became TV picture tubes, and soon there were high-fidelity recordings to be enjoyed. Radar techniques developed to detect enemy aircraft allowed commercial aviation to fly in all weather, and you could dial long distance directly and make yourself heard without shouting.

In the early 1950s, the development of the transistor inaugurated the first of several "solid-state revolutions." The integrated circuit followed and later the microcomputer on a "chip." The techniques of microelectronics seem limitless.

The idea of electronic computers was conceived before the war, but only modest applications were made even during the war. An analog computer was developed to control naval gunnery, and some simple "logic" circuits emerged but practical computers came after the war. The first digital computers were based on vacuum-tube technology and by present standards were gigantic, slow, and unfriendly to programmers. But when the idea of the digital computer merged with that of the transistor and later the integrated circuit, the development of computers became spectacular and continues so to this day.

The nature of electronics.   Electronics is a big bag of tricks. Unlike circuit theory, which submits to an orderly and logical development, electronics employs a diversity of devices, techniques, and processes. It is difficult to apply a beginning knowledge of electronics because electronic circuits are complicated. Did you ever, for example, examine the circuit diagram (the schematic) for a TV set? If one fell into your hands at this moment and you set about to apply your knowledge of Ohm's law, Kirchhoff's laws, and concepts such as impedance and Thévenin equivalent circuits, you would not make much progress in understanding such a circuit diagram.

Our goal in this book is to investigate the major themes currently driving electronics. We examine a few of the tricks in the electronics bag at the present time but only enough to impart a beginning understanding of the nature of the subject. Do not expect to be able to fix your radio or interface a microcomputer after you master this book. But you should understand how these systems work.

Electronics in this book.   This chapter explains the physical operations of diodes and transistors, and then discusses some basic circuits that use these devices. Chapter 3, "Digital Electronics," shows how these basic circuits can be adapted to process information in digital form. Chapter 4, "Analog Electronics," details how these circuits are used to process information in analog form and expands our treatment of circuit theory into signal analysis. Chapters 5 to 7 show how analog and digital circuits are integrated into systems that monitor, control, or communicate.

Ideal Diode

Circuit theory.   Electronics can be regarded as the circuit theory of nonlinear devices. The study of electronics, however, does not leave behind the circuit theory of linear devices but rather builds upon it. Electronic circuits use resistors, inductors, and capacitors, which are linear circuit elements, and the basic circuit laws are Kirchhoff's voltage and current laws, which are linear.

Nonlinear devices.   Resistors, inductors, and capacitors are still used, and certainly Kirchhoff's laws are still valid. But electronic circuits employ many devices that are nonlinear in their characteristics. Diodes, transistors, and silicon-controlled rectifiers offer examples of such nonlinear electronic devices. The nonlinearities are not undesirable hindrances to the use of these and other devices; rather, they are useful because of their nonlinear properties.

Graphical analysis.   Because of the importance of nonlinear devices, we employ more graphical analysis in electronics. Many of the circuit solution techniques we develop are based on graphical methods, and much of the information about device characteristics is given in graphical form. Figure 2.1 shows the graphical characteristics of a resistor, together with the extremes of an open circuit ($R = \infty$) and a short circuit ($R = 0$). Of course, the graphical form of Ohm's law is a straight line.

Ideal diode characteristics. Figure 2.2 shows the circuit symbol for an ideal diode, with the associated graphical characteristic, which is nonlinear. The ideal diode characteristic divides into two regions: The vertical region is called the forward-bias region and the horizontal region is called the reverse-bias region. Comparison with Fig. 2.1 suggests that the ideal diode acts as a short circuit in the forward-bias region and an open circuit in the reverse-bias region.

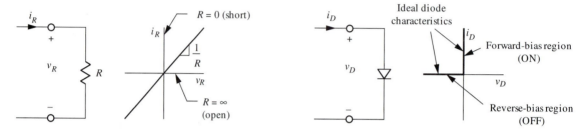

Figure 2.1 Symbol and graphical definition of a resistor.

Figure 2.2 Symbol and graphical definition of an ideal diode.

ON and OFF. The behavior of the device in terms of current flow is as follows: As long as the current is positive in the direction of the arrow of the circuit symbol, the diode acts as a short circuit and the current flows without hindrance. The diode is ON. When the voltage is positive in the direction opposite to the diode arrow, however, the diode acts as an open circuit and no current can flow. The diode is OFF. Thus, the current can flow in the direction of the arrow but cannot flow against the arrow.

Mechanical analogs. The diode can be considered as the electrical equivalent of the mechanical ratchet, such as is used to tighten the net on a tennis court. The mechanical ratchet allows motion or rotation in one direction only. Similarly, the diode allows charge motion in one direction only. Another analog would be a check valve that allows fluid flow in only one direction.

Diode switch. The diode can also be considered a voltage-actuated switch. As long as the diode voltage is positive (with the + at the top of the arrow), the switch is closed, current flows, and the diode is ON. But once the polarity of the voltage reverses, the switch opens, no current flows, and the diode is OFF.

The diode characteristic described is that of an ideal diode. Real diodes depart from this ideal characteristic, as is detailed in Sec. 2.2. But many of the common applications of the diode can be understood in terms of this ideal characteristic; hence, we present some applications before investigating the physical processes in semiconductor diodes.

LEARNING OBJECTIVE 1.

To understand how to analyze half- and full-wave rectifier circuits, with and without a filter capacitor

Rectifier Circuits

DC and electronics. Alternating current, ac, is universally used for the generation, distribution and consumption of electric power except where portability is required. However, electronic circuits are powered by direct current, dc. You might suppose, therefore, that every electronic device would contain a battery, but this is untrue. Batteries are expensive, heavy, short-lived, bulky, and filled with corrosive chemicals. Batteries are thus undesirable components and are avoided by designers except where portability is essential.

power supply

For this reason, most electronic equipment contains a power supply circuit. The function of the power supply is shown in Fig. 2.3. When you plug in and turn on your TV set, for example, the ac power enters the *power supply* section of the electronic circuit, where it is converted to dc power. From there the dc power flows to other parts of the circuit. Such power supply circuits use the nonlinear properties of diodes in circuits called rectifiers.

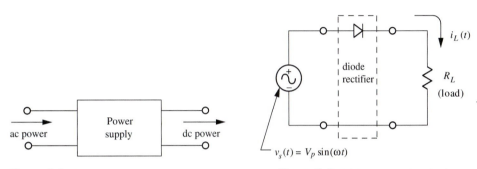

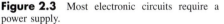

Figure 2.3 Most electronic circuits require a power supply.

Figure 2.4 Half-wave rectifier circuit.

half-wave rectifier

Half-wave rectifier. Power supplies use diodes to convert ac to dc in rectifier circuits; the diodes are said to "rectify" the ac. Figure 2.4 shows a basic rectifier circuit. This circuit is called a *half-wave rectifier* because it couples only the positive half of the input voltage to the load. The ac voltage is represented as a sine function, the rectifier is an ideal diode, and the load is represented as a resistor, although in practice the load would be the electronic circuits that require dc power.

Figure 2.5 shows the rectifying effect of the diode. When the input voltage is positive, the diode turns ON and current flows with a sinusoidal shape, but when the input voltage is negative, the diode turns OFF and no current flows. The resulting current

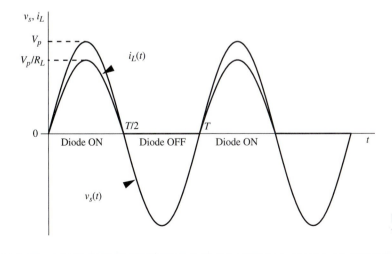

Figure 2.5 Half-wave rectifier waveforms.

flows in spurts, and the voltage across the load is that portion of the sinusoidal input that is positive. The diode blocks the negative part of the ac waveform.

DC component. Admittedly, the output of the half-wave rectifier is not pure dc power. The output, however, does contain a dc component. Indeed, if we define the dc portion of the output as the time average, we have a dc component of

$$I_{dc} = \frac{1}{T} \int_0^T i_L(t)\,dt = \frac{1}{T}\left[\int_0^{T/2} \frac{V_p}{R_L} \sin(\omega t)\,dt + \int_{T/2}^T 0\,dt \right]$$

$$= \frac{V_p}{\pi R_L}$$

(2.1)

where $\omega = 2\pi/T$. This current would be indicated by a dc ammeter in series with the load. The dc component of the load voltage is

$$V_{dc} = I_{dc} R_L = \frac{V_p}{\pi}$$

(2.2)

EXAMPLE 2.1 | **Half-wave rectifier**

A 24-V(rms) ac source is connected to a 20-Ω resistor with a diode in a half-wave rectifier circuit. Find the peak and the average current in the load.

SOLUTION:

Figure 2.4 shows the circuit. The peak current would be the peak positive voltage divided by the resistance, because the diode is ON for positive source voltage.

$$I_{\text{peak}} = \frac{V_p}{R_L} = \frac{24\sqrt{2}}{20} = 1.70 \text{ A}$$

(2.3)

The dc current is given by Eq. (2.1):

$$I_{dc} = \frac{V_p}{\pi R_L} = \frac{I_p}{\pi} = \frac{1.70}{\pi} = 0.540 \text{ A}$$

(2.4)

WHAT IF? What if the diode were reversed?[1]

---

full-wave bridge rectifier

Full-wave rectifiers. The circuit in Fig. 2.6 (a) is called a *full-wave bridge rectifier* because it inverts or rectifies the negative half of the input voltage as well as the positive half.

---

[1] Then the negative part of the ac voltage would be coupled to the load. $I_{\text{peak}} = -1.70$ A and $I_{dc} = -0.540$ A.

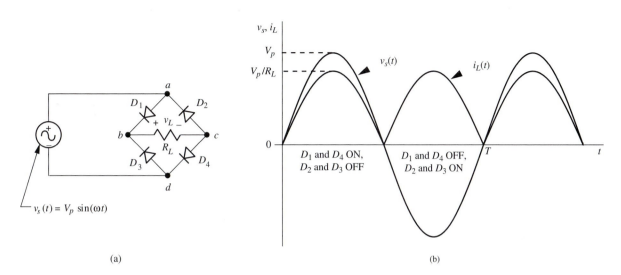

(a)

(b)

Figure 2.6 (a) Bridge full-wave rectifier circuit; (b) waveforms for the bridge full-wave rectifier.

Bridge-rectifier operation. While the source voltage is positive, the current tends to flow through the rectifier from a to d. Diode D_1 comes ON but D_2 cannot accommodate current from a to c and will turn OFF. From b, the current must flow through the resistor because D_3 will not permit current flowing directly from b to d. Finally, D_4 will turn ON and the current will flow from c to d. Thus, while the source voltage is positive, positive current flows out of the $+$ of the source through D_1, through the load resistor, and finally through D_4 returns to the source. Diodes D_2 and D_3 are OFF during this part of the cycle. While the source voltage is negative, D_2 and D_3 turn ON while D_1 and D_4 turn OFF. Thus, during the second half of the cycle, positive current flows out of the minus terminal of the source, through D_3 from d to b, through the load resistance, and back to the source through D_2. The two paths for the current are shown in Fig. 2.7. The current flows through the load from b to c during both parts of the cycle; thus, the current through the load is as shown in Fig. 2.6(b). The current again flows in spurts, but the full-wave rectifier leaves no idle time between spurts as does the half-wave rectifier.

As a power-supply circuit, the full-wave rectifier does better than the half-wave rectifier. By inverting the negative portions on the input voltage, the circuit doubles the dc component in the output; hence, the dc component in the current is

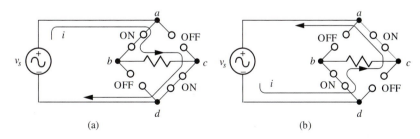

(a)

(b)

Figure 2.7 Current paths for a bridge rectifier: (a) v_s positive; (b) v_s negative.

$$I_{dc} = \frac{1}{T} \int_0^T i_L(t)\,dt = \frac{2}{T} \int_0^{T/2} \frac{V_p}{R_L} \sin(\omega t)\,dt$$

$$= \frac{2V_p}{\pi R_L}$$

(2.5)

and the dc load voltage is

$$V_{dc} = I_{dc} R_L = \frac{2}{\pi} V_p$$

(2.6)

EXAMPLE 2.2 Full-wave rectifier

A full-wave rectifier is required to provide 50 V dc and 2 A dc to a resistive load. Find the load resistance and the required rms ac voltage.

SOLUTION:

Figure 2.8 shows the bridge rectifier of Fig. 2.6(a), redrawn to put the load to the right. The load resistance is given by Ohm's law for the dc voltage and current:

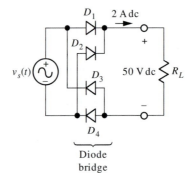

Diode
bridge

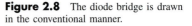

Figure 2.8   The diode bridge is drawn in the conventional manner.

$$R_L = \frac{V_{dc}}{I_{dc}} = \frac{50}{2} = 25\ \Omega$$

(2.7)

The peak ac voltage is given by Eq. (2.6):

$$V_p = \frac{\pi(50)}{2} = 78.5\ \text{V}$$

(2.8)

Hence, the rms voltage is $78.5/\sqrt{2} = 55.5$ V.

WHAT IF?

What if one of the diodes burned out and became an open circuit?[2]

---

[2] $I_{dc} = 1$ A, $V_{dc} = 25$ V.

The diode bridge as a switch. The diodes act as switches that are activated by voltage. While the source voltage is positive, D_1 and D_4 are turned ON and thus b and c are connected to the $+$ and $-$ terminals of the source, respectively. When the source voltage becomes negative, D_2 and D_3 are turned ON and hence b and c are again connected to the source, this time with b connected to the $-$ and c to the $+$ terminal. Thus, b is automatically connected to the physically positive terminal of the source while c is connected to the physically negative terminal. This automatic switching action occurs regardless of the shape of the source voltage—the shape can be sinusoidal, triangular, or an unpredictable communication signal in a radio circuit. The effect of the diode bridge is thus to produce across the load an output voltage that is the absolute value of the input voltage: $v_L(t) = |v_s(t)|$.

Full-wave rectifier with center-tapped transformer. Figure 2.9 shows a full-wave rectifier circuit that uses a transformer and two diodes. The transformer secondary is center-tapped to supply identical but opposite voltages to the two diodes. Each diode acts as a half-wave rectifier: D_1 supplies the positive part of v_{de} and D_2 supplies the positive part of v_{fe}, as shown in Fig. 2.10. The transformer gives this circuit the versatility to produce any desired dc voltage, depending on the turns ratio. It also allows both ac input and dc output to be grounded, which cannot be done with a bridge.

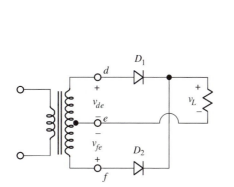

Figure 2.9 Full-wave rectifier using a center-tapped transformer.

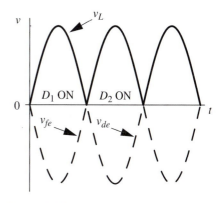

Figure 2.10 Waveform for the full-wave rectifier.

Rectifier with Filter Capacitor

filter, ripple

Filtering out the ripple. The three rectifier circuits described produce a dc component in their outputs. We may describe their outputs as a desired dc component plus an undesired *ripple*, as shown in Fig. 2.11. A *filter*, as in Fig. 2.12, is a circuit used to remove an undesirable signal in a circuit. In this case, we need a filter to eliminate, or at least greatly reduce, the ripple component from the output of the rectifier. In Chap. 4, we consider filters generally; here we introduce the simplest of filters—we merely connect a capacitor across the load, as shown in Fig. 2.13. As we will see, the capacitor stabilizes the voltage across the load resistor.

Charging the capacitor. The source is $V_p \sin(\omega t)$ and thus passes through zero volts at $t = 0$, as shown in Fig. 2.14. As the voltage increases, diodes D_1 and D_4 turn

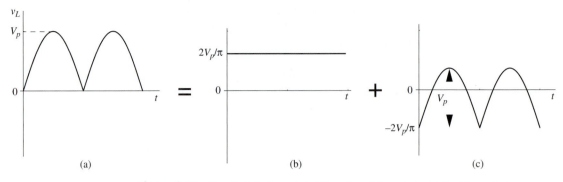

(a)   =   (b)   +   (c)

Figure 2.11 The ripple is the undesirable portion of the output. (a) Rectifier output; (b) dc component; (c) ripple component.

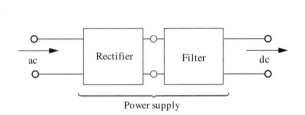

Figure 2.12 A filter is used to reduce the ripple.

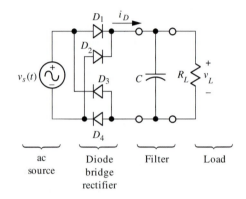

Figure 2.13 Full-wave bridge rectifier with a capacitor filter.

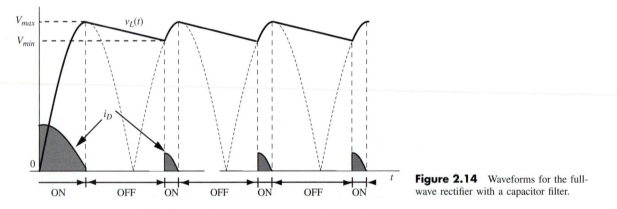

Figure 2.14 Waveforms for the full-wave rectifier with a capacitor filter.

ON and current flows through the load as before. Current also flows through the capacitor and charges it to the peak value of the input ac voltage. The first spurt of current is relatively large in Fig. 2.14 due to the initial charging of the capacitor.

Holding the charge. After its peak, the input voltage drops rapidly. If the voltage of the capacitor were to follow this voltage, a rapid discharge would have to occur. However, all diodes prevent discharge back through the input source and turn OFF when the input voltage drops below the voltage on the capacitor, because at that moment the diodes become reverse-biased. Hence, the filter capacitor and the load resistance become disconnected from the source, which continues with the negative part of its waveform.

The discharge. The capacitor discharges through the load resistor, as shown in Fig. 2.15. To write the voltage as a function of time, we need to know the time constant, the initial value, and the final value. The time constant is R_LC, the initial value is V_p, and the final value would be zero if the discharge were allowed to go on forever. Thus, the voltage of the load and the capacitor during the discharge period, while the diodes are OFF, would be

$$v_L(t') = 0 + (V_p - 0)e^{-(t'/R_LC)} = V_p\, e^{-(t'/R_LC)} \tag{2.9}$$

where t' is measured from the peak.

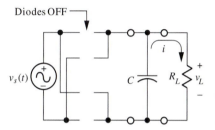

Figure 2.15 When the diodes of the bridge are OFF, the capacitor discharges through the load.

The recharge. To function well as a filter, the circuit should have a time constant, R_LC, much longer than the period of the input ac voltage, usually 1/60 s. The load voltage thus decreases only slightly between peaks. Figure 2.14 exaggerates the decrease from what it would be in practice. The diode remains OFF until the input voltage becomes equal to the decreasing load voltage. As the input voltage again exceeds the load voltage, D_2 and D_3 turn ON and current again flows through the rectifier diodes. Most of the current goes to the capacitor, replenishing the charge lost during the discharge part of the cycle. After the pulse of current that initially charges the capacitor, current flows through the diodes only during these brief recharging periods, as shown in Fig. 2.14.

The dc load voltage. Because the load voltage decreases only slightly during the period of the ac waveform, as shown in Fig. 2.14, the output voltage of the power supply remains approximately equal to the peak value of the input ac waveform. Thus, the first benefit of the filter capacitor is to increase the dc output from $(2/\pi)V_p$ to V_p. Second, the filter capacitor greatly reduces the ripple voltage. For the case we are considering, $R_LC >>$ period, the exponential decrease of the load voltage is well approximated by a straight line, given by the two leading terms in a series expansion of the exponential

$$v_L(t') = V_p\, e^{-(t'/R_LC)} = V_p\left(1 - \frac{t'}{R_L C} + \cdots\right) \tag{2.10}$$

where t' is measured from the peak. Thus from its maximum value of V_p the voltage decreases to a minimum value of approximately

$$V_{min} \approx V_p \left(1 - \frac{T/2}{R_L C}\right) = V_p \left(1 - \frac{1}{2fR_L C}\right) \qquad (2.11)$$

at the moment when the diode turns ON and permits recharging of the capacitor. In Eq. (2.11), f represents the ac frequency, the reciprocal of the period. Consequently, the peak-to-peak ripple, V_r, is

$$V_r = V_{max} - V_{min} = \frac{V_p}{2fR_L C} \qquad (2.12)$$

We note from Fig. 2.14 that a more accurate approximation to the dc component of the filtered output would be the average between the maximum and minimum voltages

$$V_{dc} \approx \frac{V_{max} + V_{min}}{2} = V_p \left(1 - \frac{1}{4fR_L C}\right) \qquad (2.13)$$

EXAMPLE 2.3 **Filtered full-wave rectifier**

Find the ripple and dc voltage out of the filtered full-wave rectifier in Fig. 2.16.

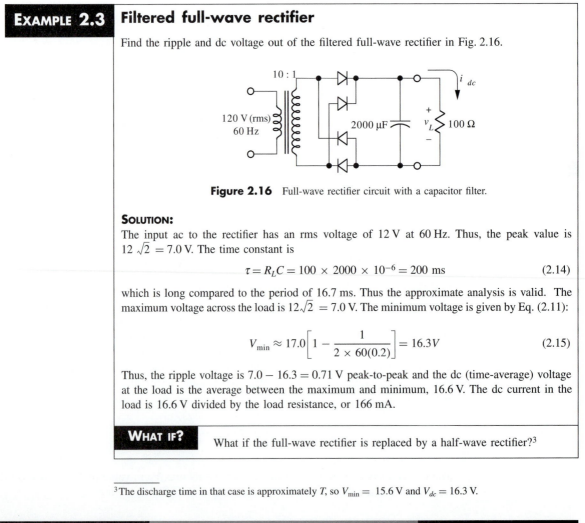

Figure 2.16 Full-wave rectifier circuit with a capacitor filter.

SOLUTION:

The input ac to the rectifier has an rms voltage of 12 V at 60 Hz. Thus, the peak value is $12\sqrt{2} = 7.0$ V. The time constant is

$$\tau = R_L C = 100 \times 2000 \times 10^{-6} = 200 \text{ ms} \qquad (2.14)$$

which is long compared to the period of 16.7 ms. Thus the approximate analysis is valid. The maximum voltage across the load is $12\sqrt{2} = 7.0$ V. The minimum voltage is given by Eq. (2.11):

$$V_{min} \approx 17.0 \left[1 - \frac{1}{2 \times 60(0.2)}\right] = 16.3 V \qquad (2.15)$$

Thus, the ripple voltage is $7.0 - 16.3 = 0.71$ V peak-to-peak and the dc (time-average) voltage at the load is the average between the maximum and minimum, 16.6 V. The dc current in the load is 16.6 V divided by the load resistance, or 166 mA.

WHAT IF? What if the full-wave rectifier is replaced by a half-wave rectifier?[3]

---

[3] The discharge time in that case is approximately T, so $V_{min} = 15.6$ V and $V_{dc} = 16.3$ V.

Better filters. We have investigated the benefits of the simplest possible filter. The performance of this filter is adequate for many applications, but high-quality power supplies employ more sophisticated filters. Some filters add inductors and additional capacitors to reduce the ripple; others employ electronic circuits to cancel the ripple.

Check Your Understanding

1. An ideal diode uses no power because either its voltage or current is zero. True or false?

2. A half-wave rectifier circuit requires at least two diodes. True or false?

3. What is the average value of a full-wave rectified sinusoid having a peak value of 10 V before rectification?

4. A battery may be used as a filter in a power supply. True or false?

5. A rectifier circuit produces pure dc. True or false?

6. What is "filtered" by a filter circuit in a power supply?

7. An unfiltered full-wave rectifier puts out 12 V dc for a sinusoidal input. If a large capacitor is added across the load, what would be the new voltage, assuming ideal diodes?

8. In a good stereo, the sound does not go away immediately when you turn off the amplifier but fades out over a period of several seconds. Explain.

Answers. **(1)** true; **(2)** false; **(3)** 6.37 V; **(4)** true—the battery acts like an infinite capacitor; **(5)** false; **(6)** the ripple, everything but the dc; **(7)** 18.8 V; **(8)** the capacitor in the power supply has to discharge before the dc voltage is zero.

2.2 THE *PN*-JUNCTION DIODE

Reasons for discussing the physical principles of diode operation. The diode finds many uses as an electronic component. We have seen how diodes are used in power supplies to convert ac to dc, but this is only one of the many uses for diodes. Diodes are used extensively in analog electronic devices such as radios and audio systems, and are even more important in digital systems such as computers and digital watches.

The diode plays an important role in the study of electronics because it is a simple electronic device; hence, it presents a good starting place in exploring the bag of electronic tricks. The diode is also important in the teaching of electronics because it offers an opportunity to investigate the physical basis of semiconductor electronics, the processes that have produced essentially all the electronic equipment we require and enjoy in our lives and work. This section describes many of these semiconductor processes.

Diode operation. We have a genuine problem in helping you understand how a diode works. It is commonly asserted, and reasonably so, that we should explain the unknown in terms of the known. But your author must explain the unknown—how a diode works—in terms of other unknowns, namely, semiconductor processes such as holes, drift currents, uncovered charges, and depletion regions. Or, put another way, we are required to consider many physical processes before we can understand how a simple *pn*-junction diode operates. Understanding of the diode merits the effort, but our investment pays further dividends because modern electronics is founded on the manipulation of charges in semiconductor materials through such physical processes.

Semiconductor Processes and the *pn* Junction

Crystalline nature of solids.   Matter exists in four states: gas, plasma, liquid, and solid. These may be understood in terms of the interaction between individual atoms or molecules comprising the matter. In a solid, the atoms remain in a fixed position relative to each other. Often the atoms exist in a regular crystalline order, held together by shared electrons in covalent bonds. Figure 2.17 shows a two-dimensional representation of such a lattice. The large circles with numbers represent the nuclei, the dots represent the valence electrons, and the curved lines represent the bonding effect of the electrons. The electrons are negatively charged and the nuclei are positively charged. Of course, there are many more electrons associated with each nucleus, but we have not represented these inner shells because these do not enter into the bonding process or the semiconductor processes we are investigating. We have represented the case where there are four electrons in the outer shell, which is typical of a semiconductor such as silicon. Because the individual atoms are electrically neutral, all the charge would be neutralized by the inner shells of electrons except for a positive charge of four times the electronic charge, which is thus the effective charge of the nucleus. This charge is indicated by the +4 in Fig. 2.17.

insulator, conductor, conduction electrons

Insulators and conductors.   At zero absolute temperature, solids are either electrical conductors or insulators. The material represented by Fig. 2.17 would be an *insulator* because all the electrons are involved in the bonding. If such a material were placed in an electric circuit, no current would flow because no electrons are free to move. In a *conductor*, each atom has at least one excess electron that is not involved in the bonding. Such excess electrons are known as *conduction electrons* and they move freely in response to electrical forces.

Semiconductors.   Thermal energy has a strong effect on the conduction properties of solids. Thermal energy is distributed throughout the electrons and nuclei of the materials and this energy is stored, among other ways, in the physical movement, or vibration, of the electrons and nuclei. The vibrations may cause some of the bonding electrons to break loose from their bonding positions to become conduction electrons.

semiconductor

Materials that have no excess electrons and retain all their electrons in bonds at normal temperatures remain good insulators. Carbon in a diamond crystalline order and many plastics and ceramics furnish examples of good insulators at normal temperatures. In other materials that are good insulators at absolute zero however, the electrons are not tightly held in the bonds, and in these materials, some of the electrons escape their bonds at normal temperatures to become conduction electrons. We have represented this condition in Fig. 2.18 with two electrons out of their bonding positions. Such a material is known as a *semiconductor* because it becomes a conductor at normal temperatures due to the electrons that have become conduction electrons.

hole

Holes and hole movement.   When an electron leaves its bonding position, it leaves behind a vacant position, which is called a *hole*. We may think of the hole as having a positive charge because the nucleus adjacent to it now has charge not neutralized by the bonding electrons. Holes are also free to move under the influence of elec-

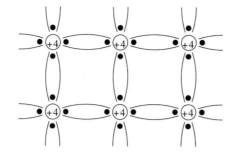

Figure 2.17 Insulator.

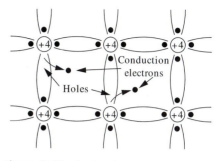

Figure 2.18 Semiconductor showing holes and electrons.

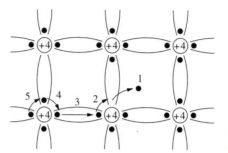

Figure 2.19 The movement of bound electrons produces hole movement.

trical forces because other bound electrons may move into the vacant location. If in Fig. 2.18 there were an electrical force tending to move electrons from left to right, the conduction electrons would move in response to such a force. But the electron next to the vacant position will also tend to move to the right and may leave one bonding position for a vacant position toward the right. Thus, we can envision the process represented in Fig. 2.19. First, the electron breaks its bond and becomes a conduction electron, creating a hole. Because there is an electric force tending to move electrons toward the right, there will also be a tendency for electrons remaining in bonds to move left to right; hence, the transitions labeled 2 through 5 are favored. The hole moves toward the left; in effect, a positive charge moves right to left because there is excess positive charge associated with the hole.

carriers

If a semiconductor were placed in a circuit where the voltage created a current, the current would be carried by both holes and electrons, which are said to be *carriers*. Note that the currents carried by the movement of holes and electrons are additive: Holes moving toward the left carry a positive current toward the left; and electrons moving toward the right carry a negative current toward the right, which is a positive current toward the left. In a typical semiconductor, the holes move about one-third as fast as the electrons and hence carry about one-fourth the total current.

Importance of thermal energy. We stated that hole–electron pairs are created because of thermal energy, and have shown how electrons changing bond positions due to thermal vibrations act like mobile positive charges. But we described these processes as if they were orderly and sedate, which they are not. For pure silicon at 300 K, there are approximately 1.5×10^{16} conduction electrons/m^3 and there are approximately 10^{22} hole–electron pairs/m^3 created and eliminated through recombination per second through thermal action.

Hence emerges a picture of violent, random thermal motion of conduction electrons in a semiconductor. Many hole–electron pairs are created and many recombinations occur during short periods of time.

intrinsic semiconductor, *n*-type doping, *n*-type semiconductor, donor

n-type doping. We can increase the concentration of carriers by adding small amounts of impurities to the pure, *intrinsic*, semiconductor. Consider that we have intrinsic silicon, which has four bonding electrons per atom. If we add a small amount of an element that has five electrons in its valence shell, such as phosphorus, the impurity nuclei will bond into the lattice with one electron left over. This electron will be a conduction electron, as shown in Fig. 2.20, and there will also be one additional positive charge fixed into the lattice of nuclei because of the additional charge of the nucleus of the impurity atom. This process of adding impurities is called *doping*; Fig. 2.20 shows *n-type doping,* so called because of the additional negative carriers. This would make an *n-type semiconductor* because it has an increased concentration of electrons. The impurity in this case would be called a *donor* atom because it donates an additional conduction electron to the semiconductor.

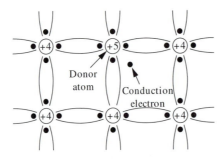

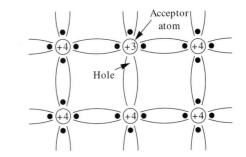

Figure 2.20 An *n*-type semiconductor has extra conduction electrons.

Figure 2.21 A *p*-type semiconductor has extra holes.

acceptor, *p*-type semiconductor

p-type doping. Similarly, if we add an impurity with three electrons in its valence shell, we would create a hole for each impurity atom, as shown in Fig. 2.21. Here we have shown the hole and we have indicated that the nucleus is deficient one positive charge with a +3. If the hole were filled by a conduction electron, that region would in effect have a negative charge built into the lattice structure of the semiconductor. Such an impurity is called an *acceptor* atom because it accepts an electron from the conduction electrons in the semiconductor and an extra hole is created. A semiconductor that is doped with acceptor atoms is called a *p-type semiconductor* because it has an excess of holes, which act like mobile positive charges.

The electrons donated by the donor atoms and the holes created by the acceptor atoms do not remain near their associated nuclei, but participate in the random thermal processes of all carriers. By doping the semiconductor, we have the ability to increase the carrier concentration and to control what type of carriers will dominate the conduction processes in the semiconductor.

drift current

Drift currents. We described what happens when a semiconductor is placed in an electric circuit. A current is formed by movement of both holes and electrons. If the semiconductor were *p*-type, most of the current would be carried by the holes. If the semiconductor were *n*-type, most of the current would be carried by the electrons. This type of current is called *drift current* because the carriers drift in a certain direction as dictated by an external voltage. Drift current is caused by an orderly process.

Diffusion currents.

However, an excess of carriers in a certain region would tend to distribute uniformly throughout the material because of their random thermal movement. Such a flow is known as *diffusion*. A diffusion flow occurs when a bottle of a smelly chemical is opened in a room. People near the open bottle would smell it first, and then those farther away. Eventually, everyone in the room would smell the chemical because the molecules would be distributed uniformly throughout the room by their thermal motion. In a similar way, carriers move away from regions of concentration due to their thermal motion. The resulting *diffusion current* is proportional to the rate of change of carrier concentration with respect to distance. Diffusion is a disorderly process because it is driven by thermal motion.

Summary.

An intrinsic semiconductor has many, but equal number of holes and electrons because thermal energy causes electrons to leave their bond positions, become conduction electrons, and leave behind a hole. We can create *p*-type or *n*-type semiconductors by doping the pure material with either acceptor or donor impurity atoms. The additional holes or electrons participate in the violent, random thermal motion of the carriers in the material. The carriers form currents either by the orderly process of a drift current or by the disorderly process of diffusion.

Bound charges.

There are also charges bound into the lattice structure associated with the acceptor or donor nuclei. The acceptor atoms act as stationary negative charges because their nuclei are deficient one positive charge relative to the other nuclei in the vicinity. The donor atoms act as stationary positive charges because their nuclei have one additional positive charge relative to the surrounding nuclei. Semiconductors are neutral in total charge because the excess mobile carriers neutralize the bound charges of the impurity atoms. All of these processes occur in a *pn* junction.

Structure.

A *pn* junction is formed at the boundary between regions of *p*-type semiconductor and *n*-type semiconductor, as shown by Fig. 2.22. When such a junction is formed, the sequence of events diagrammed in Fig. 2.23 occurs rapidly.

■ **Diffusion currents.** Electrons in the *n*-type material diffuse toward the *p*-type material, and holes in the *p*-type material diffuse toward the *n*-type material.

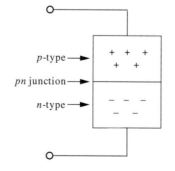

Figure 2.22 The *pn* junction.

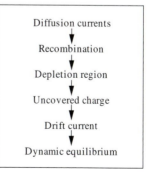

Figure 2.23 Processes involved in the formation of a *pn* junction.

These strong diffusion currents occur because the concentrations are unequal and the carriers are in violent thermal motion.

- **Recombination.** Recombinations occur immediately as the electrons that have diffused into the *p*-type material find holes to fill, and holes that have diffused into the *n*-type material are filled by conduction electrons.

depletion region, uncovered charge

- **Depletion region.** There is a region on both sides of the junction that has a deficiency of carriers due to recombinations. This region is called a *depletion region*. We might anticipate a continual pouring of carriers into this depletion region were it not for another process that occurs.

- **Uncovered charge.** On the *p*-type material side of the junction in the depletion region, the acceptor atoms bound into the lattice structure are now uncovered. Because the electrons from the *n* side have recombined with many of the holes, the deficiency of charges in the nuclei of the acceptor atoms acts as an excess of negative charges fixed in this region. Similarly, the excess positive charges of the donor nuclei act as positive charges bound into the lattice structure, now uncovered because the electrons that formerly neutralized them have recombined with holes from the *p* side. Thus, we have *uncovered charges* bound into the lattice structure in the depletion region, as shown by Fig. 2.24.

- **The battery–capacitor effect.** This charge distribution acts similar to a capacitor, as shown in Fig. 2.25. Here we show a capacitor with a charge placed on it by a battery. The bottom plate of the capacitor has positive charges and the top has negative charges. If an electron were in the region between the plates, it would move downward, attracted by the positive charges below and repelled by the negative charges above. Similarly, a positive charge between the plates would move upward. This would be an orderly process and would create a drift current in this region if there were electrons or positive charges between the plates. Similarly, in Fig. 2.24 excess carriers in the depletion region experience forces from the uncovered, bound charges.

- **Drift current.** These forces would tend to create a drift current upward as electrons are sent back toward the *n*-type material and holes are sent back toward the *p*-type material. Thus, we can envision the holes and electrons moving across the junction due to diffusion but moving the other way by the drift current caused by the uncovered charges.

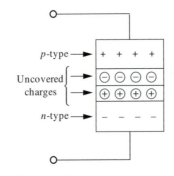

Figure 2.24 Uncovered charge.

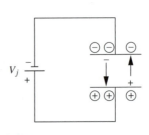

Figure 2.25 The uncovered charges act like a charged capacitor.

- **Dynamic equilibrium.** These two processes rapidly reach dynamic equilibrium. Another way to view this dynamic equilibrium is to consider that the uncovered charges constitute an internal battery–capacitor effect that automatically adjusts itself to stop the diffusion currents. Hence, we have a dynamic equilibrium between a disorderly process, the diffusion process, and an orderly process, the battery–capacitor effect.

Bound charges. Due to the uncovered charges, the *pn* junction diode has an internal voltage of approximately 0.7 V across its junction,[4] but no external voltage across the entire diode is produced. No current will flow unless the equilibrium is disturbed by an external voltage.

Forward-bias characteristic. If we add an external battery to the diode, as shown by Fig. 2.26, and if the polarity of the external battery is such as to oppose the internal battery–capacitor effect, the equilibrium will be disturbed and current will flow continually through the diode. In this case, the battery causes the holes in the *p*-type material and the electrons in the *n*-type material to move toward the junction. Because the drift process tending to prevent flow of carriers through the depletion region has been partly neutralized by the external battery, current is carried across the junction by the diffusion currents. The current increases dramatically with increases in the external voltage because it is driven by the energetic thermal motion of the carriers. If the external voltage exceeds the voltage of the internal battery–capacitor effect, large currents will flow: the diode is ON. Because this voltage is only about 0.7 V for a silicon diode, the forward-bias region of the diode approximates that of an ideal diode. A germanium diode requires about 0.2 V to cause substantial conduction and hence is more nearly ideal than a silicon diode.

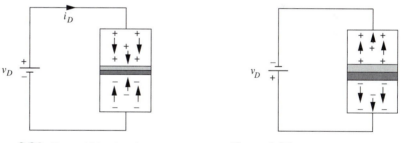

Figure 2.26 Forward-biased *pn* junction.   **Figure 2.27** Reverse-biased *pn* junction.

Reverse-bias characteristic. If the polarity of the external voltage reinforces the influence of the internal battery–capacitor effect, the holes in the *p*-type semiconductor and the electrons in the *n*-type semiconductor will tend to move away from the junction, as shown in Fig. 2.27. The external voltage merely widens the depletion region and strengthens the restraining effect of the internal battery–capacitor effect. For this polarity of the external voltage, very little current will flow through the diode and an ideal diode is well approximated: the diode is OFF. The reverse-biased *pn* junction behaves as a capacitor due to the charge separation associated with the uncovered charges. In the

---

[4]For a silicon-based *pn* junction. Other semiconductors give slightly different voltages.

next section we give the equation of a *pn*-junction diode and see how well an ideal diode approximates a semiconductor *pn* junction.

Physical Properties of Real Diodes

The *pn*-junction equation. The voltage–current characteristic of the *pn*-junction diode is well described by

$$i_D = I_0\left[\exp\left(\frac{qv_D}{\eta kT}\right) - 1\right] \tag{2.16}$$

where

$\exp(x) = e^x$

i_D = diode current, A

v_D = diode voltage, V

I_0 = a constant called the reverse saturation current, which depends on the semiconductor materials, manner of junction formation, and junction size

$q = |e|$, the magnitude of the electronic charge, 1.60×10^{-19} C

k = Boltzmann's constant, 1.38×10^{-23} J/K

T = absolute temperature, K

η = a constant between 1 and 2, called the *ideality factor*, which depends on junction materials and method of formation

Figure 2.28 shows a plot of Eq. (2.16) for the parameters $\eta = 1.5$ and $I_0 = 10^{-9}$ A. These would be typical of a silicon diode at room temperature. If you wish to plot manually the *pn*-junction equation, it is useful to have v_D as a function of i_D.

$$v_D = \eta V_T \ln\left(\frac{i_D}{I_0} + 1\right) \tag{2.17}$$

voltage equivalent of temperature

where $V_T = kT/q$ is called the *voltage equivalent of temperature* and has a magnitude of about 25.9 mV at $T = 300$ K.

If you compare the characteristics of a *pn*-junction diode, Fig. 2.28, with those of an ideal diode, Fig. 2.2, you might be disappointed, for the properties of a real diode appear quite nonideal. That impression is created by our plotting the *pn*-junction characteristic on an expanded voltage scale. Figure 2.29 shows the diode characteristics on the same scale as a 1-kΩ resistor.

threshold voltage

Junction properties. The properties of the *pn*-junction diode may be summarized in this fashion: Essentially, zero current flows in the reverse-bias region. Negligible current flows in the forward-bias region until a small *threshold voltage* is reached, after which the magnitude of the current rises rapidly. Once the current begins to rise, the voltage remains fairly constant. The threshold voltage is about 0.7 V for a silicon diode. The *pn*-junction equation, Eqs. (2.16) and (2.17), gives the precise characteristic, but the concept of a threshold voltage is used to simplify analysis for many diode applications. For the rectifier circuits presented earlier in this chapter, we may assume that the diodes will have a voltage of about 0.7 V when they are ON and thus the rectifier output volt-

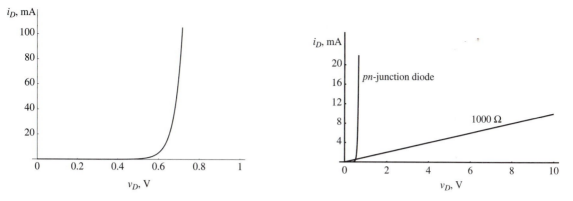

Figure 2.28 Typical current–voltage characteristic for a silicon diode.

Figure 2.29 A *pn*-junction diode characteristic compared with a 1-kΩ resistor.

ages and currents will be reduced accordingly. When modeled by a threshold voltage, the voltage drop of the diode is easy to include in design and analysis of diode circuits.

Power limits and heat transfer. Temperature strongly affects the semiconductor processes upon which diode operation depends. In Eq. (2.16), temperature appears explicitly in the exponential term, but the reverse saturation current, I_0, is also strongly affected by temperature. For this reason, the electronics designer must prevent the diode from getting too hot.

Heat is produced by the electrical power given to the diode. The power into a device is

$$p = vi \qquad (2.18)$$

The ideal diode requires no power because, in the forward-bias region, the voltage is zero, and in the reverse-bias region, the current is zero; hence, the product of voltage and current is always zero. A real diode inherently receives little power because it also has small voltage when the current is high in the forward-bias region and low current when the voltage is high in the reverse-bias region. Even so, the small amount of power that is given to the diode is important because the heat is generated in the junction region, which is physically small. Thus, a small power can cause a significant rise in the junction temperature and affect diode performance. For this reason, diodes in power supplies are designed to have good heat conduction between the junction and the outer case of the diode, and the diode is mounted in such a way as to enhance heat transfer to the ambiance. Often, heat exchangers[5] and even fans are used to improve cooling of the diode. Occasionally, water cooling is used on large power supplies. Figure 2.30 shows a power semiconductor mounted on a heat sink.

heat sink

reverse-breakdown voltage

Breakdown. A physical diode is also limited by the amount of voltage it can withstand in the reverse-bias region. Conduction is prevented in the reverse-bias region, where an external voltage reinforces the effect of the uncovered charges in the depletion region, as suggested in Fig. 2.22. When too much voltage appears across the depletion region, however, the forces on the bound electrons in that region become so great that

---

[5] Also called *heat sinks*.

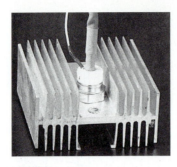

Figure 2.30 Power semiconductor mounted on a heat sink.

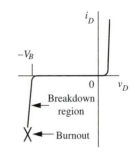

Figure 2.31 When reverse-bias voltage exceeds the breakdown voltage, the current increases rapidly.

they can be stripped from their bonds. In this condition, a rapid buildup of current occurs, as shown in Fig. 2.31. The rapid buildup of current at the *reverse-breakdown voltage*, $-V_B$, increases the power into the junction region, and the diode fails.

peak inverse voltage, PIV

Breakdown in power supplies. Reverse breakdown must be avoided in power supply operation. Designers have available diode types with breakdown voltages[6] in excess of -1000 V. In the unfiltered half-wave rectifier shown in Fig. 2.4, the maximum diode voltage is $-V_p$, which occurs when the diode is OFF and the source at its maximum negative value. For successful operation in such a power supply, the *PIV* of the diode must exceed this voltage.

In the power supply in Fig. 2.9 with a capacitor, the maximum voltage across the diode is approximately $-2V_p$, as shown in Fig. 2.32. This maximum occurs because the

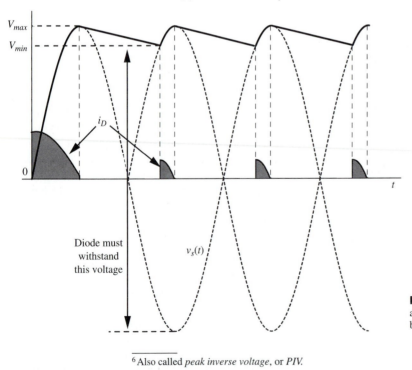

Diode must withstand this voltage

$v_s(t)$

Figure 2.32 To operate successfully in a rectifier, a diode must never have its breakdown voltage exceeded.

---

[6] Also called *peak inverse voltage*, or *PIV*.

capacitor holds the load voltage at approximately $+V_p$, whereas the voltage source swings negative to $-V_p$. To operate successfully in such a power supply, the diode must be able to withstand $-2V_p$.

Check Your Understanding

1. Does an intrinsic semiconductor have an excess of holes, conduction electrons, or neither?

2. In a *pn* junction, does the *p* stand for semiconductor material that has an excess of holes or electrons?

3. Which of the following are nonlinear devices: resistor, *pn*-junction diode, capacitor, short circuit, ideal diode?

4. In the ON state, does current cross a *pn* junction under the influence of a drift or diffusion process?

Answers. (**1**) Neither; (**2**) holes; (**3**) *pn*-junction diode and ideal diode; (**4**) diffusion.

2.3 BIPOLAR JUNCTION TRANSISTOR (BJT) OPERATION

Importance of the Transistor

From your experience with circuit theory, you will recall that one can do a fair amount of work solving a two- or three-loop ac circuit. That being true, do you wonder how electrical engineers can design circuits containing literally thousands of loops and nodes? For example, if you ever examine the circuit diagram for a relatively simple piece of electronics, say a TV set or an FM radio, you would see hundreds of resistors and capacitors, not to mention diodes and transistors and other components.

One-way property of the transistor. The design of such complicated circuits is made possible by the unidirectional properties of transistors. The transistor is a three-terminal device connected normally as shown in Fig. 2.33. The transistor has an input and an output, suggesting that the cause–effect relationship goes from left to right. The input affects the output but the output has little effect on the input. This one-directional causality is indicated by marking an arrow from left to right and crossing out the arrow from right to left. We might say that the transistor is a "one-way street" to the electrical signal. This one-directional property of the transistor isolates the output from the input and allows electrical engineers to build complex circuits. This complexity can be mas-

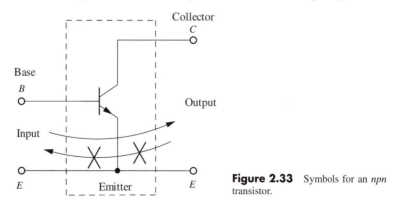

Figure 2.33 Symbols for an *npn* transistor.

tered because circuits can be designed (or analyzed) one part at a time, unlike the two- or three-loop circuits in circuit theory, which must be analyzed all at once.

Other transistor properties. Transistors can give signal gain, thus allowing small signals, such as a voltage induced in a radio antenna, to be amplified by stages until large enough to power a radio speaker. Transistors also can be used to couple circuits of greatly differing impedance levels, allowing more efficient transfer of signals between them. Transistors are used in digital circuits as electrically controlled switches. Finally, transistors offer a variety of nonlinear effects that are used in communication circuits for manipulating signals in the frequency domain. In this section, however, we limit our attention to the isolating, amplifying, and switching properties of the transistor.

**LEARNING
OBJECTIVE 3.**

**To understand the
semiconductor
processes in the
npn bipolar–
junction
transistor and
the resulting input
and output
characteristics**

Types of transistors. All transistors accomplish the purposes stated before. There are *npn* and *pnp* bipolar-junction transistors (BJTs) and there are *p*-channel and *n*-channel field-effect transistors (FETs). FETs can be either junction field-effect transistors (JFETs) or they can be metal-oxide semiconductor field-effect transistors (MOSFETs). In this section, we deal with the *npn* bipolar-junction transistor, which is symbolized in Fig. 2.33. In the next section, we deal with field-effect transistors.

BJT Characteristics

As shown in Fig. 2.33, the transistor is a three-terminal device that is connected with one terminal in common between the input and output circuits. The parts of the transistor are the emitter (*E*), the base (*B*), and the collector (*C*), each with appropriately labeled terminals. Our goal in this section is to describe the input and output characteristics of a typical *npn* transistor.

BJT transistor structure. Figure 2.34 shows a simple diagram of the physical structure of an *npn* transistor in the common-emitter configuration. The structure is that of a sandwich, with the *p* material like a very thin piece of bologna between two thick pieces of *n*-material bread. The external connections are made through wires that are bonded to the three regions of the transistor. Depletion regions form at the two *pn* junctions, and, in the absence of external applied voltages, the orderly and disorderly processes rapidly come to equilibrium at both junctions.

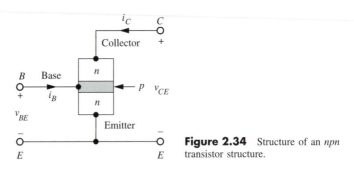

Figure 2.34 Structure of an *npn* transistor structure.

Transistor input characteristics. First we consider the effect of placing a voltage at the base–emitter (*B-E*) junction, the input part of the transistor. Ignoring for the present the effect of the collector–base junction, we note that the base–emitter junction forms a *pn*-junction diode. We conclude that the input *i*–*v* characteristic is like that of a

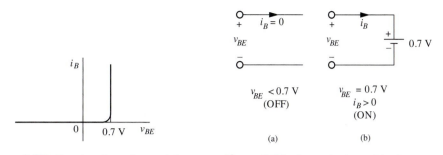

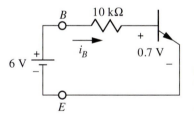

Figure 2.35 Transistor input characteristic.

Figure 2.36 Input circuit model of an *npn* transistor.

diode, as shown in Fig. 2.35. The base current is very small until sufficient voltage exists across the junction to turn it ON, about 0.7 V for a silicon transistor. Once the junction is turned ON, the base current increases rapidly, with the base–emitter voltage remaining constant at about 0.7 V. Therefore, we can model the base–emitter characteristic as either an open circuit (for $v_{BE} < 0.7$) or else a constant voltage of 0.7 V once the input voltage tries to go above that value. This model is shown in Fig. 2.36. In justifying this model, we ignored the state of the output circuit (v_{CE} and i_C); but as we stressed before, the output has negligible effect on the input.

EXAMPLE 2.4 | **Base current**

A 6-V battery in series with a 10-kΩ resistor is connected to the base circuit of an *npn* transistor, with the positive voltage connected to the base of the transistor. Find the base current.

Figure 2.37 The base–emitter circuit. We represent the base–emitter junction with a voltage of 0.7 V because it is ON for the input.

SOLUTION:

Figure 2.37 shows the base–emitter circuit. The *pn* junction between base and emitter will be ON, so the base–emitter voltage is approximately 0.7 V. Thus, KVL around the base–emitter loop is

$$-6 + i_B \times 10 \text{ k}\Omega + 0.7 = 0 \implies i_B = \frac{6 - 0.7}{10 \text{ k}\Omega} = 0.530 \text{ mA} \tag{2.19}$$

WHAT IF?   What if the required base current were 0.70 mA? Find the required base voltage if the 10-kΩ resistor is unchanged.[7]

---

[7] 7.7 V.

Output Characteristics

Effect of doping. Like the input characteristic, the output characteristics depend on whether the collector–base *pn* junction is forward-biased or reverse-biased. We assume that the input current to the base has been fixed at i_B, which requires approximately 0.7 V at the base region, considering the reference node to be the emitter. Now we increase the collector–emitter voltage, beginning at zero volts, and observe the collector current.

Consider first the holes and electrons in the base region. When we discussed the *pn* junction earlier in describing diode operation, we implied that the same density of holes exists in the *p* region as electrons in the *n* region. In this case, forward biasing the *pn* junction causes roughly as many holes to diffuse into the *n*-type material as electrons to diffuse into the *p*-type material. But if the *n*-type material were more heavily doped than the *p*-type material, the electron density would greatly exceed the hole density. For this case, a forward bias would produce many more electrons diffusing into the *p*-type material than holes diffusing into the *n*-type material. The diode would still work; however, the current would be carried across the junction largely by the electrons, and most of the recombinations would occur in the *p*-type material.

Emitter doping. When a transistor is made, the emitter is doped more heavily than the base; hence, for an *npn* transistor, the conditions are those described earlier—excess electrons diffuse into the base region. For zero volts between collector and emitter, the base–collector junction is also forward-biased. Hence, electrons also tend to diffuse into the base from the collector region. Although there is a buildup of excess electrons in the base region, the only current that flows is that permitted by recombination of electrons in the base: This current is the i_B assumed for the base–emitter bias circuit.

saturation region

Saturation region. If we now increase the collector–emitter voltage, we reinforce the orderly process forcing a drift current of electrons from the base to the collector. The result is a rapid buildup of collector current as excess electrons are permitted to pass into the collector region, where they flow through the collector–emitter circuit. This region of rapid increase in collector current is the *saturation region* labeled in Fig. 2.38.

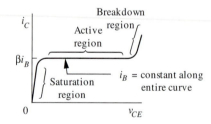

Figure 2.38 Output current–voltage characteristic with base current held constant.

active region

Active region. By the time the collector–base voltage is about 0.2 to 0.4 V, all the excess electrons in the base region are being drawn into the collector region and the curve levels off, resulting in the active region in Fig. 2.38. In the *active region,* the collector current is controlled by the number of electrons injected into the base region by the emitter. This is controlled by the base–emitter voltage, which is controlled in turn by the amount of base current the base circuit allows. Thus, in the active region, the collector current is controlled by the base current.

CHAPTER 2 SEMICONDUCTOR DEVICES AND CIRCUITS

Let us review the principal processes that occur in the active region. The external bias of about 0.7 V that is applied to the base–emitter junction diminishes the orderly process that might hold back the electrons in the heavily doped emitter region. These electrons diffuse into the base region, where a few, say, 1%, combine with holes and create the base current. The remaining 99% diffuse into the collector–base depletion region, where the orderly process of the uncovered bound charges forces them into the collector region. Thus, in the active region, the collector current is controlled by the base current.

alpha, beta, current gain

Current gain of the transistor. If we call α (alpha) the fraction of electrons that diffuse across the narrow base region, the fraction of electrons that recombine with holes in the base region to create the base current is $1 - \alpha$. The ratios of the base, collector, and emitter currents are thus

$$i_C = \alpha i_E \quad \text{and} \quad i_B = (1 - \alpha) i_E \qquad (\alpha < 1) \tag{2.20}$$

IDEA 1
Conservation of Charge

The value of α is fairly constant throughout the active region for a given transistor, and α characterizes the current gain of the device. Usually, the current gain is described in terms of the β (beta) of the transistor, defined as

$$\beta = \frac{i_C}{i_B} = \frac{\alpha}{1 - \alpha} \tag{2.21}$$

In terms of β, the ratios of the currents become

$$i_C = \beta i_B, \quad i_E = \frac{\beta + 1}{\beta} i_C = (\beta + 1) i_B \tag{2.22}$$

We stress that Eqs. (2.20) to (2.22) are valid only in the active region.

EXAMPLE 2.5 **Transistor currents**

A transistor has a β of 150. Find the collector and emitter currents if $i_B = 10 \, \mu A$.

SOLUTION:
Assuming the active region, we find from Eq. (2.22)

$$i_C = 150 \times 10 \, \mu A = 1.5 \, mA$$
$$i_E = (150 + 1) \times 10 \, \mu A = 1.51 \, mA \tag{2.23}$$

WHAT IF? What if the transistor is saturated?[8]

---

[8] We then know that $i_C < 1.5$ mA and $i_E < 1.51$ mA.

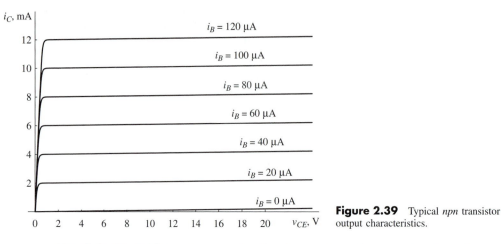

Figure 2.39 Typical *npn* transistor output characteristics.

Breakdown region. Figure 2.38 also shows a breakdown region where the current increases rapidly with increasing collector–emitter voltage. In this region, the power into the transistor becomes excessive and thermal failure often occurs.

Normally, the output characteristics are shown for many values of base current, as in Fig. 2.39. Here we have given typical characteristics for a transistor with a β of approximately 100 ($\alpha \approx 0.99$).

Transistor Amplifier-Switch Circuit Analysis

common-emitter connection

Problem statement. Figure 2.40 shows an important transistor circuit. This circuit uses an *npn* transistor in the *common–emitter connection*, meaning that the emitter of the transistor provides the common terminal between the input and output circuits. The circuit has an input voltage, v_{in}, and an output circuit with its output voltage, v_{out}. A dc voltage source, V_{CC}, in the output circuit supplies the energy required by the circuit, and two resistors, R_C and R_B, control the currents and voltages applied to the transistor. The transistor input characteristics are those of a silicon *pn* junction, Fig. 2.35, and the output characteristics are shown in Fig. 2.39. Our goal is to determine how the output voltage depends on the input voltage.

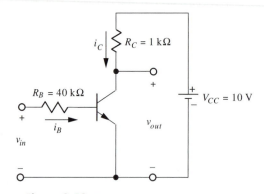

Figure 2.40 Transistor amplifier-switch circuit.

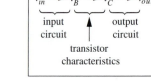

Figure 2.41 Causal relationships in the amplifier-switch circuit go from input to output.

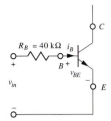

Figure 2.42 Input circuit.

Causality. Figure 2.41 shows the causal relationships for this circuit. The input voltage controls the base current through KVL in the input circuit. In the active region, the base current controls the collector current, as described by the transistor characteristics in Fig. 2.39. The collector current controls the output voltage, as determined by KVL in the output circuit. Our analysis proceeds therefore from input to output.

Input-circuit analysis. We begin at the input circuit, shown in Fig. 2.42. We first determine the base current, i_B, because this current controls the state of the transistor. Let us start with a negative value of v_{in} and increase this voltage to positive values. Comparing the input circuit with the model of the input characteristics in Fig. 2.36, we note that no current flows until the input voltage becomes at least $+0.7$ V because the *pn* junction is OFF. This is known as *cutoff*, for no current flows in the base or in the collector. When the input voltage exceeds $+0.7$ V, the base–emitter junction turns ON and the second model circuit in Fig. 2.36 applies. Because in this case v_{BE} is constant at 0.7 V, KVL requires the base current to be

$$-v_{in} + i_B R_B + 0.7 = 0 \Rightarrow i_B = \frac{v_{in} - 0.7}{R_B} \quad v_{in} > 0.7 \text{ V} \quad (2.24)$$

cutoff

This current is graphed in Fig. 2.43. This completes our analysis of the input circuit.

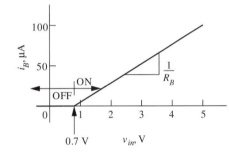

Figure 2.43 Once the base–emitter junction is ON, the base current is limited by the base resistor, R_B.

EXAMPLE 2.6	**Ideal transistor**

Consider the base–emitter *pn* junction an ideal diode. What changes in Eq. (2.24) are required under this assumption?

SOLUTION:
The 0.7 represents the threshold voltage of the base–emitter *pn* junction. For an ideal diode, the threshold voltage is zero. Thus, the formula becomes

$$i_B = \frac{v_{in}}{R_B} \quad v_{in} \geq 0 \text{ V} \quad (2.25)$$

WHAT IF? What if the base voltage is negative and so large in magnitude as to cause reverse breakdown of the junction?[9]

_____
[9] Good-bye transistor.

Output-circuit analysis. The output characteristic of the transistor is controlled by the input circuit. In the active region the base current controls the collector current, which in turn determines v_{out}.

Load-line analysis. The concept of a load line offers an excellent way to understand the interaction of the transistor with the rest of the output circuit. In Fig. 2.44, we have shown the output circuit broken at the transistor collector and emitter connections. When we stand at that break and look to the left, we see the i–v output characteristics of the transistor, which are given in Fig. 2.39. If we look to the right, we see the i–v characteristic of the external circuit, V_{CC} in series with R_C. Using KVL and Ohm's law, we find the collector circuit i–v characteristic to be

$$-V_{CC} + iR_C + v = 0 \implies i = \frac{V_{CC} - v}{R_C} \tag{2.26}$$

load line

When we plot current versus voltage for Eq. (2.26), Fig. 2.45, we observe a straight line with a voltage intercept of V_{CC}, a current intercept of V_{CC}/R_C, and a slope of $-1/R_C$. This line, known as the *load line*, represents the output circuit external to the transistor.

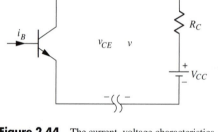

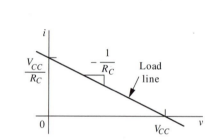

Figure 2.44 The current–voltage characteristics of the left and right parts of the output circuit may be considered separately.

Figure 2.45 The right half of the circuit is characterized by a load line.

Figure 2.46 repeats Fig. 2.39 with the load line drawn for our specific values of $V_{CC} = 10$ V and $R_C = 1$ kΩ. The mental break in the output circuit was made for the purpose of examining independently the characteristics of the two parts of the circuit. When we mentally reconnect the circuit, the two voltages must be the same and the two currents must be the same. These requirements determine the current and voltage in the output circuit for a prescribed value of i_B.

Input–output characteristic. Consider now that the base current is 50 µA. According to the transistor output characteristics, the transistor must operate on the line corresponding to that base current. On the other hand, the collector circuit must operate on the load line. Both requirements are satisfied at the intersection of the two lines. For 50-µA base current, the intersection occurs at a collector current of 5.3 mA and a collector–emitter voltage of 4.7 V, as indicated in Fig. 2.46. This result demonstrates the method for finding the output voltage from the input voltage. For each input voltage, we can determine the base current, Eq. (2.24) and Fig. 2.43, and for each value of base current, we can determine the output voltage from the intersection with the load line. We

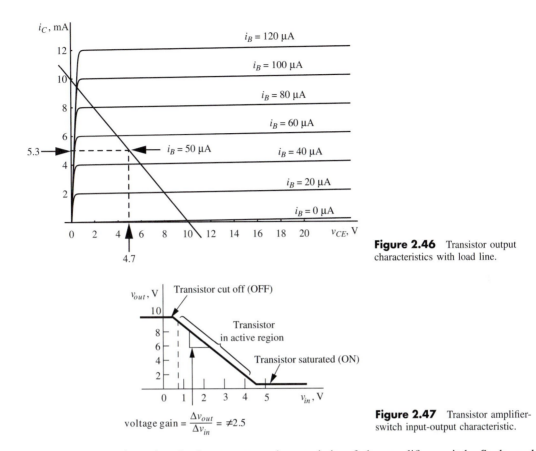

Figure 2.46 Transistor output characteristics with load line.

voltage gain $= \dfrac{\Delta v_{out}}{\Delta v_{in}} = \neq 2.5$

Figure 2.47 Transistor amplifier-switch input-output characteristic.

can thus plot the input–output characteristic of the amplifier–switch. Such a plot is shown in Fig. 2.47.

EXAMPLE 2.7 **Different base current**

If the base current is $30\,\mu\text{A}$, find the input and output voltages.

SOLUTION:
The input voltage is given by Eq. (2.24):

$$v_{in} = i_B R_B + 0.7 = 0.03\ \text{mA} \times 40\ \text{k}\Omega + 0.7 = 1.9\ \text{V} \qquad (2.27)$$

The output voltage is given by the intersection of the $i_B = 30\,\mu\text{A}$ transistor characteristic with the load line in Fig. 2.46, about $i_C = 3.1$ mA and $v_{CE} = v_{out} = 6.9$ V.

WHAT IF? What if you want the input voltage corresponding to an output of 2 V?[10]

---

[10] 3.74 V.

Cutoff. Several features of Fig. 2.47 merit close attention. When the input voltage is less than +0.7 V, no base current flows and hence no collector current flows. The transistor is cut off. This cutoff condition fixes the output voltage at V_{CC}, 10 V in this case. Because no current flows in R_C, the full power supply voltage must appear across the transistor. This is like turning OFF a valve in a water pipe; the full pressure must be supported by the valve in the absence of flow.

gain

Active region. As the input voltage increases beyond +0.7 V, the base current begins to flow and the transistor moves out of cutoff into the active region. In the active region, the transistor amplifies small changes in the input voltage. The incremental *gain* of the amplifier is the slope of the input–output characteristic in the active region. The characteristic in Fig. 2.43 shows an incremental voltage gain of

$$A_v = \frac{\Delta v_{out}}{\Delta v_{in}} = \frac{-\Delta i_C R_C}{\Delta i_B R_B} = -\beta \frac{R_C}{R_B} = \frac{-100 \times 1\,\text{k}\Omega}{40\,\text{k}\Omega} = -2.5 \qquad (2.28)$$

The minus sign of the gain signifies the inversion of the incremental changes. That is, a small positive *change* in the input voltage produces a larger negative *change* in the output voltage.

Saturation. As the input voltage continues to increase, the base current eventually saturates the transistor. This is like opening fully a valve in a water pipe; the valve relinquishes control of the flow rate to the capacity of the supply. With the transistor saturated, the output voltage remains small, 0.2 to 0.4 V, and the collector current remains at approximately $V_{CC}/R_C = 10$ mA, even though the base current continues to increase as v_{in} continues to increase. Thus in the saturation region the output voltage will remain small for increasing values of the input voltage.

EXAMPLE 2.8	**Saturation**

What is the base current required to saturate the transistor in Fig. 2.46?

SOLUTION:

The load line intersects the saturation region where the base current is approximately 95 μA.

WHAT IF?	What if the collector resistor, R_C, is changed from 1 to 2 kΩ? What is the base current required to saturate the transistor?[11]

Transistor Applications

ideal switch, real switch

Ideal and real switches. An *ideal switch* is either an open circuit (OFF) or a short circuit (ON). A *real switch*, like the wall switch for the lights, has a large resistance when OFF and a small resistance when ON. To function properly as a switch, the de-

---

[11] About 0.047 mA.

Impedance Level

vice must have a large OFF resistance compared with the impedance level of the load such that almost all the voltage appears across the switch and very little voltage appears across the load. Likewise, in the ON state, the switch must have a small resistance compared with the impedance level of the load so that almost all the voltage appears across the load with very little voltage across the switch.

The transistor as a switch. Thus, the transistor can function as an electronic switch in the circuit shown in Fig. 2.40 if its OFF resistance is much larger than its load resistor, R_C, and its ON resistance is much smaller than its load resistor.

If the input voltage is less than 0.7 V, the transistor is cut off, that is, the electronic switch is OFF. If R_C represented a light bulb, no current would flow through the bulb and it would not glow. If, for example, the input voltage exceeded about 4.0 V, the transistor would be saturated and our electronic switch would be ON. If R_C were a light bulb, it would glow. Thus, we can use the transistor as a voltage-controlled switch to turn the bulb on and off.

The importance of transistor switches. The switching action of the transistor is one of its most valuable properties. This is true because digital circuits—computers, calculators, digital watches, and digital instrumentation—utilize transistors in switching operation. Some of the important applications of transistors in digital circuits are explored in Chaps. 3 and 5.

large signal amplifier

Large-signal amplifiers. An amplifier is called a *large-signal amplifier* when the signal levels require the full transistor characteristics, from near cutoff to near saturation. Such amplifiers can furnish moderate amounts of power to transducers such as loudspeakers or control motors. The amplifier–switch circuit in Fig. 2.40 is limited to an output voltage of 10 V, peak to peak, but this limitation can be overcome by increasing the power supply voltage or the circuit complexity. As shown in Chap. 4, feedback techniques can be used to reduce distortion and generally improve the characteristics of large-signal amplifiers.

EXAMPLE 2.9 | **Large-signal amplifiers**

What is the maximum peak-to-peak output voltage (v_{CE}) and the corresponding peak-to-peak input voltage?

SOLUTION:
Figures 2.46 and 2.47 show the maximum output voltage to be 10 V when the transistor is cut off and about 0.3 V when saturated; hence, the peak-to-peak is 9.7 V. The corresponding range at the input is shown by Fig. 2.47 to be 0.7 to 4.06 V.

WHAT IF? | What if you compute the large-signal gain from these peak-to-peak values?[12]

---

[12] $A_V = -2.49$.

Small-Signal Amplifiers

stages of
amplification

The importance of small-signal amplifiers. Most amplifiers are small-signal amplifiers. Consider, for example, a radio that receives a signal of 10 mV and produces an output voltage of 10 V. Such a radio would require a voltage gain of 1000. This gain would be accomplished in *stages*, each transistor amplifier stage taking as its input the output of the previous stage. If each stage had a voltage gain of $\sqrt{10}$, six stages of amplification would provide the necessary gain. Of these, all but the last stage would be small-signal amplifiers.

dc bias

Converting the amplifier switch to a small-signal amplifier. A small-signal amplifier must have a *dc bias* circuit for placing the transistor in its amplifying region, a means for introducing the input signal, and a means for supplying its output signal to the next stage. The circuit shown in Fig. 2.48 is the basic amplifier in Fig. 2.40 with a voltage divider added to provide bias at the input and with coupling at input and output.

IDEA 6 Equivalent Circuits

We symbolized the power supply connection with the terminal marked $+V_{CC}$, which appears as an open circuit but actually is connected through a dc power supply to ground. The two resistors R_1 and R_2 comprise a voltage divider to supply dc current to the transistor base to bias the transistor into its amplifying region. At the input, we have a Thévenin equivalent circuit of the signal source; this represents the previous stage of the amplifier or the origin of the signal such as a microphone or a radio antenna. The load, R_L, represents the input impedance of the next stage of the amplifier.

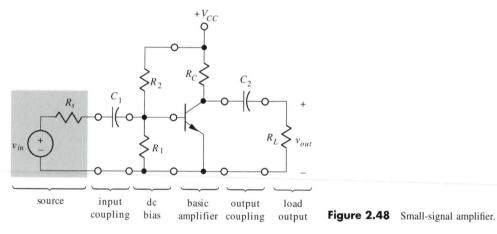

Figure 2.48 Small-signal amplifier.

coupling capacitors

Coupling capacitors. Capacitors C_1 and C_2 in Fig. 2.48 are coupling capacitors. Large capacitors block dc current but pass ac current. The infinite impedance of the capacitor at dc allows the dc state of each amplifier stage to be independent of the adjoining stages. If the impedance of the capacitors is small relative to the impedance level of the circuit, the time-varying signals will pass through the capacitors undiminished. Thus, the stages are isolated for dc but coupled for ac signals by the *coupling capacitors* at the input and output.

IDEA 5 Impedance Level

Bias voltage divider. A dc current is supplied to the transistor base by the voltage divider, R_1 and R_2. Because the coupling capacitors act as blocks to the dc current, the equivalent circuit at dc is as shown in Fig. 2.49 (a). Although there is only one power

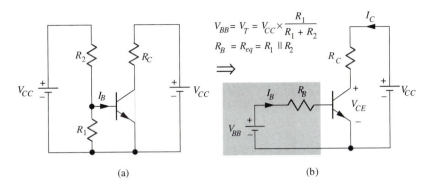

(a)   (b)

Figure 2.49 (a) DC bias circuit; (b) Thévenin equivalent circuit of the input portion of bias circuit.

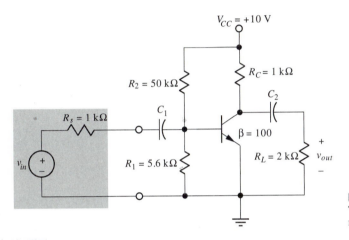

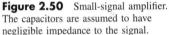

Figure 2.50 Small-signal amplifier. The capacitors are assumed to have negligible impedance to the signal.

supply, we have replaced the $+V_{CC}$ symbol with two voltages sources. We draw the circuit this way for two reasons: (1) to emphasize that the power supply acts independently on the bias and collector circuits, and (2) to help you recognize the voltage divider. The base circuit in Fig. 2.49 (a) can be reduced to the familiar circuit in Fig. 2.40 by converting the voltage divider to a Thévenin equivalent circuit, as shown in Fig. 2.49 (b). We used the symbol V_{BB} for the open-circuit voltage at the base of the transistor. This plays the role of v_{in} in Fig. 2.40 and Eq. (2.24).

Bias analysis. We continue this development for the specific circuit shown in Fig. 2.50. The open-circuit base dc bias voltage is

$$V_{BB} = 10 \times \frac{5.6 \text{ k}\Omega}{50 \text{ k}\Omega + 5.6 \text{ k}\Omega} = 1.0 \text{ V} \tag{2.29}$$

and the dc output impedance of the bias network, R_B, is

$$R_B = 5.6 \text{ k}\Omega \parallel 50 \text{ k}\Omega = 5.0 \text{ k}\Omega \tag{2.30}$$

Assuming a silicon transistor that requires a threshold voltage of $V_{BE} = 0.7$ V to turn ON the base–emitter junction, we can use the equivalent circuit in Fig. 2.36. Thus, the dc base current by Eq. (2.24) is

$$I_B = \frac{V_{BB} - 0.7}{R_B} = \frac{1.0 \text{ V} - 0.7 \text{ V}}{5.0 \text{ k}\Omega} = 60 \text{ μA} \tag{2.31}$$

Assuming the transistor to be in the active region, we calculate the collector dc current, I_C, to be

$$I_C = \beta I_B = 100 \times 60 \text{ μA} = 6.0 \text{ mA} \tag{2.32}$$

To confirm that the transistor is in the active region, not saturated, we must determine the dc collector–emitter voltage, V_{CE}. We could use a load line, but KVL around the collector–emitter loop in Fig. 2.49 (b) will do as well.

$$-V_{CC} + I_C R_C + V_{CE} = 0 \quad \Rightarrow \quad V_{CE} = 10 - (6.0 \text{ mA})(1 \text{ k}\Omega) = 4.0 \text{ V} \tag{2.33}$$

Thus, the transistor is operating near the middle of its active region. If there were no input signal, the transistor would have a steady voltage of +4.0 V across it, but the voltage across the 2-kΩ load resistor would remain zero because the dc voltage would be blocked by the output coupling capacitor, C_2.

EXAMPLE 2.10 **Changing R_B**

What if R_B were 2 kΩ but V_{BB} were the same? What would be V_{CE}?

SOLUTION:
Equation (2.24) is still valid:

$$I_B = \frac{1.0 \text{ V} - 0.7 \text{ V}}{2 \text{ k}\Omega} = 150 \text{ μA} \tag{2.34}$$

But Eq. (2.32) is no longer valid because 150 μA of base current saturates the transistor, Fig. 2.46. Thus $V_{CE} \approx 0.3$ V and $I_C = I_{C \text{ (sat)}} \approx 9.7$ mA.

WHAT IF?
What if V_{BB} and R_B have their original values but R_C is changed to 2 kΩ? Find V_{CE}[13]

---

small-signal component

DC and small signals. We now consider the effect of an input signal. In general, all voltages and currents in the circuit will have two components, a dc component and a time-varying component, which we call the *small-signal component*. For example, the base current is

$$i_B(t) = I_B + i_b(t) \tag{2.35}$$

where $I_B = 60$ μA and $i_b(t)$ is the small-signal component, with $i_b \ll I_B$.

---

[13] $V_{CE} = 0.3$ V, still saturated.

Equivalent Circuits

Small-signal equivalent circuit. What is the equivalent circuit that the signals "see," in the sense that Fig. 2.49 (a) is the circuit seen by the dc voltages and currents? Beginning at the input and moving toward the output, we now justify the circuit shown in Fig. 2.51 as the appropriate small-signal equivalent circuit. The input source and its resistance, R_s, are now coupled to the amplifier with a short circuit representing the negligible impedance of C_1 to the signal. The base bias resistor R_1 appears as expected, but R_2 now connects to the *signal ground*[14] because of the low impedance of the power supply to the signal. The power supply circuit, not shown but represented by the V_{CC} symbol, is connected to the circuit ground by a filter capacitor, as shown, for example, in Fig. 2.13. This filter capacitor not only filters the power supply output, but also provides a low-impedance path to ground for the signal.

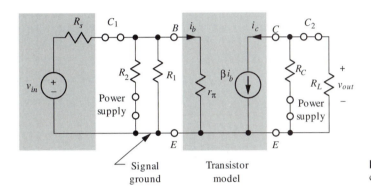

Signal ground   Transistor model

Figure 2.51   Small-signal equivalent circuit.

Transistor small-signal model. The input impedance of the transistor is represented by a resistor, r_π, and the output by a dependent current source. Representing the base–emitter by a constant 0.7 V, as shown in Fig. 2.36 (b), is adequate for the dc solution but is not accurate for the signal: r_π is a relatively low resistance in the range 300 to 1000 Ω that may be determined from theory,[15] from measured transistor input characteristics, or from published specifications of the transistor. The *dependent current source*,[16] $i_c = \beta i_b$, represents the current gain of the transistor.

In the output circuit, the collector bias resistor, R_C, is shown connected to the signal ground for the reason given before for R_2. Finally, the collector is connected to the load resistor, R_L, through the short circuit that represents the output coupling capacitor.

Analyzing the small-signal circuit. The small-signal equivalent circuit in Fig. 2.51 may be analyzed for the voltage gain, $A_v = v_{out}/v_{in}$, by the standard methods of circuit theory. Our analysis begins at the input with the calculation of i_b, the base signal current. From the base current, we can calculate the collector small-signal current, i_c, and then the small-signal output voltage.

---

[14] The "signal ground" is the reference node for the signals in the circuit. Here it is identical to the power "ground." These "grounds" may not correspond to actual earth ground in the circuit.

[15] See Problem 2.31.

[16] A dependent current source is a current source whose value depends on a voltage or current elsewhere in the circuit. Here we have a current-controlled current source because the collector current depends on the base current in the active region.

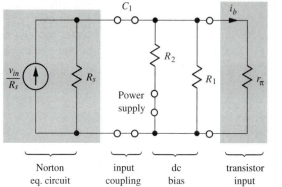

Figure 2.52 Analysis of the input circuit is simplified through use of a source transformation.

Norton eq. circuit   input coupling   dc bias   transistor input

The small-signal base current is easily calculated if the input source is converted to a Norton circuit, as shown in Fig. 2.52. We thus have a four-way current divider. The base current is the base voltage divided by r_π:

$$i_b = \frac{v_{in}}{R_s} \times \frac{R_s \| R_2 \| R_1 \| r_\pi}{r_\pi} \tag{2.36}$$

For the values in Fig. 2.50 and $r_\pi = 450\ \Omega$, Eq. (2.36) leads to $i_b = v_{in}/1.54\ \text{k}\Omega$.

The collector current is βi_b and the output voltage is thus

$$v_{out} = -\beta i_b(R_C \| R_L) = -100\, \frac{v_{in}}{1.54\ \text{k}\Omega}(1\ \text{k}\Omega \| 2\ \text{k}\Omega) = -43.3 v_{in} \tag{2.37}$$

Thus, the voltage gain of the amplifier is $A_v = v_{out}/v_{in} = -43.3$. The minus sign represents inversion of the signal.

EXAMPLE 2.11   **Small signal amplifier**

What value of R_L gives a gain of -30?

SOLUTION:
The only equation that changes is Eq. (2.37), which becomes

$$-30 v_{in} = -100\, \frac{v_{in}}{1.54\ \text{k}\Omega}(1\ \text{k}\Omega \| R_L') \tag{2.38}$$

where R_L' is the new load resistor. Equation (2.38) yields $R_L' = 858\ \Omega$.

WHAT IF?   What if r_π was 740 Ω, but R_L was the original value of 2 kΩ? Find the new voltage gain.[17]

---

[17] $A_v = -35.3$.

Summary. The calculation of the small-signal gain of the amplifier requires analysis of a dc equivalent circuit and a small-signal equivalent circuit. In the latter, the transistor is represented by its input resistance and a dependent current source, coupling capacitors by short circuits, and the power supply by a short circuit to signal ground. Because of the one-way property of the transistor, the analysis proceeds from input to output.

Check Your Understanding

1. For a transistor, the conditions in the output (*C–E*) circuit have little influence on the input (*B–E*) circuit. True or false?

2. What type of semiconductor material is the base region for an *npn* transistor?

3. Is the input (*B–E*) current–voltage characteristic of a transistor similar to that of a resistor, an ideal diode, a *pn*-junction diode, or an open circuit?

4. The load line intersects the current axis at the current that would flow if the transistor were replaced by a short circuit. True or false?

5. In the saturation region, does the transistor act as a switch that is ON or OFF?

6. To be used as an amplifier, the transistor must be biased into its cutoff, active, breakdown, or saturation region?

2. If the voltage gain of a transistor amplifier is negative, it means that the signal is diminished by the amplifier. True or false?

Answers. (1) True; (2) *p*-type; (3) a *pn*-junction diode; (4) true; (5) ON; (6) active; (7) false.

2.4 | FIELD-EFFECT TRANSISTORS

> **LEARNING OBJECTIVE 6.**
>
> **To understand FET operation and applications**

The bipolar-junction transistor (BJT) was invented in 1947 and developed to usable form in the late 1950s. The field-effect transistor (FET) was also invented in the late 1940s but was made practical by manufacturing developments in the 1970s. Today, virtually all digital electronic systems such as watches and computers use integrated circuits of FETs operating as electronic switches.

Like the BJT, the FET comes in two polarity types: The *n*-channel corresponds to the *npn* and the *p*-channel corresponds to the *pnp*. We limit our study to *n*-channel devices. Unlike the BJT, FETs come in two varieties: the junction FET (JFET) and the metal-oxide semiconductor FET (MOSFET). We consider the JFET in detail and then deal briefly with the MOSFET because its properties are similar.

Junction Field-Effect Transistors

The physical construction of an *n*-channel JFET is shown in Fig. 2.53(a). The transistor has three terminals: gate (*G*), drain (*D*), and source (*S*). The input signal is the voltage applied between gate and source, v_{GS}, and the output signal is the current from drain to source, i_D. The gate is connected to the *p*-type semiconductor, and the drain and source are connected to the channel, which is *n*-type material. For proper operation, the *pn* junction between the gate and channel regions must be reverse-biased, $v_{GS} < 0$, so that a depletion region is formed, as shown. The depletion region acts as a nonconductor, but the remainder of the channel acts as a conductor and thus forms a "resistor" with pecu-

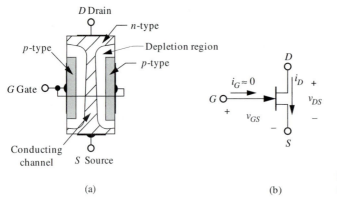

(a)

(b)

Figure 2.53 (a) JFET construction; (b) JFET circuit symbol.

liar properties. The width of the depletion region and thus the width of the conducting channel depends on the gate voltage. The drain is operated at a positive voltage relative to the source, and hence the gate–drain end of the *pn* junction is more strongly reverse-biased than the gate–source end. For this reason, we show the conducting channel to be more narrow near the drain. The circuit symbol for the JFET is shown in Fig. 2.53(b). Notice that the arrow points from *p* to *n*, as for a diode.

IDEA 5 **Impedance Level**

Input characteristics. Because the *pn* junction between gate and channel is reverse-biased, very little current flows into the gate. Thus, we show $i_G \approx 0$ in Fig. 2.53(b), and the drain current flows through the channel and out the source connection. The gate controls the current in the channel through an electric field[18] that affects the depletion region. Because very little current flows into the gate, the input impedance level of the device is extremely high, up to $10^{12}\,\Omega$, and very little energy is required to control the device.

ohmic region, pinchoff, saturation region

Output characteristics. Like the BJT, the FET has several regions of operation, which we may examine by fixing the input voltage, v_{GS}, and observing the output current as we vary the output voltage, v_{DS}. Figure 2.54(a) shows the circuit for the experiment, and Fig. 2.54(b) shows the results. Consider first the top curve, where $v_{GS} = 0$ V. For small values of drain-source voltage, the current increases as for a resistor. This is the *ohmic region*. However, as v_{DS} increases, the current begins to level off because the channel narrows at the drain end of the device. At the negative of the *pinchoff* voltage, $-V_P$, the conducting channel reaches a minimum size and the current becomes constant, independent of further increases in v_{DS}. This is the *saturation region*.[19] In the saturation region, the drain current is controlled totally by the input signal, and thus the saturation region for the FET corresponds to the active region for the BJT. Off the graph, to the right, is a breakdown region, where the current increases rapidly.

For smaller values of v_{GS} (negative values), currents fall below the curve corresponding to $v_{GS} = 0$, and the device changes from the ohmic region to the saturation re-

---

[18] Think of an electric field as something to move the carriers.

[19] Warning: This "saturation region" is totally different from the saturation region for the BJT. Using the same term for both is unfortunate and confusing, but customary.

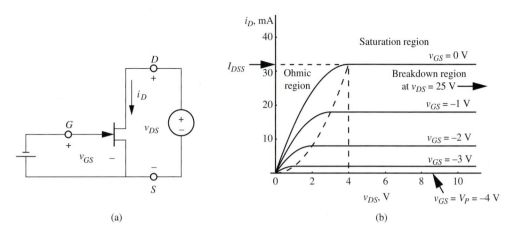

(a)   (b)

Figure 2.54 (a) Circuit for determining the output characteristic; (b) output characteristics for $V_P = -4$ V and $I_{DSS} = 32$ mA.

gion at lower values of v_{DS}. For $v_{GS} < V_P$, the FET is cut off, meaning that no current flows, regardless of v_{DS}.

Theoretical model. In the ohmic region, the drain current is given by

$$i_D = \frac{2I_{DSS}}{V_P^2}\left[(v_{GS} - V_P)v_{DS} - \frac{v_{DS}^2}{2}\right], \tag{2.39}$$

$$0 < v_{DS} < (v_{GS} - V_P),\ v_{GS} > V_P$$

where I_{DDS} is the saturation current for $v_{GS} = 0$, as indicated in Fig. 2.54(b). Equation (2.39) describes a parabola passing through the origin and tangent to the point of pinchoff. After pinchoff, the current is constant at the value

$$i_D = \frac{I_{DSS}}{V_P^2}(v_{GS} - V_P)^2,\quad v_{GS} > V_P,\quad v_{DS} > v_{GS} - V_P \tag{2.40}$$

EXAMPLE 2.12 **FET characteristics**

What value of v_{GS} corresponds to $i_D = 15$ mA and $v_{DS} = 6$ V?

SOLUTION:
This falls in the saturation region. Equation (2.40) yields

$$15 = \frac{32}{(-4)^2}[v_{GS} - (-4)]^2 \Rightarrow v_{GS} = -1.26\ \text{V} \tag{2.41}$$

WHAT IF? What if you are in the ohmic region with $i_D = 15$ mA and $v_{DS} = 1.5$ V?[20]

---

[20] $v_{GS} = -0.75$ V.

Summary. The JFET has four regions of operation. In the cutoff region, no current flows. In the ohmic region, the device behaves like a nonlinear resistor, with the resistance controlled mainly by the gate–source voltage. In the saturation region, the device behaves as a current source, with the current controlled by the gate–source voltage. Finally, in the breakdown region, the drain current increases rapidly. With the exception of the breakdown region, which is to be avoided, all the regions are useful.

Determining the region of operation. The region of operation for a JFET can be determined by the simple rules given in Table 2.1. We consider that the transistor has two ends: a gate–source end and a gate–drain end. The ends can be ON or OFF, depending on the voltage between the gate and the source or drain. For example, in the device shown in Fig. 2.54(b), the pinchoff voltage is -4 V. Therefore, if the gate–source voltage is greater than -4 V, the gate–source end of the transistor will be ON, and if the gate–source voltage is less than (more negative than) -4 V, the gate–source end of the transistor will be OFF. Likewise, if the gate–drain voltage is greater than -4 V, the gate–drain end of the transistor will be ON; otherwise, it is OFF. With these definitions of ON and OFF, the transistor has the four states shown in Table 2.1.

TABLE 2.1   Determination of Transistor Operating Region

Gate-Source End	Gate-Drain End	Condition
OFF	OFF	Cutoff region
OFF	ON	Reverse saturation
ON	OFF	Saturation region
ON	ON	Ohmic region

We have already discussed rows 1, 3, and 4 in Table 2.1. Row 2 deals with a condition not previously discussed. For the gate–source end to be OFF and the gate–drain end to be ON, the drain voltage must be negative relative to the source voltage; and hence source and drain exchange roles. In this case, the current will flow from source to drain if we keep the same labels. This condition is useful also but will not be discussed further.

EXAMPLE 2.13 **Finding the region of operation**

An n-channel JFET has a pinchoff voltage of -3 V. The gate–source voltage is -1 V and the drain–source voltage is $+2$ V. What is the region of operation?

SOLUTION:

The gate–source voltage is greater than the pinchoff voltage, so the gate–source end is ON. The gate-drain voltage may be determined from KVL and the law of subscripts:

$$v_{GD} = v_{GS} + v_{SD} = v_{GS} - v_{DS} = -1 - (+2) = -3 \text{ V} \tag{2.42}$$

This is equal to the pinchoff voltage, and hence the gate–drain end of the transistor is borderline between ON and OFF. Therefore the transistor is on the boundary between the ohmic region (row 4) and saturation (row 3).

JFET Applications

Variable resistance. In the ohmic region, the JFET acts as a resistor between drain and source whose resistance is controlled by v_{GS}. For small signals of v_{DS}, the resistance, r_{DS}, can be determined from Eq. (2.39):

$$r_{DS} \approx \frac{1}{\partial i_D / \partial v_{DS}} = \frac{V_p^2}{2 I_{DSS} (v_{GS} - V_P)}, \qquad v_{DS} \ll v_{GS} - V_P \qquad (2.43)$$

Hence, the resistance is smallest for large v_{GS}.

EXAMPLE 2.14 | **FET resistance**

Find the small-signal resistance of the transistor in Fig. 2.54(b) for $v_{GS} = 0$.

SOLUTION:
From Eq. (2.43),

$$r_{DS} \approx \frac{(-4)^2}{2(32 \times 10^{-3})[0 - (-4)]} = 62.5 \ \Omega \qquad (2.44)$$

WHAT IF? What if you want $r_{DS} = 1000 \ \Omega$?[22]

Small-signal amplifier. Like the BJT, the JFET may be used as a small-signal amplifier. In this application, the JFET is biased into its saturation region, and small changes in v_{GS} produce changes in the drain current that, passed through a resistor in the drain circuit, amplify the signal. To represent the JFET for this application, we require a small-signal model for the transistor, Fig. 2.55. The voltage and current variables are signal components and must remain much smaller than the dc bias voltages and current. The gate circuit is shown as an open circuit because the gate draws no current. The drain–source circuit is shown as a voltage-controlled current source.[23] The *transconductance*, g_m, may be derived from Eq. (2.40):

transconductance, voltage-controlled current source

---

[21] Both ends are OFF. The transistor is cut off and $i_D = 0$.

[22] $v_{GS} = -3.75$ V.

[23] A *voltage-controlled current source* is a current source whose current depends on a voltage elsewhere in the circuit. The constant relating the signal current in the drain to the signal voltage between gate and source has units of siemens, or mhos (reciprocal ohms).

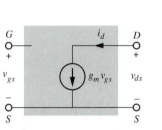

Figure 2.55 Small-signal model for the JFET. The voltages and current are signal components of the total voltages and current. The gain-related parameter g_m is the transconductance.

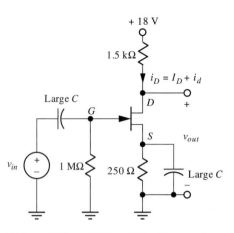

Figure 2.56 A JFET amplifier. The capacitors block dc but act as short circuits for the signal.

$$g_m = \frac{\partial i_D}{\partial v_{GS}} = \frac{2I_{DSS}}{V_P^2}(v_{GS} - V_P)\big|_{v_{GS} = V_{GS}} \tag{2.45}$$

where V_{GS} is the dc bias voltage between gate and source. The use of Eq. (2.45) and the small-signal equivalent circuit are illustrated in the following analysis.

JFET-amplifier bias analysis. We will determine the small-signal gain of the amplifier shown in Fig. 2.56. The transistor has the output characteristics shown in Fig. 2.54(b). Our first task is to determine the dc conditions in the circuit because the transconductance, g_m, depends on the dc conditions. To establish the bias condition, we must solve simultaneously for the dc gate–source voltage and the drain current. The source is not grounded but is biased to a positive voltage by the current passing through the 250-Ω resistor. The gate is grounded through the 1-MΩ resistor, but because no current flows in this resistor, the gate is at zero volts. Consequently, the gate–source voltage is

$$v_{GS} = v_G - v_S = 0 - 0.250 i_D \tag{2.46}$$

where current is expressed in milliamperes and the reference node is the signal ground. We assume that the JFET is saturated and hence we obtain another equation relating the gate-source voltage and the drain current from Eq. (2.40).

$$i_D = \frac{I_{DSS}}{V_P^2}(v_{GS} - V_P)^2 = \frac{32}{(-4)^2}[v_{GS} - (-4)]^2 \tag{2.47}$$

When we eliminate v_{GS} between Eqs. (2.46) and (2.47), we obtain a quadratic equation for the drain current. The two solutions are $i_D = 8.0$ and 32.0 mA. The second value is unrealistic for several reasons.[24] The $I_D = 8.0$ mA can be accepted tentatively, but we

---

[24] See Problem 2.44.

must verify our assumption that the transistor is in the saturation region. With a drain current of 8 mA, the source voltage is $+2$ V and hence $v_{GS} = -2$ V. The gate–source end of the transistor is ON, as required for saturation. The drain voltage can be obtained from KVL as $+18$ V $-$ (8 mA $\times$ 1.5 kΩ) $= +6$ V. Hence, the gate–drain voltage is -6 V and the gate–drain end of the transistor is OFF. Therefore, the transistor is operating in row 3 of Table 2.1, in the saturation region as assumed.

Small-signal analysis. Figure 2.57 shows the small-signal equivalent circuit for the amplifier in Fig. 2.56. Because the capacitors have a small impedance to the signal, we replaced them with short circuits. The power supply also has a small impedance to the signal, and hence we have connected one end of the 1.5-kΩ resistor to the signal ground. The transistor had been replaced by its equivalent circuit from Fig. 2.55. The transconductance of 8×10^{-3} mhos has been determined from Eq. (2.45) for the dc operating conditions established before.

The analysis of the circuit proceeds from input to output. The gate–source signal voltage is equal to the input voltage because the source is connected to the signal ground. Thus, the signal current in the drain–source circuit is $(8 \times 10^{-3}) v_{in}$. This current flows upward through the 1.5-kΩ resistor, and hence the output voltage is

$$v_{out} = -(8 \times 10^{-3}) v_{in} \times 1.5 \times 10^3 = -12.0 v_{in} \tag{2.48}$$

Thus, the voltage gain of the amplifier is -12.0. The minus sign indicates inversion of the signal.

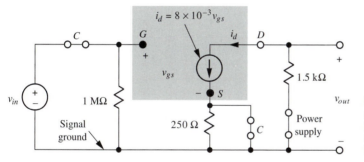

Figure 2.57 Small-signal equivalent circuit. The transistor has been replaced by its small-signal model, and the capacitors and the power supply have been replaced by short circuits.

EXAMPLE 2.15 | **Grounding the source**

What is the drain current if the source is grounded in Fig. 2.56?

SOLUTION:
Because $v_G = 0$ and $v_S = 0$, then $v_{GS} = 0$. To find i_D, we must assume either the ohmic region, Eq. (2.39), or the saturation region, Eq. (2.40). Because the latter is simpler, we will see if it produces consistent results.

$$i_D = \frac{32}{(-4)^2} [0 - (-4)]^2 = 32 \text{ mA} \tag{2.49}$$

If this is the drain current, then the drain voltage is

$$v_D = 18 \text{V} - (32 \text{ mA} \times 1.5 \text{ k}\Omega) = -30 \text{ V} \tag{2.50}$$

This would turn ON the gate–drain junction and thus the transistor cannot be in the saturation region; hence, Eq. (2.50) cannot be valid. This inconsistency shows that the transistor must be in the ohmic region.

WHAT IF?   What if you try it with Eq. (2.39)? What is i_D? [25]

Impedance Level

IDEA 5

JFET application as a switch.   Figure 2.58 (a) shows a circuit that uses a JFET as a voltage-controlled switch. Figure 2.58(b) shows the load line. With v_{GS} less than -4 V, the transistor is cut off and the output voltage is $+10$ V. With $v_{GS} = 0$ V, the transistor is in the ohmic region and, as shown in Example 2.14 on FET resistance, has a resistance of about 63 Ω. Thus, the FET switch must operate in circuits with high-impedance levels. In series with 2 kΩ, this drops the output voltage to about 0.3 V. Thus, the JFET can act as a voltage-controlled switch.

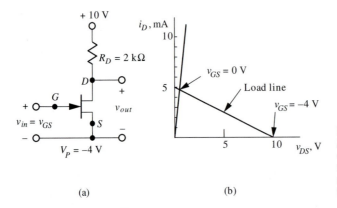

(a)   (b)

Figure 2.58   (a) In this circuit, the JFET can operate as a switch; (b) by changing the input voltage from -4 to 0 V, the JFET can be changed from OFF to ON and the output voltage changed from 10 V to near zero.

Summary.   In this section, we investigated the physical construction, input and output characteristics, and equivalent circuits appropriate to the various regions of operation of the n-channel JFET. We investigated its applications as a switch, a voltage-controlled resistor, and a small-signal amplifier. In the next section, we introduce briefly a class of devices very similar to the JFETs.

Metal-Oxide Semiconductor Field-Effect Transistors (MOSFETs)

Depletion-mode MOSFET.   Figure 2.59(a) shows the physical structure of an n-channel depletion-mode MOSFET. The substrate of p material is maintained at a voltage equal to or less than any voltage anticipated at either source or drain so that a depletion region forms at the pn junction and isolates the channel from the substrate. Normally, the drain is operated positive relative to the source, and hence the substrate can be connected to the source to accomplish this purpose. On the circuit symbol, Fig. 2.59(b), the arrow in the substrate connection points from p to n.

---
[25] 6.98 mA.

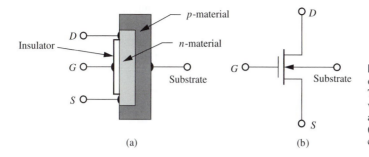
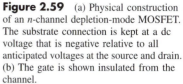

Figure 2.59 (a) Physical construction of an *n*-channel depletion-mode MOSFET. The substrate connection is kept at a dc voltage that is negative relative to all anticipated voltages at the source and drain. (b) The gate is shown insulated from the channel.

threshold voltage (MOSFET)

The gate is separated from the channel by an insulating layer and hence no current flows into the gate. If the gate has a large negative voltage relative to the channel, the carrier electrons are driven from the channel, which becomes a nonconductor and the channel is cut off. Like the JFET, there exists a critical voltage to cut off the channel. This is called the *threshold voltage*, V_T, and has the same effect as the pinchoff voltage for the JFET. Indeed, the depletion-mode MOSFET has similar characteristics to the JFET, except that the gate–source voltage can be positive. It has regions of cutoff, saturation, and ohmic operation, as described by Table 2.1. For example, if a depletion-mode MOSFET has a threshold voltage of -5 V, the gate–source voltage is -2 V, and the drain–source voltage is $+2$ V, then both ends of the device are ON (voltage is greater than the threshold) and hence operation falls in the ohmic region (row 4 of Table 2.1).

The depletion-mode MOSFET differs from the JFET in its physical construction and operating principles, but otherwise it functions similarly. Indeed, the equations describing operation are identical to Eqs. (2.39) and (2.40) except for a change in notation and the permitting of positive gate–source voltage. It can operate as a voltage-controlled resistor, small-signal amplifier, or switch.

Enhancement-mode MOSFET.
Figure 2.60(a) shows the physical structure of an enhancement-mode MOSFET, which differs from the depletion-mode device by having no *n*-type channel. For proper operation, the substrate must be connected to a voltage more negative than any voltage anticipated at either source or drain.

Both the drain-substrate and source–substrate *pn* junctions are reverse biased, and no channel exists until external voltage is applied to the gate. Thus, the device is cut off if the gate-source voltage is less than a positive threshold voltage. When the gate-source voltage exceeds the threshold voltage (V_T), the resulting electric field attracts electrons and repels holes in the *p*-type substrate, and a channel forms. The circuit symbol in Fig. 2.60(b) shows a broken line for a "channel" to indicate that a channel does not ex-

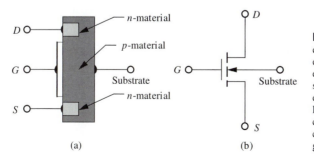

Figure 2.60 (a) Physical structure of an enhancement-mode MOSFET. No channel exists until the gate-source voltage is large enough to create a channel out of the *p*-type substrate. (b) The circuit symbol for an enhancement-mode MOSFET. The broken line for the "channel" indicates that a channel does not exist naturally but must be created by the electric field between the gate and source.

ist naturally. As usual, the arrow on the substrate connection points from p to n, where n refers to the n channel formed by enhancement. The depletion- and enhancement-mode devices have identical characteristics except V_T is negative for the former and positive for the latter. Equations describing operation in the ohmic and saturation regions of both types of MOSFETs are identical to those of the JFET except for a change in notation. In the ohmic region, the drain current is given by

$$i_D = \frac{K}{V_T^2}\left[(v_{GS} - V_T)v_{DS} - \frac{v^2{}_{DS}}{2}\right], \qquad v_{GS} > V_T, \qquad v_{DS} < v_{GS} - V_T \tag{2.51}$$

where K is a constant, and in the saturation region, we have

$$i_D = \frac{K}{2V_T^2}(v_{GS} - V_T^2) \qquad v_{DS} > v_{GS} - V_T \tag{2.52}$$

For $K = 16$ mA and $V_T = +4$ V (enhancement mode), we have the characteristics shown in Fig. 2.61.

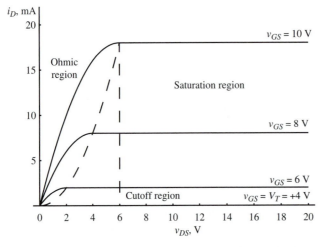

Figure 2.61 Device characteristics for an enhancement-mode n-channel MOSFET for $K = 16$ mA and $V_T = +4$ V.

Cutoff, ohmic, and saturation regions. In the cutoff region, the drain–source circuit acts as an open circuit or a switch that is OFF. In the ohmic region, the enhancement-mode MOSFET acts as a voltage-controlled resistor, and in the saturation region it acts as a voltage-controlled current source. The region of operation may be determined from Table 2.1. For example, consider the top curve given in Fig. 2.61. With $v_{GS} = +10$ V, the source end of the transistor is ON because the gate–source voltage exceeds the threshold voltage of $+4$ V. If the gate-drain voltage is less than $+4$ V, then the gate–drain end of the transistor will be OFF and the transistor will be in the saturation region. This occurs for $v_{DS} > 10 - 4 = 6$ V. Hence, the transistor will be in the saturation region for drain–source voltages greater than 6 V and in the ohmic region for voltages less than 6 V.

Circuit models for operation in the ohmic and saturation regions. In the ohmic region, the transistor may be modeled as a voltage-controlled resistance. We may determine the transistor resistance in the ohmic region from Eq. (2.51) in the manner used to derive Eq. (2.43). For small-signal operation in the saturation region, the circuit model shown in Fig. 2.55 is valid. We may determine the transconductance for the model from Eq. (2.52) in the same manner as Eq. (2.45) was derived.

MOSFET resistance

Find the resistance of the MOSFET in Fig. 2.61 for small v_{DS} and $v_{GS} = 10$ V.

SOLUTION:

From Eq. (2.51),

$$r_{DS} \approx \frac{1}{\partial i_D / \partial v_{DS}} = \frac{V_T^2}{K(v_{GS} - V_T)} = \frac{(+4)^2}{0.016(10-4)} = 167 \ \Omega \qquad (2.53)$$

WHAT IF? What if a resistance of $100 \ \Omega$ were required?[26]

Summary. The JFET and both types of MOSFETs have similar characteristics. The main difference between them is that the enhancement-mode MOSFET has a positive threshold voltage and the JFET and the depletion-mode MOSFETs have negative threshold voltages (called the pinchoff voltage for the JFET). Operation in the ohmic or saturation region may be determined in all cases from Table 2.1. These FETs can function as switches, voltage-controlled resistors, or small-signal amplifiers.

Circuit designers also have available *p*-channel JFETs and MOSFETs. These devices have identical characteristics except that all polarities of voltages and currents are reversed.

Check Your Understanding

1. Why is the channel shown as a dashed line for the enhancement-mode MOSFET, yet shown as a solid line for the depletion-mode device?

2. For a JFET with $V_P = -5$ V and $v_{GS} = -3$ V, what range of drain–source voltage corresponds to operation in the ohmic region? Assume $v_{DS} > 0$ V.

3. For the device described in the previous problem and $I_{DS} = 20$ mA, determine the resistance of the device for small values of drain–source voltage.

4. An *n*-channel MOSFET has a threshold voltage of -3 V. Is this a depletion- or enhancement-mode MOSFET?

5. For the device described in the previous problem, assume the gate–source voltage is -1.5 V. What range of drain–source voltages corresponds to operation in the saturation region (positive voltages only)?

Answers. (**1**) Because the channel exists naturally for the depletion-mode device but must be created by the gate voltage for the enhancement-mode device; (**2**) $0 < v_{DS} < +2$ V; (**3**) $313 \ \Omega$; (**4**) depletion mode; (**5**) $v_{DS} > 1.5$ V.

---

[26] $v_{GS} = 14$ V.

CHAPTER SUMMARY

Electronics applies the circuit theory of nonlinear devices. We begin with the ideal diode and its applications in rectifier circuits. Semiconductor phenomena are introduced to describe the *pn* junction for understanding diodes, bipolar-junction transistors, and the various types of field-effect transistors. One-stage transistor circuits are analyzed to show the isolation, switching, and amplifying properties of transistors.

Objective 1: To understand how to analyze half- and full-wave rectifier circuits, with and without a filter capacitor. After introducing the role of power supplies in electronic circuits, we study several rectifier configurations. The advantages of a filter capacitor are shown.

Objective 2: To understand the physical processes and the *i–v* characteristic of a *pn* junction. The variety of semiconductor processes at work in the *pn* junction are described, and the characteristic of the ideal *pn* junction is presented.

Objective 3: To understand the semiconductor processes in the *npn* bipolar-junction transistor and the resulting input and output characteristics. The roles of the two *pn* junctions of the BJT are discussed. The input characteristic is that of a *pn*-junction diode. The output characteristic depends on the state of the input circuit and includes cutoff, saturation, breakdown, and active regions.

Objective 4: To understand how to analyze a common-emitter amplifier-switch circuit. We derive the input–output characteristic of the common-emitter circuit. The effects of cutoff, saturation, and active regions are noted, and the amplifier gain in the active region is shown graphically and determined analytically.

Objective 5: To understand how to analyze a common-emitter small-signal amplifier for operating point and gain. The dc bias and output circuits are added to the basic common-emitter circuit to give a small-signal amplifier. The bias conditions are analyzed. The small-signal equivalent circuit is developed, including the equivalent circuit for the transistor, and analyzed for small-signal gain.

Objective 6: To understand FET operation and applications. Junction and MOS field-effect transistors are described. Large- and small-signal circuits are analyzed.

Chapter 3 shows how transistor circuits in the switching mode are used to store and process information in digital form. Chapters 4-7 show the roles of amplifier gain in analog and communication electronics.

GLOSSARY

Active region, p. 92, the region where the collector current of a BJT is controlled by the base current. Also called the amplifying region.

Carriers, p. 81, holes and electrons that are free to move in a conductor or semiconductor.

Common-emitter connection, p. 94, a transistor connection in which the emitter of the transistor provides the common terminal between the input and output circuits, frequently used in both switching and amplifying applications.

Conductor, p. 80, a material that has at least one excess valence electron that is not involved in the bonding. Such excess electrons are known as conduction electrons and they move freely in response to electrical forces.

Coupling capacitors, p. 100, capacitors that couple signals in and out of transistor amplifiers and hence isolate dc conditions in adjoining stages of amplification.

Current gain, p. 93, the current gain of a BJT, the β (beta) of the transistor, is defined as the collector current divided by the base current, which is roughly constant in the active region.

Cutoff, p. 95, the transistor state where no current flows in the output circuit; thus the transistor acts as an open switch.

DC bias circuit, p. 100, a circuit for placing the transistor in its amplifying region.

Dependent current source, p. 103, a current source that depends on a voltage or current elsewhere in the circuit.

Depletion region, p. 84, a region on both sides of a *pn* junction that has a deficiency of carriers due to recombinations.

Diffusion current, p. 83, a current formed by the random motion of carriers from a region of high concentration to a region of low concentration.

Drift current, p. 82, a current formed by movement of carriers responding to an electric field.

Filter, p. 75, a circuit used to remove an undesirable signal.

Full-wave rectifier, p. 72, a circuit that inverts or rectifies the negative half of the input voltage and corrects both halves to the load.

Half-wave rectifier, p. 71, a circuit that couples either the positive half or the negative half of the input voltage to the load.

Heat sink, p. 87, a heat exchanger used to improve cooling of an electronic device.

Hole, p. 80, when a valence electron leaves its bonding position, it leaves behind a vacant position, called a hole, which acts as a positive charge.

Ideal switch, p. 98, a device that acts as an open circuit (OFF) or a short circuit (ON).

Incremental gain, p. 98, small changes in the output voltage or current divided by small changes in the input voltage or current.

Insulator, p. 80, a material with all the valence electrons involved in the bonding.

Intrinsic semiconductor, p. 82, a pure, undoped semiconductor.

Large-signal amplifier, p. 99, an amplifier in which the signal levels require the full transistor characteristics, from near cutoff to near saturation.

Load line, p. 96, the constraint presented to the transistor by the output circuit, usually represented as a line superimposed on the output characteristics of the transistor.

***n*-type semiconductor**, p. 82, a semiconductor doped to have an excess of electronic carriers.

Ohmic region, p. 106, the region where an FET behaves as a voltage-controlled resistance.

***p*-type semiconductor**, p. 82, a semiconductor doped to have an excess of holes as carriers.

Pinchoff voltage, p. 106, the voltage that cuts off the channel of a JFET.

pn junction, p. 83, the boundary between regions of *p*-type semiconductor and *n*-type semiconductor.

Power supply, p. 71, an electronic circuit that converts ac power to dc power.

Real switch, p. 98, a device that has an OFF resistance that is large compared with the impedance level of the load and an ON resistance that is small compared with the impedance level of the load.

Reverse-breakdown voltage, p. 87, the voltage in a reversed-biased *pn* junction at which a rapid buildup of current occurs, usually leading to failure of the device.

Saturation region (BJT), p. 92, for a BJT, the region in which the transistor is ON and the current is controlled by the collector circuit and not the base current.

Saturation region (FET), p. 106, the region where an FET behaves as a voltage controlled current source. This is the active region of the FET and does not correspond to the saturation region of the BJT.

Semiconductor, p. 80, an insulator at absolute zero temperature that becomes a conductor at normal temperatures due to the electrons that have escaped their bonding positions.

Signal ground, p. 103, the reference node for the signals in the circuit. It may or may not be identical to the power "ground." In electronics these "grounds" may not correspond to actual earth ground.

Signals, p. 68, patterns of electrical voltage or current that convey information.

Small signals, p. 102, information-bearing variations in voltage and current that are much smaller than the dc voltage or current.

Threshold voltage, p. 86, a critical voltage in a semiconductor device, usually the voltage at which the magnitude of the current rises rapidly. For a silicon diode, the threshold voltage for forward bias is about 0.7 V.

Threshold voltage, p. 113, the voltage that separates the cutoff region for a MOSFET from the conducting regions.

Transconductance, p. 109, for an FET, changes in the output current divided by changes in the input voltage, a measure of the gain of the device.

Voltage equivalent of temperature, p. 86, a measure of the thermal energy of carriers, expressed in electrical units. The voltage equivalent of temperature has a value of 25.9 mV at 300 K.

PROBLEMS

Section 2.1: Rectifiers and Power Supplies

2.1. Figure P2.1 shows a sinusoidal voltage source, a diode, and a load resistor, *R*.

(a) A dc voltmeter measures 12 V when connected between *A* and *G*. What would an ac voltmeter measure between *B* and *G*? Assume an ideal diode.

(b) If a very large capacitor were connected between *A* and *G*, what then would the dc meter read? Assume that *RC* >> *T*, where *T* is the period of the sinusoid.

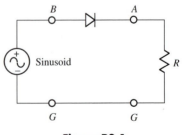

Figure P2.1

2.2. Using the unfiltered half-wave rectifier circuit shown in Fig. 2.4, design a power supply to produce 50 mA of dc current into a 600-Ω load. Give the circuit and the value that an ac voltmeter would indicate for the input voltage.

2.3. **(a)** Derive the equations for the dc voltage and ripple voltage for the half-wave rectifier shown in Fig. 2.4 with a capacitor across the load to reduce the ripple. Assume that the time constant of the *RC* part of the circuit is large compared with the period of the input sinusoid. Have *f* continue to represent the frequency of the input.

(b) Using your results from part (a), design a 60-Hz filtered power supply to provide 60 V dc, 40 mA dc to a load, with a ripple voltage of 0.3 V. You must specify the rms value of the input voltage and the value of the capacitor to be used. Assume ideal diodes. *Hint:* The dc voltage and current define an equivalent resistance for the load.

2.4. Draw the circuit of a 60-Hz filtered half-wave rectifier that will deliver 30 V dc and 270 mA dc to a resistive load. Specify the value of all components, including the equivalent load resistor. Assume an ideal diode. The peak-to-peak ripple voltage should not exceed 0.5 V. Also specify the rms input voltage.

2.5. **(a)** Draw the circuit of an unfiltered half-wave rectifier, with a generator, an ideal diode, and a 10-kΩ load.

(b) Consider now that the generator puts out the square wave shown in Fig. P2.5. What would be the dc current in the load?

(c) Now add a 10-μF capacitor across the load. In this case, what would be the current in the load?

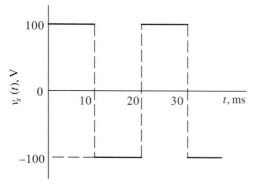

Figure P2.5

2.6. A simple rectifier circuit has the unsymmetric square waveform for an input as shown in Fig. P2.6. What is the average load voltage, with the polarity markings shown? Assume an ideal diode.

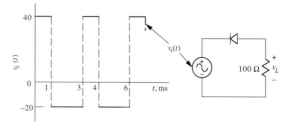

Figure P2.6

2.7. For the circuit shown in Fig. P2.7, the diode is an ideal diode.

(a) What fraction of the time does current flow in the load?

(b) Find the dc current through the load.

(c) If a 600-μF capacitor is added across the load to improve performance, find the dc current in that case.

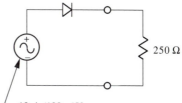

$12 \sin(120\pi t)$V

Figure P2.7

2.8. The load for the rectifier shown in Fig. P2.8 draws 10 mA dc regardless of the voltage.

(a) Name the type of rectifier (half-wave bridge, half-wave center-tapped transformer, full-wave bridge, full-wave center-tapped transformer)?

(b) Find the peak-to-peak ripple voltage across the load.

(c) Estimate the dc voltage across the load.

2.9. For the circuit shown in Fig. P2.9, assume ideal diodes.

(a) Is the circuit a half- or full-wave rectifier?

(b) Is the circuit a bridge or center-tapped rectifier?

(c) What is the maximum load voltage?

(d) What is the minimum load voltage?

(e) What is the dc load voltage?

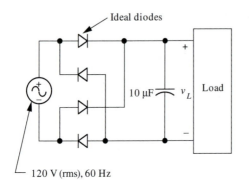

Ideal diodes

10 μF

v_L

Load

+

−

120 V (rms), 60 Hz

Figure P2.8

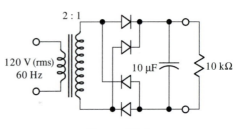

2 : 1

120 V (rms)
60 Hz

10 μF

10 kΩ

Figure P2.9

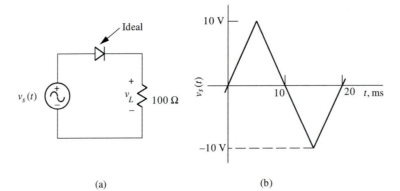

Ideal

$v_s(t)$

+

v_L

−

100 Ω

10 V

$v_s(t)$

10

20 t, ms

−10 V

(a)

(b)

Figure P2.10

2.10. The rectifier circuit in Fig. P2.10(a) has a "sawtooth" input voltage as shown in Fig. P2.10(b). Assume an ideal diode.
  (a) Determine the maximum and average current through the load.
  (b) If a 5000 μF capacitor is placed across the load, determine the average and ripple voltage across the load.

2.11. In Fig. P2.11, the load requires 10 W of dc power. The capacitor has a value of 600 μF. Assuming an ideal diode, find the following:
  (a) DC load voltage.
  (b) DC load current.
  (c) Ripple voltage at the load.
  (d) Maximum reverse voltage across the diode.

2.12. The circuit for a power supply is shown in Fig. P2.12. Consider the diodes ideal except in (d).
  (a) Is this a half-wave or full-wave rectifier?
  (b) Is this a raw or filtered rectifier?
  (c) Find the dc current in the load.

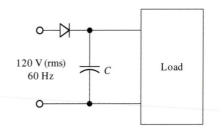

120 V (rms)
60 Hz

C

Load

Figure P2.11

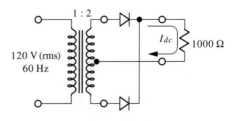

1 : 2

120 V (rms)
60 Hz

I_{dc}

1000 Ω

Figure P2.12

(d) Find the dc current in the load if one of the diodes burned out and became an open circuit.

2.13. For the power-supply circuit shown in Fig. P2.13, an ac voltmeter measures 80 V between A and B. Find the following:

(a) Maximum load voltage.

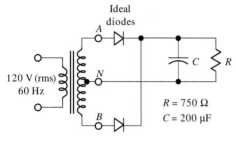

Figure P2.13

Section 2.2: The *pn*-Junction Diode

2.15. (a) A *pn*-junction diode has a characteristic well described by Eqs. (2.16) and (2.17). Plot the $i–v$ characteristic in the region $0 < i_D < 100$ mA, assuming $I_0 = 10^{-10}$ A and $\eta = 1.4$. On a linear scale, plot both i_D and v_D. Assume $T = 300$K.

(b) If the diode in part (a) were placed in series with a 12-V source and a 500-Ω resistor, with the battery polarity such as to forward bias the diode, what current would flow? You may use trial and error, an analytic, or graphical technique.

(c) What current would flow in part (b) if the battery were reversed so as to reverse bias the diode?

2.16. Figure P2.16 shows a piece of silicon that is doped with donor and acceptor atoms to form a *pn*-junction diode. Assume an ideality factor of 1.4.

(a) Where does there exist (1) more carrier electrons than holes; (2) more holes than carrier electrons; and (3) relatively few carriers?

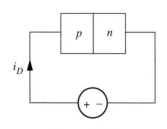

Figure P2.16

(b) Minimum load voltage.
(c) DC load current.
(d) DC voltage across the load if C were removed.

2.14. The power-supply circuit shown in Fig. P2.14 converts 120-V (rms), 60-Hz voltage to 24 V dc, with 20 mA of dc current. Find the following:

(a) Turns ratio of the transformer.
(b) Equivalent load resistance.
(c) Peak-to-peak ripple voltage.

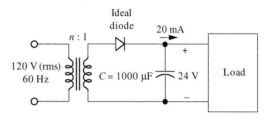

Figure P2.14

(b) If you measured $i_D = 15$ mA of current for $v_D = 0.75$ V, what would you expect for $v_D = 0.80$ V? Assume a temperature of 300 K.

(c) What current would you expect for $v_D = -0.3$ V?

2.17. (a) The usual symbol for a *pn*-junction diode is shown in Fig. P2.17 together with a corresponding piece of silicon. Indicate which half is *p*-type material and which is *n*-type material. Which contains the donor and which the acceptor atoms?

(b) Indicate with an arrow the direction in which holes cross the junction due to diffusion (thermal motion) when the diode is ON.

(c) With the diode reverse-biased, the current is determined to be 10^{-11} A. How much voltage would it take to turn ON the diode to a current of 10 mA? Assume an ideality factor of 1.5 and $T = 300$K.

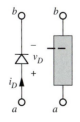

Figure P2.17

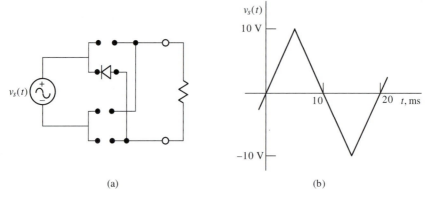

(a)

(b)

Figure P2.18

2.18. The circuit in Fig. P2.18(a) shows a bridge rectifier with only one diode in place. The input voltage source is the "sawtooth" wave shown in Fig. P2.18(b).
  (a) Put the other three diodes in place such that the rectifier is functional.
  (b) Find the time-average voltage across the load resistor with the polarity shown, assuming the diodes are ideal.
  (c) Assuming real pn-junction diodes, what is the peak inverse voltage the diodes have to withstand to function satisfactorily in the circuit?

2.19. If the diodes in P2.12 were pn-junction diodes instead of ideal diodes, answer the following true or false questions.
  (a) The dc load voltage would decrease by approximately 0.7 V. True or false?
  (b) The reverse voltage might burn out one diodes.True or false?
  (c) A large filter capacitor across the load would make little difference in the dc current. True or false?
  (d) Changing to a bridge rectifier would require the addition of two more diodes. True or false?

2.20. The circuit shown in Fig. P2.20 is a battery charger. The current flows when the instantaneous voltage out of the transformer exceeds the battery voltage plus the diode threshold voltage. As the battery is charged, its voltage increases until the current stops flowing. For this problem, however, consider the battery voltage constant at 12.6 V. Consider that it takes 0.7 V to turn ON the diode.
  (a) Find the peak value of the secondary voltage, V_p, such that the peak current in the battery is 10 A.
  (b) With this value for V_p, what percent of the time is the diode conducting?
  (c) What is the peak inverse voltage that the diode must withstand?

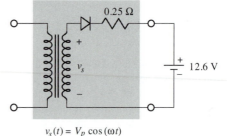

$v_s(t) = V_p \cos(\omega t)$

Figure P2.20

Section 2.3: Bipolar Junction-Transistor (BJT) Operation

2.21. (a) For the transistor output characteristics shown in Fig. P2.21, determine the approximate value of β in the active region.
  (b) What value of α does this imply?
  (c) If the collector current is saturated at a value of 28 mA, what is the collector–emitter voltage that results, and what is the base current

required to saturate the transistor at this value of collector current?

2.22. For the transistor amplifier shown in Fig. 2.40, change $V_{CC} = 12$ V and $R_B = 20$ kΩ.
  (a) Draw the new load line on the transistor characteristics in Fig. 2.46.
  (b) Find the input and output voltages for the

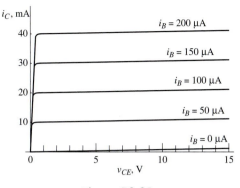

Figure P2.21

following base currents: 0, 20, 40, 60, 80, 100, and 120 µA.

(c) From part (b), plot v_{out} vs. v_{in}.

(d) Find the incremental gain from the slope, $A_v = \Delta v_{out}/\Delta v_{in}$.

2.23. A transistor circuit is shown in Fig. P2.23(a), along with the transistor characteristics in Fig. P2.23(b). The base–emitter voltage is 0.7 V when ON.

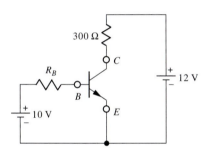

(a)

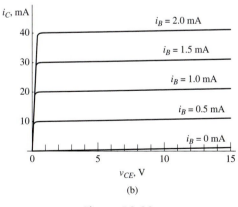

(b)

Figure P2.23

(a) What is the beta of the transistor?

(b) Place a load line on the characteristics.

(c) Find R_B to give $v_{CE} = 10$ V.

(d) What is the power out of the 12-V power supply with $v_{CE} = 10$ V?

(e) Find R_B to saturate the transistor.

(f) Find R_B to cut off the transistor.

2.24. A transistor circuit is shown in Fig. P2.24. The transistor output characteristics are shown in Fig. P2.21.

(a) Draw the load line on the characteristics. Find the required current into the base to give a collector–emitter voltage of 6 V.

(b) What value of the base voltage, V_{BB}, is required to give this amount of base current? Assume 0.7 V between base and emitter.

(c) What value of V_{BB} is required to saturate the transistor for this circuit? What is the collector current at saturation?

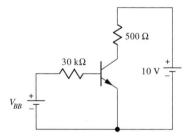

Figure P2.24

2.25. A transistor amplifier–switch circuit in Fig. P2.25(a) has a load line as shown on the transistor characteristics in Fig. P2.25(b). Assume that $v_{BE} = 0.7$ V when ON.

(a) What are V_{CC}, R_C, and β for the transistor?

(b) What value of v_{in} is required to saturate the transistor?

(c) What is the output voltage with the transistor $v_{in} = $ one-half the value in part (b)?

2.26. An amplifier-switch circuit is shown in Fig. P2.26. Assume that it takes 0.7 V to turn ON the base-emitter junction.

(a) The collector–emitter voltage $v_{CE} = 4.5$ V. Find the collector current, i_C.

(b) The current gain of the transistor is $\beta = 50$. Find the input voltage for the condition in (a).

(c) Find v_{CE} if $v_{in} = 0$ V.

2.27. A transistor circuit is shown in Fig. P2.27. Transistor characteristics are those shown in Fig. P2.21.

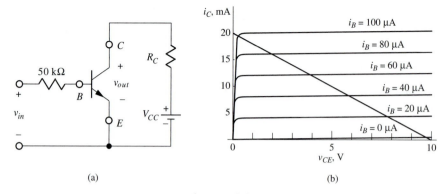

(a)

(b)

Figure P2.25

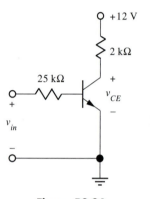

+12 V

2 kΩ

25 kΩ

v_{CE}

v_{in}

Figure P2.26

Assume 0.7 V for the base–emitter junction when ON.
(a) Draw the load line. What value of R_B will give a dc collector current of 15 mA?
(b) Find the voltage between collector and emitter for this value of R_B.

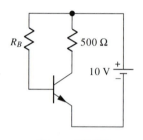

R_B

500 Ω

10 V

Figure P2.27

2.28. The BJT amplifier-switch circuit is shown in Fig. P2.28(a), with the transistor characteristics shown in Fig. P2.28(b). The input voltage is fixed at $v_{in} = +2.5$ V. The turn-ON voltage of the base–emitter junction is 0.7 V.
(a) Estimate β for the transistor.
(b) Find the base current.
(c) Find V_{CC} to give $v_{CE} = 3$ V for this value of v_{in}.

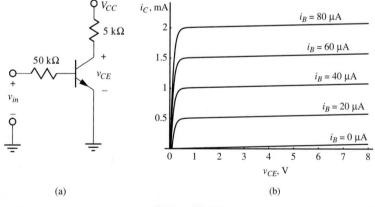

(a)

(b)

Figure P2.28

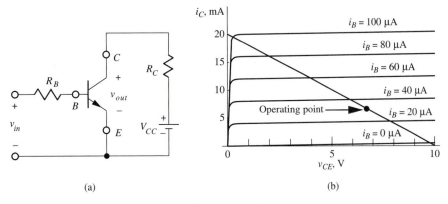

(a)

(b)

Figure P2.29

(d) With this value of V_{CC} fixed, what value of R_C ($\neq 5$ kΩ) will put the transistor barely in saturation?

2.29. The circuit shown in Fig. P2.29(a) is a standard amplifier-switch circuit. The transistor output characteristics are shown, with the circuit load line in place. The operating point for $v_{in} = 2$ V is shown. Assume $V_{BE(ON)} = 0.7$ V.
  (a) Find the power supply voltage, V_{CC}.
  (b) Find the base resistor, R_B.
  (c) Find the collector resistor, R_C.
  (d) Estimate the β of the transistor.

2.30. For the transistor amplifier shown in Fig. P2.30, find the following:
  (a) What is the collector–emitter voltage, v_{CE}, if the transistor is cut off?
  (b) What is the collector current, i_C, if $v_{CE} = 8$ V?
  (c) What is the minimum base current required to saturate the transistor?

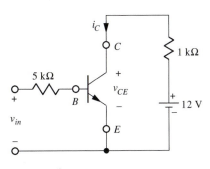

$\beta = 100$
$v_{BE} = 0.7$ V to turn ON
$v_{CE} = 0.3$ V at saturation

Figure P2.30

(d) What is the minimum value of v_{in} to place the transistor in the active region?

2.31. The input characteristic of a transistor is that of a *pn* junction, Eq. (2.17). For a transistor, we would associate the diode current with the base current, i_B, and the diode voltage with the base–emitter voltage, v_{BE}. The appropriate input resistance for the transistor in a small-signal equivalent circuit would be the slope of the input characteristic at the dc base current level: $r_\pi = dv_{BE}/di_B$ at $i_B = I_B$. Using Eq. (2.17) and ignoring the +1 term in the ln term, derive an expression for the input resistance to the transistor base. Also, evaluate with the result r_π for the circuit of Fig. 2.50 and confirm that a value of 450 Ω is reasonable for $\eta \approx 1$.

2.32 Figure P2.32(a) shows an amplifier-switch circuit and Fig. P2.32(b) shows the output characteristics of the transistor. The input voltage consists of a dc and an ac source in series. Assume 0.7 V for the base-emitter junction when ON.
  (a) Draw the load line.
  (b) Find V_{BB} to give an output voltage (with $V_p = 0$) of 8 V.
  (c) With the value of V_{BB} found in part (b), what is the largest value of V_p that can be amplified without serious distortion?

2.33. For the transistor in Fig. P2.33, 0.7 V is required to turn ON the base–emitter junction, and β is 100. Find R_C and R_B to make the collector current 5 mA and the collector–emitter voltage 4 V.

2.34. Figure P2.34(a) shows a transistor amplifier–switch circuit and Fig. P2.34(b) the transistor output characteristics.

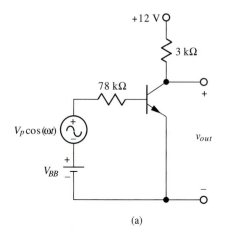

(a)

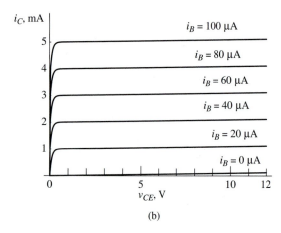

(b)

Figure P2.32

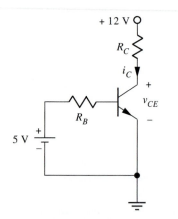

Figure P2.33

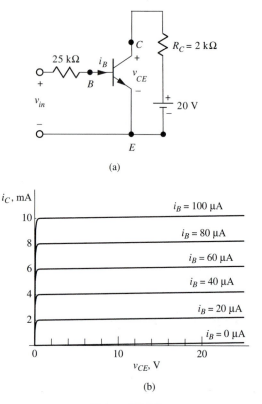

(a)

(b)

Figure P2.34

(a) What is β for the transistor?
(b) Draw the load line.
(c) Estimate the base current required to saturate.
(d) Find the collector current for $v_{CE} = 12$ V.
(e) What is the voltage across R_C if the transistor is cut off?

2.35. The transistor amplifier-switch in Fig. P2.35(a) has the input–output characteristic shown in Fig. P2.35(b). When the transistor is saturated, the power out of V_{CC} is 110 mW. Find V_{CC}, R_B, and R_C.

2.36. For the transistor in Fig. P2.36, $\beta = 150$, $v_{CE(\text{sat})} = 0.3$ V, and $v_{BE(\text{ON})} = 0.7$ V. Find the following:
(a) What are the values of v_{CE} and i_C at which the load line will cross the voltage and current axes on the transistor output characteristics?
(b) What are the collector current and base current at saturation?
(c) What is the range (from ? to ?) of input voltages to have the transistor in the active region?

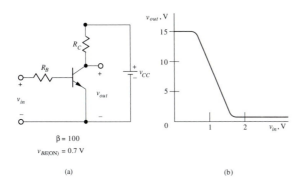

$\beta = 100$
$v_{BE(ON)} = 0.7$ V

(a)   (b)

Figure P2.35

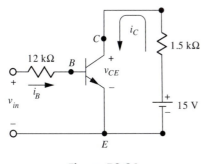

Figure P2.36

2.37. Figure P2.37 shows a transistor amplifier biased to operate as a small-signal amplifier.
  (a) Draw the dc bias circuit and solve for the base and collector dc currents and the dc voltage from collector to emitter.
  (b) Draw the small-signal equivalent circuit and solve for the voltage gain of the amplifier, v_{out}/v_s. Consider the capacitors as short circuits at the signal frequency. Let $r_\pi = 750\ \Omega$.
  (c) Find the input impedance of the amplifier as seen by the input generator. This does not include the 1.5-kΩ source resistance.

2.38. In Fig. P2.38(a), we show a transistor amplifier-switch circuit, with the output characteristics of the transistor in Fig. P2.38(b). The base–emitter voltage of the transistor is 0.7 V when ON.
  (a) Draw the load line on the characteristics.
  (b) What is the beta of the transistor?
  (c) What would have to be the power rating of the 5-kΩ resistor to operate satisfactorily in all conditions of the input voltage?
  (d) If the input voltage were 1.9 V, what would be the output voltage?

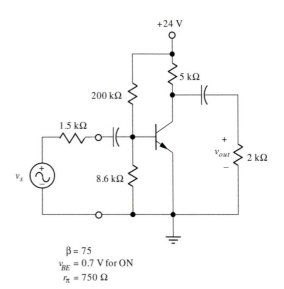

$\beta = 75$
$v_{BE} = 0.7$ V for ON
$r_\pi = 750\ \Omega$

Figure P2.37

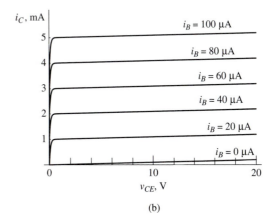

(a)

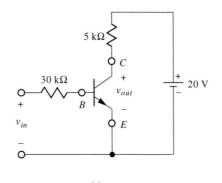

(b)

Figure P2.38

Section 2.4: Field-Effect Transistors

2.39. For a JFET with $V_P = -2.5$ V and $v_{DS} = 2$ V, what range of gate–source voltage corresponds to operation in the ohmic region? Assume v_{GS} and v_{DS} both < 0 V.

2.40. For the JFET amplifier circuit in Fig. 2.56 with characteristics shown in Fig. 2.54, find the following:
   (a) Redesign the circuit to have $i_D = 7$ mA by changing the 250-Ω resistor to a different value. Find the corresponding value of v_{DS}.
   (b) Determine the small-signal gain of the amplifier at the new operating point.

2.41. For the JFET amplifier circuit in Fig. 2.56, the designer wishes to increase the gain of the amplifier by changing the 1.5-kΩ resistor to a larger value.
   (a) How large a resistor value can be used and still have the transistor in the saturation region?
   (b) What is the small-signal gain with this limiting resistance?

2.42. The JFET in Fig. P2.42 has $I_{DSS} = 32$ mA and $V_P = -4$ V, and hence has the output characteristics shown in Fig. 2.54(b). Assume that the transistor is in the saturation region throughout this problem.
   (a) Find v_{GS} to give $i_D = 4$ mA.
   (b) Find the required value of R_S to give $i_D = 4$ mA.
   (c) Find the drain voltage v_D relative to ground, the reference node.
   (d) Determine the voltage gain of the amplifier stage, $A_v = v_{out}/v_{in}$.

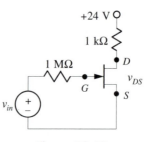

Figure P2.43

transistor is in the saturation region throughout this problem.
   (a) Find i_D to give $v_{DS} = 12$ V.
   (b) What value of v_{in} gives $v_{DS} = 12$ V?

2.44. In solving for the dc conditions for the JFET amplifier circuit in Fig. 2.56 we referred to a quadratic equation derived from Eqs. (2.46) and (2.47).
   (a) Derive the quadratic equation and show that the solution is $I_D = 8.0$ and 32.0 mA.
   (b) Show that 32.0 mA is not a realistic solution because it violates the assumed device characteristics.

2.45. An n-channel MOSFET has a threshold voltage of -2 V, and the drain–source voltage is $+3$ V. What range of gate–source voltages corresponds to operation in the ohmic region?

2.46. Show that Eqs. (2.51) and (2.52) match each other in both current and slope at the transition between the ohmic and saturation regions.

2.47. The MOSFET in Fig. P2.47 has the characteristics described by Eqs. (2.51) and (2.52), with $K = 16$ mA

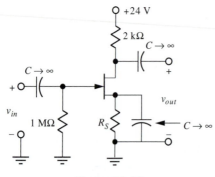

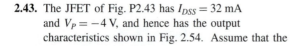

Figure P2.42

2.43. The JFET of Fig. P2.43 has $I_{DSS} = 32$ mA and $V_P = -4$ V, and hence has the output characteristics shown in Fig. 2.54. Assume that the

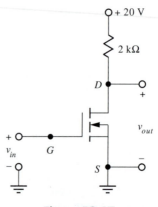

Figure P2.47

and $V_T = 4$ V, and displayed in Fig. 2.61. Determine the input–output characteristic, v_{out} vs; v_{in} for the range $0 < v_{in} < 10$ V. *Hint:* A load line on the characteristics would be a good place to start.

2.48. The MOSFET in the small-signal amplifier in Fig. P2.48 has the characteristics described by Eqs. (2.51) and (2.52), with $K = 16$ mA and $V_T = 4$ V, and displayed in Fig. 2.61. The capacitors may be treated as dc open circuits and signal short circuits.

   (a) Determine the operating point for the dc drain current and drain–source voltage.

   (b) Calculate the mutual conductance for the transistor, g_m, and draw the small-signal equivalent circuit for the amplifier, replacing the transistor with its small-signal equivalent circuit.

   (c) Determine the small-signal gain of the amplifier, v_{out} / v_{in}.

General Problems

2.49. Find the current in the diode in Fig. P2.49 by the following method: Plot the diode $i_D(v_D)$ characteristics; draw a load line; and find the intersection.

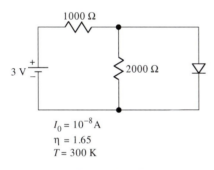

$I_0 = 10^{-8}$ A
$\eta = 1.65$
$T = 300$ K

Figure P2.49

2.50. Repeat Problem 2.7 if the diode is a *pn*-junction diode that requires 0.7 V to turn ON.

2.51. The power supply circuit in the dashed-line-box in Fig. P2.51 can be represented by a Thévenin equivalent circuit for moderate values of I_{dc}. The diode requires 0.7 V to turn ON. Find V_T and R_{eq}.

2.52. A three-phase system with a neutral is used in a rectifier configuration, as shown in Fig. P2.52. The time-domain line-to-neutral voltages are shown.

   (a) If a voltmeter measures 460 V between A and B, what is the peak value of the voltage, V_p?

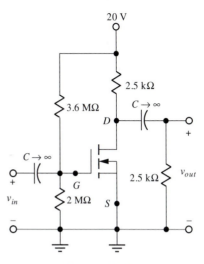

Figure P2.48

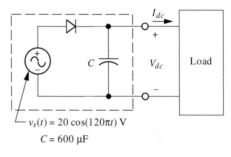

$v_s(t) = 20 \cos(120\pi t)$ V
$C = 600$ μF

Figure P2.51

   (b) Sketch the current in the load, $i_L(t)$.

   (c) Find the dc current through the load.

2.53. A transistor circuit is shown in Fig. P2.53. The β for the transistor is 125. The base–emitter voltage is 0.7 V when ON and the collector–emitter voltage is 0.2 V when the transistor is saturated.

   (a) What range of input voltages corresponds to the transistor's being cut off? What is the collector current of the transistor when cut off?

   (b) What range of input voltages corresponds to the transistor's being saturated? What is the collector current of the transistor when saturated?

2.54. The circuit shown in Fig. P2.54 will operate as an ac/dc voltmeter, depending on the switch setting. The meter movement responds to dc, 0 to 100 μA full scale.

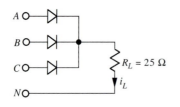

(a)

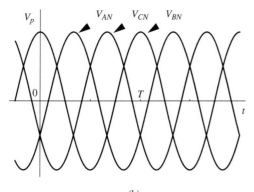

(b)

Figure P2.52

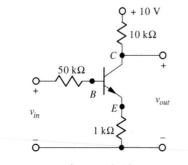

Figure P2.53

(a) Explain how the circuit works on dc. Mention what would happen if the input polarity were reversed.

(b) Explain how the circuit works on ac. What would happen in this case if the input leads were reversed?

(c) Determine R_1 so that the dc meter indicates full scale with 150 V dc input. Assume ideal diodes.

(d) Determine R_2 for the meter to read full scale with 150 V ac (rms) input. The meter movement registers full scale with 100 μA of dc

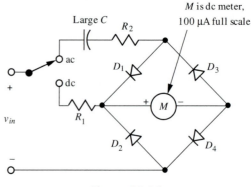

Figure P2.54

current through it. Assume ideal diodes and that the capacitor is large.

2.55. A calculator "charger," shown in Fig. P2.55, is actually a simple transformer. The remainder of the power supply is located in the calculator, as shown. The battery pack has a voltage ranging from 3.0 V (discharged) to 3.6 V (charged). The current limit for the diode is 50 mA and its ON voltage is 0.7 V. No current should flow into the fully charged battery. Determine the turns ratio required for the "charger" and the minimum value of R.

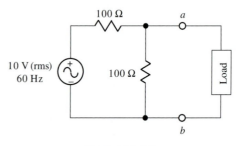

Figure P2.55

2.56. For Fig. P2.56, find the dc current from a to b for the following conditions:

(a) The load is a short circuit.

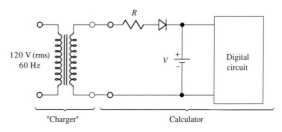

Figure P2.56

(b) The load is a 5-V battery, with + connected to *a*.

(c) The load is an ideal diode, with the allowed direction from *a* to *b* through the load.

2.57. A transistor circuit is shown in Fig. P2.57(a). The transistor characteristics are shown in Fig. P2.57(b). Determine R_2 to make the collector–emitter voltage +5 V, as shown on the circuit.

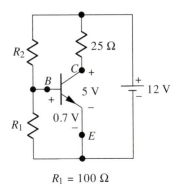

$R_1 = 100\ \Omega$

(a)

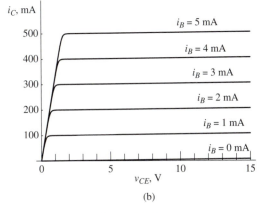

(b)

Figure P2.57

2.58. The transistor circuit shown in Fig. P2.58 has a constant voltage of +5 V at the base. The voltage at the emitter is 4.3 V because the ON voltage of the base-emitter junction is 0.7 V. The β for the transistor is 100.

(a) Is the transistor in the active, cutoff, or saturation region?

(b) What is the base current if *a* and *b* have no load?

(c) If *a* is shorted to *b*, what is the current that would flow in the short circuit?

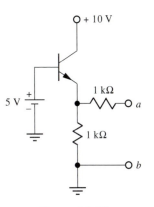

Figure P2.58

(d) Draw a Thévenin or Norton circuit with *a* and *b* as the output terminals. *Note:* Although the transistor is a nonlinear device, here it operates in a linear region of its characteristics. Hence, we may use an equivalent circuit to represent circuit output characteristics in that linear region.

2.59. The transistor in the circuit of Fig. P2.59 has a β of 60, requires a base-emitter voltage of 0.7 V to turn ON, and saturates at a voltage of 0.3 V. The input voltage v_{in}, is zero for a long time, and then at $t = 0$, changes to a value V_{in}.

(a) Find the minimum value of V_{in} to saturate the transistor for all positive times.

(b) Sketch the output voltage for positive time.

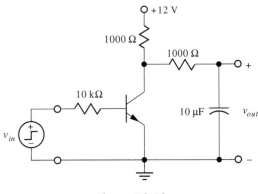

Figure P2.59

2.60. The circuit in Fig. P2.60 shows a standard switch-amplifier circuit, except that the collector and base currents (rather that collector-emitter voltage) are monitored with dc ammeters, *A*. An experiment is performed in which collector and base currents are

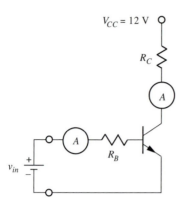

Figure P2.60

measured with dc ammeters as the input voltage is changed, with the results shown in the table.

v_{in} (V)	I_B (μA)	I_C (mA)
0	0	0
1	12	1.7
2	52	2.3
3	92	8.7
4	132	8.7

Assume that 0.7 V is required to turn ON the base–emitter junction and that the collector–emitter saturates at 0.3 V.

(a) Find β for the transistor.
(b) Find the base resistor, R_B.
(c) Find the collector resistor, R_C.
(d) What is the input voltage that first saturates the transistor?

Answers to Odd-Numbered Problems

2.1. **(a)** 26.7 V; **(b)** 32.7 V.

2.3. **(a)** $V_{dc} = V_p(1 - 1/(2fRC))$, $V_r = V_p/fRC$;
(b) 42.5 V, rms, 2230 μF.

2.5. **(a)**

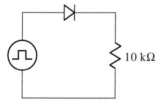

(b) 5.0 mA; **(c)** 9.76 mA.

2.7. **(a)** 50%; **(b)** 15.3 mA; **(c)** 45.5 mA.

2.9. **(a)** full wave; **(b)** bridge; **(c)** 84.9 V; **(d)** 72.8 V; **(e)** 81.3 V.

2.11. **(a)** 169 V; **(b)** 59.2 mA; **(c)** 1.65 V; **(d)** 339 V.

2.13. **(a)** 56.6 V; **(b)** 53.4 V; **(c)** 73.3 mA; **(d)** 36.0 V.

2.15. **(a)**

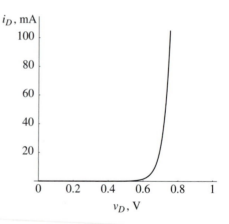

(b) 22.6 mA at 0.6975 V; **(c)** -10^{-10} A.

2.17. **(a)** n top, p bottom; **(b)** holes diffuse upward;
(c) 0.805 V.

2.19. **(a)** True; **(b)** true; **(c)** false; **(d)** true.

2.21. **(a)** 200; **(b)** 200/201; **(c)** $\approx$ 0.3 V, base current about 140 μA.

2.23. (a) 20; (b) intercepts at 12 V and 40 mA; (c) ≈ 26.6 kΩ; (d) 80 mW; (e) 4.77 kΩ; (f) ∞ Ω to cut off.

2.25. (a) 200, $V_{CC} = 10$ V, $R_C = 500$ Ω; (b) 5.7 V; (c) 5.3 V.

2.27. (a) 124 kΩ; (b) ≈ 2.5 V.

2.29. (a) 10 V; (b) 43.3 kΩ; (c) 500 Ω; (d) ≈ 200.

2.31. exact is 432 Ω.

2.33. $R_C = 1.6$ kΩ, $R_B = 86$ kΩ.

2.35. (a) 15 V, 12.7 kΩ, 2.05 kΩ.

2.37. (a) 35.1 μA, 2.63 mA, 10.8 V; (b) -44.9; (c) 687 Ω.

2.39. (a) $-0.5 < v_{GS} < 0$ V.

2.41. (a) 1.75 kΩ; (b) -14.0.

2.43. (a) 12 mA; (b) -1.55 V.

2.45. $-2 < v_{GS} < -1$ V.

2.47. starts at 20 V, drops to about 3 V in the range from 6 to 10 V.

2.49. ≈ 2.2 mA at voltage 0.51V.

2.51. 19.3 V, 27.8 Ω.

2.53. (a) $v_{in} < 0.7$ V, 0 A; (b) $v_{in} > \approx 1.95$ V, 9.8 mA.

2.55. (a) 39.5:1; (b) 12 Ω.

2.57. 1.15 kΩ.

2.59. (a) 4.60 V; (b) decays from 12 V to 0.3 V with a time constant of 10 ms.

3

Digital Electronics

1. To understand how information is coded in digital form
2. To understand how to perform NOT, OR, NOR, AND, NAND, and XOR operations on binary variables
3. To understand how digital information is represented and manipulated with electronic circuits
4. To understand how to use Boolean algebra to simplify and manipulate logic expressions
5. To understand how to efficiently implement logic expressions using standard logic gates
6. To understand how flip flops are used to store and process digital information
7. To understand how a computer uses digital circuits to process information

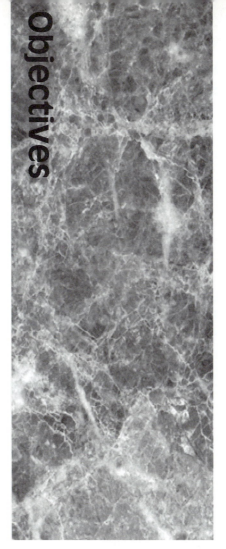

objectives

The basic idea of digital electronics is simple enough: represent information by signals that are either **ON** or **OFF**, and then use electronic circuits to store or process the information. But this idea has affected not only technology, but most aspects of modern living—think "computer." Digital techniques increasingly dominate electronic systems and are replacing many mechanical systems such as clocks.

What Is a Digital Signal?

An historical example. "Listen my children and you shall hear/Of the midnight ride of Paul Revere…." According to Longfellow's poem, Paul Revere was sent riding through the New England countryside by a signal from the bell tower of the Old North Church in Boston. "One if by land and two if by sea." Thus, one light was to be displayed if the British forces were advancing toward Concord by the road from Boston, and two lights were to be displayed if they were crossing the Mystic River to take an indirect route. The message received by the patriot was coded in digital form. We would say today that two "bits" of information were conveyed by the code.[1]

Digital Information

Information can be communicated in digital form if a message is capable of being defined by a series of yes/no statements. There can be only two states of each variable used in conveying the information. Reducing information to a series of yes/no statements might appear to be a severe limitation, but the method is in fact quite powerful. Numbers can be represented in base 2, and the alphabet by a digital code. Indeed, any situation with a finite number of outcomes can be reduced to a digital code. Specifically, n digital bits can represent 2^n states or possible outcomes.

binary variables

Binary variables. *Binary variables* are unusual mathematical variables because each can have only two values. We may call those two values by any names we wish: yes/no, true/false, ONE/ZERO, high/low, or black/white. When such variables were used primarily for analysis of philosophical arguments through symbolic logic, the values of the variables were called true or false. Recently, the names ONE/ZERO have come to be preferred by engineers and programmers dealing with digital codes. These names have the obvious advantage of fitting with the binary (base-2) number system for representation of numerical information.

Numerical and nonnumerical codes. Although other codes can be used, the most common digital code for numbers is in base 2. Table 3.1 shows the binary code for the numerals 0 to 9 and a few letters of a common code for the alphabet that is used in computers and communication.

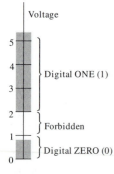

Figure 3.1 Typical ranges of voltage represent digital ONEs and ZEROs.

logic levels

Representing digital information electrically. In digital electronics, digital variables are represented by *logic levels*. At any given time, a voltage is expected to have one value or another, or more precisely to lie within one region or another. In a typical system, a voltage between 0 and 0.8 V would be considered a digital ZERO, a voltage above 2 V would be considered a digital ONE, and anything between 0.8 and 2 V would be forbidden; That is, if the voltage fell within this range, you would know that the digital equipment needs repair. These definitions are shown in Fig. 3.1.

Amplifier-switch logic levels. As an illustration of a digital circuit, we analyze the BJT amplifier switch we studied in Sec. 2.3 as a NOT circuit. The output of a *NOT*

---

[1] Strictly speaking, two bits could indicate four possible messages and would require distinguishable lights, say, one red and one white.

TABLE 3.1	Some Common Digital Codes	
Symbol	Digital Code	Type of Code
0	0000	Binary number
1	0001	Binary number
2	0010	Binary number
3	0011	Binary number
4	0100	Binary number
5	0101	Binary number
6	0110	Binary number
7	0111	Binary number
8	1000	Binary number
9	1001	Binary number
A	1000001	ASCII
B	1000010	ASCII
C	1000011	ASCII

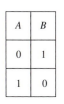

A	B
0	1
1	0

$B = \text{NOT } A = \overline{A}$

Figure 3.2 NOT binary function.

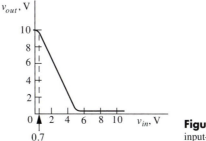

Figure 3.3 Amplifier-switch input–output characteristic.

complement, NOT circuit, truth table

circuit is the digital *complement*, or the opposite, of the input. First we represent the definition of the NOT circuit with the *truth table*[2] shown in Fig. 3.2: A represents the input, which may be either 1 or 0; B represents the output, which may also be 1 or 0 but depends on the input. The NOT, or logical complement, operation is indicated algebraically by the equation under the truth table, $B = \overline{A}$.

We now define logic levels for the amplifier–switch we studied in Chap. 2 such that it performs the NOT function. The input–output characteristic of the circuit is repeated in Fig. 3.3. Clearly, we wish 10 V to be in the region for a 1 and 0.7 V to be in the region for a 0. That is, if the input were 10 V (digital 1), the output should be less than 0.7 V (digital 0), and vice versa. Hence, we might consider making the region for a digital 0 to be from 0 to 1 V, and the region for a digital 1 to be, say, from 8 to 10 V. This works but leaves insufficient range for a working digital system. We broaden the range of values in the regions for 1 and 0 to allow for variations in transistors or power supply voltage, noise that might get mixed with the signal, and other uncertainties. In the present case, we can determine by trial and error that the region 0 to 1.5 V as a digital 0 works well with 5 to 10 V as a digital 1; with these definitions, the circuit operates as a NOT circuit. These logic levels are shown in Fig. 3.4.

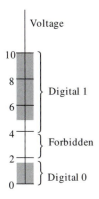

Figure 3.4 Digital definitions for the amplifier switch.

---

[2] A truth table systematically enumerates all possible states of the system. Here there are two states.

Digital Representation of Information

Elevator door controller. Having explained the nature of digital signals, how digital information is represented electrically, and how transistor circuits have the possibility of performing digital operations, we turn to a more complete example showing how to represent a situation in digital form. Our purpose is to introduce the AND and OR digital functions and to illustrate further the language and mathematics of the digital approach.

The door on a typical elevator closes from two sources: (1) a passenger who pushes a button or (2) a timer that closes the door automatically if empty. For safety, it also has an "electric eye" to prevent the door from closing on a passenger. Let us represent a command to close the door with the binary variable D ($D = 1$ if the door is to close). The state of the door, D, will be controlled by three binary variables: T represents the state of the timer ($T = 1$ means that the timer is running, time has not yet elapsed); B represents someone's pushing a button for another floor ($B = 1$ means that a button has been pushed); and S represents the state of the safety device ($S = 1$ means that someone is in the door). We see that D is the dependent variable and is a function of three independent variables (T, B, and S). Keep in mind that these are all binary variables and hence can be only 1 or 0.

$$D = f(T, B, S) \qquad (3.1)$$

Logic circuit

T	B	S	D
0	0	0	1
0	0	1	0
0	1	0	1
0	1	1	0
1	0	0	0
1	0	1	0
1	1	0	1
1	1	1	0

Figure 3.5 Truth table for an elevator door controller.

logic circuit

Truth-table representation. One useful method for describing a binary function is a truth table, as shown in Fig. 3.5. Here we have enumerated all possible combinations of the independent variables and shown the appropriate value of the dependent variable. In general, when there are n independent variables, each having two possible states, there will be 2^n possible combinations, 2^3 in this case, which may be enumerated to define the function. The truth table offers a systematic form for displaying such an enumeration. The name *truth table* originated historically from the application of this type of representation to the systematic investigation of logical arguments. Because of this association, digital circuits are often called *logic circuits*.

Truth table for elevator door. Figure 3.5 presents the truth table for the elevator door function. The 1's and 0's in the first three columns result from a systematic counting of the eight combinations. We call it "counting" because the pattern counts in the base-2 number system. The pattern is clear: We alternated the 1's and 0's fastest for S, slower for B, and slowest for T, thus covering all possible combinations. These represent all values of our independent variables. For filling out the 1's and 0's in the last column, we looked first at the S column, which represents the safety switch. We do not want the door to close when the safety switch indicates that someone stands in the door ($S = 1$), so we put 0 in the D column ($D = 0$ means do not close the door) for every 1 in the S column. This accounts for four of the eight states. The other four states depend on the button and the timer. If $S = 0$ (nothing blocking the door), the door should close if either the button is pushed ($B = 1$) or the timer expires ($T = 0$). We examine the remaining four states and put a 1 in column D if there is a 1 in the B column or a 0 in the T column, or both.

Interpretation of the truth table. Three combinations close the door. The first, $TBS = 000$, represents the timer running out to close the door. The second, 010, represents a button being pushed and the timer running out simultaneously, and the third, 110, represents a button being pushed before the timer runs out. The other five combinations leave the door open.

EXAMPLE 3.1

Electric fan

An electric fan will cool ($C = 1$) if it is plugged in ($P = 1$) and the switch is ON ($S = 1$). Show the truth table for the operation of the fan.

P	S	C
0	0	0
0	1	0
1	0	0
1	1	1

Figure 3.6 Truth table for the fan.

SOLUTION:

There are two independent variables, P and S, and one dependent variable, C. Thus, $n = 2$ and there are $2^n = 4$ states. The truth table, given in Fig. 3.6, shows that the fan must be plugged in and turned on to cool.

WHAT IF? What if we consider the possibility of a faulty cord ($F = 1$ means bad cord). How many rows will there be now?[3]

NOT function

LEARNING OBJECTIVE 2.

To understand how to perform NOT, OR, NOR, AND, NAND, and XOR operations on binary variables

OR function inclusive OR

AND function

NOT function. The truth-table method is a brute-force way for describing a binary function. The same information can be represented algebraically through the AND, OR, and NOT binary functions. Consider first the NOT function, the logical complement, described in the truth table of Fig. 3.2. The NOT function is involved in this problem because NOT S allows the door to close, and NOT T prompts the closing of the door by the timer. The NOT function is represented algebraically by an overscore added to the variable or expression to be NOTed: $\bar{S}$ means NOT S. We need the NOT when a 0 is to trigger the OR or AND combinations because these trigger on a 1.

OR function. The OR binary function is defined in Fig. 3.7. The dependent variable $C = A$ OR B is 1 when either A or B (or both) is (are) 1. This is thus the *inclusive* OR because it includes the case where both A and B are 1. The OR *function* is involved in our elevator problem in describing the combined effect of the timer and the button. We wish the door to close when the timer elapses ($T = 0$) OR the button is pushed ($B = 1$) or both. The way to express this algebraically is $\bar{T}$ OR B. The truth table for this function is shown in Fig. 3.8. In constructing the truth table in Fig. 3.8, we added a NOT T column. Then we put a 1 in the last column wherever there is a 1 in either of the previous two columns because these are the variables we are ORing.

AND function. Next we need to account for the safety switch. The AND function is required because we must express the simultaneous occurrence of an impulse to close the door and the lack of an obstacle in the door. The truth table for the AND function appears in Fig. 3.9. Here we get a 1 only when A AND B is[4] 1. To complete the truth table for our door-closing variable, D, we must AND S with the last column in Fig. 3.8 to cover all the possibilities. Thus, we may state the door-closing function as

---

[3] Eight rows for the eight states.

[4] This seems like bad grammar, but A AND B is a singular subject.

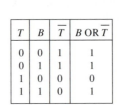

A	B	C
0	0	0
0	1	1
1	0	1
1	1	1

$C = A$ OR B

Figure 3.7 OR function.

T	B	$\overline{T}$	B OR $\overline{T}$
0	0	1	1
0	1	1	1
1	0	0	0
1	1	0	1

Figure 3.8 Binary function $\overline{T}$ OR B.

A	B	C
0	0	0
0	1	0
1	0	0
1	1	1

$C = A$ AND B

Figure 3.9 AND function.

$$D = (\overline{T}\,\text{OR}\,B)\,\text{AND}\,\overline{S} \tag{3.2}$$

When interpreted as a binary or logical function, Eq. (3.2) states algebraically the same information as the truth table in Fig. 3.5, which we worked out by considering all possible combinations. We have now introduced a method of representing information in digital form and we have defined some basic logic relationships. We turn next to describing how electrical circuits can perform logical operations such as AND and OR.

EXAMPLE 3.2 ## Fan motor again

In Example 3.1, express C in terms of P, S, and F and the logical NOT, OR, and AND functions.

SOLUTION:

To operate, we must have power ($P = 1$) AND the switch must be ON ($S = 1$) AND the cord must NOT be faulty ($\overline{F} = 1$); hence the function is

$$C = P\,\text{AND}\,S\,\text{AND}\,\overline{F} \tag{3.3}$$

WHAT IF? What if we include the possibility of a burned-out motor ($G = 1$ means good motor)?[5]

 ### Check Your Understanding

1. A truth table having five independent and three dependent binary variables would have how many rows?

2. How many bits of information (binary variables) are required to specify one of the 50 states in the United States?

3. If a signal in a digital system is between the regions for a 1 and a 0, you should assume 1, assume 0, or repair the circuit. Which?

Answers. **(1)** 32; **(2)** 6 bits; **(3)** repair the circuit.

---

[5] $C = P$ AND S AND $\overline{F}$ AND G.

NOT Circuit

Circuit improvements. In Sec. 3.1, we showed that the transistor amplifier-switch circuit performs the digital NOT function, provided that we define a digital 1 as any voltage between 5 and 10 V and a digital 0 as any voltage between 0 and 1.5 V. We propose here three modifications of the circuit to improve its performance as a NOT circuit. Specifically, we add two diodes to the base circuit, lower the resistance in the base circuit to 10 kΩ, and lower the power-supply voltage to 5 V, all shown in Fig. 3.10.

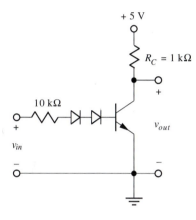

Figure 3.10 Improved NOT circuit.

Analysis of the modified circuit. We assume the diodes and the base–emitter *pn* junction are OFF when the voltage across each is less than 0.7 V. After they are turned ON and current begins to flow, their voltage will remain at 0.7 V each. This is the simple model of the *pn* junction shown in Fig. 2.36.

Base-circuit analysis. To derive the input–output characteristic of the circuit in Fig. 3.10, we increase the input voltage, beginning at zero volts. No current will flow into the base of the transistor until both diodes and the base–emitter junction turn ON, which requires about 3×0.7 or 2.1 V at the input. Once the voltage at the input rises to 2.1 V, therefore, the three *pn* junctions will turn ON, and the transistor will move out of cutoff. Once this occurs, Kirchhoff's voltage law in the base–emitter loop takes the form

$$i_B = \frac{v_{in} - 3 \times 0.7}{R_B} = \frac{v_{in} - 2.1}{10 \text{ k}\Omega}, v_{in} > 0 \tag{3.4}$$

Collector-circuit analysis. The base current controls the current in the collector–emitter loop, the output part of the circuit. While the transistor is cut off, no collector current flows and the output voltage remains at +5 V. As the base current begins to flow, however, the transistor moves into the active region and the collector (output) voltage begins to fall. This moves the operating point from cutoff toward saturation, as shown on the load line in Fig. 3.11. Because we changed the power-supply voltage, the load line now goes from $V_{CC} = +5$ V on the voltage axis to 5 V/1 kΩ = 5 mA on the

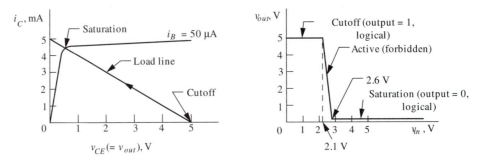

Figure 3.11 Load line for improved circuit.

Figure 3.12 Input–output characteristic of improved NOT circuit.

current axis. Because the transistor beta (β) is about 100, a base current of about $5 \text{ mA}/100 = 50 \text{ μA}$ is required for saturation; hence, we put the 50-μA characteristic on Fig. 3.11. Base currents between 0 and 50 μA put the transistor in the active region. Equation (3.4) requires, therefore, that input voltages between 2.1 and 2.6 V correspond to the active region. After the input voltage exceeds 2.6 V, the transistor is saturated; that is, the collector voltage (v_{out}) falls to about 0.4 V and further increases in base current cause little change in the output. Thus, the input–output characteristic of the modified amplifier–switch is shown in Fig. 3.12.

Benefits of changing the circuit. If you compare the characteristic in Fig. 3.12 with that in Fig. 3.3, you will note several differences. The slope is greater, higher gain, in the active region, a result of decreasing the resistance in the base circuit. We want higher gain, so that the transistor passes through the active region, the forbidden region in digital operation, with a smaller range of input voltage and hence passes through with greater speed. Next, the modified circuit remains in cutoff until the input exceeds 2.1 V, in contrast to 0.7 V for the unmodified circuit. Thus, two diodes in the base circuit increase the range for a digital 0 and enhance symmetry between the allowed ranges of the digital 0 and the digital 1. Finally, we note that the output voltage for a 1 is decreased from +10 to +5 V. We lowered the value of V_{CC} for two reasons, to save diodes and to save power. We added two diodes to raise the threshold of the active region to roughly half of the +5 V. If we had to add enough diodes to raise the active region up to one-half of 10 V, we would have had to use five or six diodes; thus, we save diodes by lowering the power-supply voltage. We also save power by lowering the voltage because the saturation current is lowered correspondingly. The power used by the circuit when saturated is approximately V_{CC} times the saturation current; hence, we use one-fourth the power with the smaller V_{CC}.

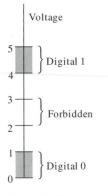

Figure 3.13 Logic levels for improved NOT circuit.

Logic levels for the modified NOT circuit. Appropriate digital levels for the modified amplifier–switch are shown in Fig. 3.13. The range 0 to 1.0 V defines a digital 0, 4.0 to 5.0 V defines a digital 1, and the range from 1.0 to 4.0 V is forbidden. We desire these broad, symmetrical regions for several reasons. Digital equipment is reliable and inexpensive because the exact voltages do not matter, as long as they lie in the ranges for a 0 or 1. Furthermore, broad well-separated regions for the logic levels immunize digital circuits to noise and false signals to some degree. We use these voltage ranges as our logic levels for the remainder of this section. Thus, a "1," meaning a digital 1, a symbol and not a number, means a voltage between 4.0 and 5.0 V, and a "0" means a voltage between 0 and 1.0 V.

EXAMPLE 3.3 | **Other circuit changes**

If three diodes are used in the base circuit and the collector resistor is changed to 2 kΩ, do these logic levels still work?

SOLUTION:
The 2-kΩ collector resistor changes the base saturation current to 25 µA. Now the transition begins when $v_{in} = 2.8$ V and ends when $v_{in} = 2.8 + 25$ µA $\times 10$ kΩ $= 3.05$ V. Thus, we lose a bit of the symmetry but the logic levels still work.

WHAT IF? How about three diodes and a 500-Ω collector resistor?[6]

BJT Gates

gates, NOR gate

NOR gate. Digital *gates* are circuits that pass or block signals moving through a logic circuit. We now examine the NOR *gate*, a circuit that combines the OR function with the NOT function. A simple NOR circuit is shown in Fig. 3.14(a). The inputs are *A* and *B*, and the output is *C*.

NOR truth table. We will justify the truth table shown in Fig. 3.14(b). Because there are two inputs, there will be 2^n ($n = 2$) or 4 possible input combinations for *A* and *B*, which we have listed systematically in the truth table. The first of these (00) corresponds to having voltages below 1.0 V at both inputs. Although the three *pn* junctions (two diodes and the base–emitter junction) are slightly forward-biased, insufficient voltage is present to turn them ON and in particular the transistor base–emitter junction will not turn ON; hence, the transistor remains OFF. This cutoff condition causes the output to be +5 V, a digital 1; hence, we place 1 in the *C* column, first row. The next row has a digital 1 at *B* and a digital 0 at *A*. The voltage at *B* exceeds 4 V, which turns ON diodes D_2, D_3, and the base–emitter junction, leading to saturation and a digital 0 at the output. Notice that the voltage at *P* is at least $4 - 0.7 = 3.3$ V and the voltage at *A* is at most 1.0 V; hence, D_1 is OFF. This diode, acting as an open circuit in its OFF condition, prevents the signal at *B* from coupling back into the source of *A*. The last two rows in the truth table follow from similar considerations: Clearly, if either (or both) of the inputs is (are) a 1, the transistor will turn ON, current will flow in the saturated transistor, and the output voltage will drop into the range for a digital 0. The diodes perform the OR operation and the transistor gives the NOT. Incidentally, we are not limited to two inputs; we can have three, four, or more inputs coming into the OR part of the circuit.

---

[6] Now the transition ends about 3.8 V; not enough margin.

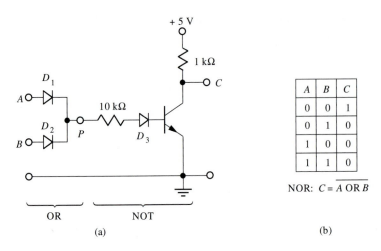

A	B	C
0	0	1
0	1	0
1	0	0
1	1	0

NOR: $C = \overline{A \text{ OR } B}$

OR   NOT

(a)   (b)

Figure 3.14   (a) NOR circuit; (b) truth table for the NOR function.

Role of transistor.   The output C is isolated by the transistor from affecting the inputs. The transistor firms up the output of the diode OR logic and isolates the input circuit from the output circuit. In the process of offering these benefits, the transistor inverts the digital signal and we thus pick up the NOT operation. If we require an OR circuit, we would put C into a NOT circuit; this would NOT the NOR to give the OR operation.

NAND gate

NAND gate.   The NAND circuit shown in Fig. 3.15(a) combines the AND and the NOT operations. Ignore for the moment the two inputs with their diodes and think of the circuit as a NOT circuit with the input (to the 10-kΩ resistor) connected to the +5-V power supply. Without action at the A and B inputs, the circuit would be a NOT circuit with the input locked at a digital 1 and the output locked at 0.

NAND truth table.   Now let us consider the effect of the inputs. The first row of the truth table has both A and B at low voltage, below 1 V. This causes both D_1 and D_2 to turn ON and current flows through the 10-kΩ resistor, through the diodes, and to ground through whatever is controlling A and B. The voltage at P is at most

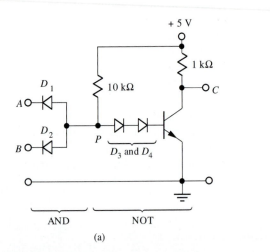

A	B	C
0	0	1
0	1	1
1	0	1
1	1	0

NAND: $C = \overline{A \text{ AND } B}$

AND   NOT

(a)   (b)

Figure 3.15   (a) NAND circuit; (b) truth table for the NAND function.

$1 + 0.7 = 1.7$ V, not enough to turn ON diodes D_3 and D_4 and the base–emitter junction. Hence, the transistor is cut off and the output voltage is $+5$ V, a digital 1. The next state (01) leads to similar operation, the only difference being that the voltage at B is now at least 4 V and hence D_2 is OFF. But the voltage at P remains no higher than 1.7 V and the output remains at 1. Only if both A and B are a digital 1 does the current through the 10-kΩ resistor return to the transistor base, turn ON the base–emitter junction, saturate the transistor, and drop the output voltage to a digital 0. The input diodes perform the logical AND function and the transistor the NOT function, giving the entire circuit a NAND function. If we require the AND function, we can invert C with a NOT circuit. If we add other inputs in parallel with A and B, all would have to be at a digital 1 to give a digital 0 at the output.

MOSFET Gates

A number of features make MOSFETs attractive as logic gates: They are small and easily fabricated in integrated circuits, they have a high input impedance, and they require very little power to hold and change logic states. In Sec. 2.4, we showed how a MOSFET can operate as a switch when used in series with a load resistor. The circuits used in practice replace the load resistor with another FET.

driver transistor, load transistor

Using an *n*-channel depletion-mode MOSFET for a load.   Figure 3.16(a) shows a NOT gate that uses an enhancement-mode MOSFET as a switch, or *driver*, transistor, and a depletion-mode MOSFET as a *load* transistor. When the input voltage is less than the threshold voltage for the driver transistor, it will be OFF and no current will flow through the transistor. The load transistor will be in the ohmic region because the output voltage will be very near V_{DD}.

When the input voltage to the driver transistor is much greater than its threshold voltage, it allows current to flow, the output voltage drops, and the driver transistor enters the ohmic region. The load transistor moves into its saturation region and thus acts as a current source having a high output impedance. Because a large current would require high power, the load transistor is fabricated to have a relatively small current when in saturation with $v_{GS} = 0$.

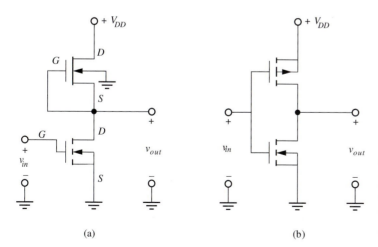

(a)   (b)

Figure 3.16   (a) A NOT gate using an enhancement-mode *n*-channel FET for a driver transistor and a depletion-mode FET for a load transistor; (b) a NOT gate using two enhancement-mode MOSFETs, an *n*-channel driver and a *p*-channel load transistor.

Using a _p_-channel enhancement-mode MOSFET for a load. The NOT gate shown in Fig. 3.16(b) uses a _p_-channel enhancement-mode MOSFET for a load transistor. Here a low input voltage causes the driver transistor to be cut off and the load transistor to be in the ohmic region; hence, the output voltage is nearly V_{DD}. Likewise, an input voltage near V_{DD} puts the driver transistor into the ohmic region and the load transistor into cutoff; hence, the output voltage is small. Note that in both states, one of the transistors is cut off and no dc current flows through the inverter. Current flows in the gate only during transitions, and hence this circuit uses very little power.

NMOS, CMOS

NMOS and CMOS logic families. The type of circuit shown in Figure 3.16(a) is called NMOS because the logic circuits are constructed entirely of _n_-channel MOS transistors. The type of circuit in Fig. 3.16(b) is called CMOS because it uses _p_-channel and _n_-channel, or complementary, transistors. Although we have shown only the NOT gate, families of logic elements including NAND and NOR gates are constructed of these transistors without use of resistors or capacitors.

The NOR, NAND, and NOT circuits are the basic building blocks from which digital systems are constructed. We will consider how large systems are formed after we have refined our mathematical language.

Check Your Understanding

1. Do the BJTs in a digital watch spend the least time in the cutoff, saturation, or active region?
2. A transistor, among other things, produces the complement (or NOT) of a binary signal. True or false?
3. In the NAND circuit in Fig. 3.15, is the "AND" part done by the diodes or the transistor?
4. If the collector voltage is used to represent a digital signal, would the output of a transistor in saturation be 1 or 0?
5. The diodes in the base circuit of Fig. 3.10 are there to increase the gain. True or false?
6. In a three-input NAND gate, how many input states correspond to 1 at the output?
7. Which type of NOT gate would use less power, the BJT or the CMOS?

Answers. **(1)** Active; **(2)** true; **(3)** by the diodes; **(4)** 0; **(5)** false; **(6)** seven; **(7)** CMOS.

3.3 ▊ THE MATHEMATICS OF DIGITAL ELECTRONICS

Need for a Mathematical Language

realize a logic function

We showed earlier how to express a digital function with a truth table. We termed this a brute-force method, for a truth table lists all possible states of the independent variables (the inputs) and lists the corresponding values of the dependent variables (the outputs). This way of expressing a digital function has several limitations. The truth table offers little guidance about how to _realize_ the logic function, how to assemble the required logical operations with digital circuits to produce the desired output. In our example about the elevator door, for example, we were able to translate the problem description into a logical function by common sense rather than through examination of the truth table. The techniques we develop in this section not only suggest a realization of the logic function but also permit manipulation of the function into different forms, thus offering alternative realizations. Also, we demonstrate ways to simplify logical expressions, thus permitting simpler realizations.

Boolean algebra

Origins of Boolean algebra. The algebra of two-valued variables is called *Boolean algebra* after George Boole, an English mathematician who first investigated this type of mathematics. Boole was interested in symbolic logic, the formal examination of logical arguments to establish their soundness or expose their fallacies. As we have already remarked, this early application of Boolean algebra has influenced the language of digital electronics.

A warning. Whereas the ideas, theorems, and applications of Boolean mathematics are relatively simple, the nomenclature and language can be confusing. One difficulty is that this mathematics uses some of the same symbols as ordinary algebra, but with different meanings. Thus, $A + B = C$ is a meaningful equation in both systems but has totally different meaning and is read differently as a binary expression.[7] Another difficulty is that common English words such as "and" are given technical meanings. Until you become accustomed to this new usage, many of the statements about Boolean variables sound like double talk.

Common Boolean Theorems

Boolean variables. A Boolean, or digital, or binary variable has two values, which we call 1 and 0. The values are *defined* to satisfy the definitions of OR, AND, and NOT in Fig. 3.17. The $+$ sign is used for OR, the "$\bullet$" symbol for AND, and the overscore for NOT or logical complement.

One-variable theorems. Theorems involving one variable, here A, are shown in Fig. 3.18. All of these may be verified by testing the validity of the expression for both states of A and comparing with the definitions in Fig. 3.17.

OR	AND	NOT
$1 + 1 = 1$	$1 \bullet 1 = 1$	$\overline{1} = 0$
$1 + 0 = 1$	$1 \bullet 0 = 0$	$\overline{0} = 1$
$0 + 1 = 1$	$0 \bullet 1 = 0$	
$0 + 0 = 0$	$0 \bullet 0 = 0$	

Figure 3.17 Basic definitions of OR, AND, and NOT functions.

OR	AND	NOT
$1 + A = 1$	$1 \bullet A = A$	$\overline{\overline{A}} = A$
$0 + A = A$	$0 \bullet A = 0$	
$A + A = A$	$A \bullet A = A$	
$A + \overline{A} = 1$	$A \bullet \overline{A} = 0$	

Figure 3.18 Boolean theorems for one variable.

EXAMPLE 3.4 | **Proof with truth table**

Verify $A \bullet \overline{A} = 0$ using a truth table.

SOLUTION:

Because A has two states, the truth table has two rows, Fig. 3.19. The first column lists the

---

[7] For example, the equation $1 + 1 = 1$ is correct in Boolean mathematics but is incorrect in ordinary algebra.

two values of A. The second substitutes these into the expression to be proved. The third uses the definitions under NOT in Fig. 3.17, and the fourth uses definitions under AND in Fig. 3.17. The theorem is proved because the last column shows $A \bullet \overline{A} = 0$ for both states of A.

A	$A \bullet \overline{A}$	Definition	Definition
0	$0 \bullet \overline{0}$	$\overline{0} = 1$	$0 \bullet 1 = 0$
1	$1 \bullet \overline{1}$	$\overline{1} = 0$	$1 \bullet 0 = 0$

Figure 3.19 Proof of the theorem $A \bullet \overline{A} = 0$ with a truth table.

WHAT IF? What if you have to simplify $A + (A \bullet \overline{A})$?[8]

Two or three variables. Some useful theorems and properties involving two or three binary variables are shown in Fig. 3.20. Many of these are deceptively similar to the familiar properties of algebra. Note that we show the expressions involving AND operations both with and without the "$\bullet$" symbol; thus, AB means the same as $A \bullet B$. Writing the AND without any symbol is common, even though the confusion with ordinary multiplication is compounded.

Commutation:	$A + B = B + A; A \bullet B = B \bullet A; AB = BA$
Association:	$A + (B + C) = (A + B) + C; A \bullet (B \bullet C) = (A \bullet B) \bullet C; A(BC) = (AB)C$
Absorption:	$A + (A \bullet B) = A; A \bullet (A + B) = A; A(A + B) = A$
Distribution:	$A \bullet (B + C) = (A \bullet B + A \bullet C); A(B + C) = AB + AC$
	$A + (B \bullet C) = (A + B) \bullet (A + C); A + BC = (A + B)(A + C)$
De Morgan's Theorems:	$\overline{A + B} = \overline{A} \bullet \overline{B}; \overline{A + B} = \overline{A}\overline{B}; \overline{A \bullet B} = \overline{A} + \overline{B}; \overline{AB} = \overline{A} + \overline{B}$

Figure 3.20 Some useful theorems and properties involving two or three binary variables.

EXAMPLE 3.5

Absorption rule

Verify the first absorption rule in Fig. 3.20.

SOLUTION:
Figure 3.21 shows the truth table. The last column is identical to the column for A, validating the theorem. Examination of the truth table shows how B is "absorbed" by A: If $A = 1$, B does not matter; and if $A = 0$, B does not matter. Hence, B is absorbed.

A	B	AB	$A + AB$
0	0	0	0
0	1	0	0
1	0	0	1
1	1	1	1

Figure 3.21 Truth-table proof of the absorption theorem $A + (A \bullet B) = A$.

De Morgan's theorems. *De Morgan's theorems* reveal how to distribute the NOT over variables that are ANDed or ORed. Figure 3.22 gives a proof for the first theorem.

---

[8] $A + (A \bullet \overline{A}) = A + 0 = A$.

A	B	$A + B$	$\overline{A + B}$	$\overline{A}$	$\overline{B}$	$\overline{A} \cdot \overline{B}$
0	0	0	1	1	1	1
0	1	1	0	1	0	0
1	0	1	0	0	1	0
1	1	1	0	0	0	0

Figure 3.22 Truth-table proof of one of De Morgan's theorems, $\overline{A + B} = \overline{A} \cdot \overline{B}$.

De Morgan's theorems

The fourth and seventh columns are identical, thus proving the theorem. The form of De Morgan's theorems is that on the left side of the identities, the NOT covers two variables that are either ANDed or ORed, and on the right side, the NOT has been distributed to the individual variables. Both rules are summarized by the following statement: A NOT can be distributed in a logical expression involving two variables provided that ANDs are changed to ORs, and vice versa. This rule can be applied to expressions involving more than two variables, provided care is taken with the grouping of variables.

Importance of De Morgan's theorems. A transistor inverts the output signal relative to its input(s). We saw before in examining NOR and NAND gates that the diodes at the input perform the OR or AND logic and the transistor firms up the decision, isolates the input from the output, and inverts the signal in the process, thus adding the NOT unavoidably to the logical operation. This "NOT" makes De Morgan's theorem useful in digital electronics. We explore the usefulness of De Morgan's theorems in digital electronics after we have introduced symbols for the logic gates.

EXAMPLE 3.6

Simplifying logic expressions

Simplify the expression $\overline{(\overline{AB} + A)}$ using the theorems in Fig. 3.20.

SOLUTION:

$$\overline{(\overline{AB} + A)} = \overline{(\overline{A} + \overline{B} + A)} = \overline{1 + \overline{B}} = 0 \tag{3.5}$$

Equation (3.5) shows that a logical expression that appears to depend on two input variables is in fact constant and remains in the 0 state, regardless of the values of A and B.

WHAT IF? What if you simplify $\overline{A \cdot B} + \overline{\overline{A} + B}$?[9]

EXCLUSIVE OR, XOR

EXCLUSIVE OR (XOR) and the equality function. The OR operation is an *inclusive* OR, meaning that it includes the case where both variables are 1. An EXCLUSIVE OR, or XOR, function is a logic function that takes the value 1 if either variable is 1 but takes the value 0 if both variables are either 1 or 0. The truth table for such a

---

[9] $\overrightarrow{A \cdot B}$.

A	B	$A \oplus B$	$\overline{A \oplus B}$
0	0	0	1
0	1	1	0
1	0	1	0
1	1	0	1

Figure 3.23 EXCLUSIVE OR and EQUALITY functions.

function is given in Fig. 3.23. The third column contains the definition of the EXCLU-SIVE OR, symbolized by a + inside a circle, $\oplus$. The bottom row of the table shows the exclusion of the state where both inputs are 1. The EXCLUSIVE OR indicates inequality between the variables, since the output 1 indicates that A and B are unequal. It follows that the complement of the EXCLUSIVE OR, $\overline{A \oplus B}$, also called XNOR, indicates equality, as the last column of Fig. 3.23 shows.

Check Your Understanding

1. Evaluate $D = \overline{B} + B(C + \overline{A})$ if $A = 1$, $B = 1$, and $C = 0$.

2. A digital system has three inputs, A, B, and C, and one output, $\overline{AB} + C$. How many of the possible input states correspond to a 1 at the output?

3. Under what condition is the Boolean expression $A + A = 1$ valid?

4. Simplify $\overline{\overline{A}(1 + \overline{A}) + \overline{B}}$.

5. As a Boolean equation, $C + C = C$ is valid. True or false?

Answers. **(1)** 0; **(2)** 7 states; **(3)** if $A = 1$; **(4)** AB; **(5)** true.

3.4 COMBINATIONAL DIGITAL SYSTEMS

Logic Symbols and Logic Families

Logic symbols. Digital systems consist of vast numbers of AND, NAND, OR, NOR, XOR, and NOT gates, plus memory and timing circuits that we discuss later, all interconnected to perform some useful task such as count and display time, measure a voltage, or perform arithmetic. If we were to draw a circuit diagram for such a system, including all the resistors, diodes, transistors, and interconnections, we would face an overwhelming task. And the task would be unnecessary because anyone who read the circuit diagram would group the components together into standard circuits and think in terms of the "system" functions of the individual gates. For this reason, we design and draw digital circuits with standard logic symbols, as shown in Fig. 3.24.[10]

The small circle at the output of the logic symbol indicates the inversion of the signal. Thus, without the small circles, the triangle of the NOT would represent an amplifier (or buffer) with *output = input*, and the second symbol would indicate an OR gate. As stated earlier, however, the common circuits are those that invert. These logic symbols show only the input and output connections. When wired into a digital circuit, the gates would have a power-supply voltage, V_{CC}, and grounding connections as well.

---

[10]The + (OR) and • (AND) markings in the gates are optional. The shape of the gate symbol defines its function.

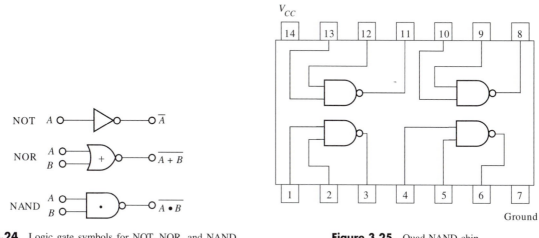

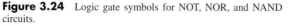

NOT $\quad A \circ\!\!-\!\!\triangleright\!\!\circ-\!\!\circ \overline{A}$

NOR $\quad \begin{matrix} A \circ\!\!-\!\! \\ B \circ\!\!-\!\! \end{matrix} \!\!+\!\!\circ-\!\!\circ \overline{A + B}$

NAND $\quad \begin{matrix} A \circ\!\!-\!\! \\ B \circ\!\!-\!\! \end{matrix} \!\!\cdot\!\!\circ-\!\!\circ \overline{A \cdot B}$

Figure 3.24 Logic gate symbols for NOT, NOR, and NAND circuits.

Figure 3.25 Quad-NAND chip.

Figure 3.25 shows the connections for a quadruple, two-input NAND gate. The supply voltage is applied between pins 14 and 7, with 7 grounded.

Logic families. If you wished to construct a digital circuit, you would not assemble a pile of diodes, resistors, and transistors and proceed to wire them together, first into standard gate circuits, and then into larger functions. You would purchase the gates already fabricated on an integrated circuit (IC), and packaged in a plastic capsule, as suggested by Fig. 3.26.[11] This commercial IC chip includes four NAND gates packaged together. An early and important step in the design would be to select a particular logic family, depending on the nature and working environment of your eventual product. The logic families are composed of a large selection of compatible circuits that can be connected together to make digital systems. The logic families differ in the details of the circuits used to perform the logical operations. Here are some of the logic families.

Figure 3.26
Physical appearance of a logic chip.

1. Diode–transistor logic (DTL) circuits are similar to the circuits in Figs. 3.14 and 3.15. These circuits are now obsolete. We used this simple type of logic only to illustrate the principles of logic gates.

2. High-threshold logic (HTL) circuits are similar to DTL gates but include a special diode in place of the two series diodes in Figs. 3.14 and 3.15, and use a larger power-supply voltage, V_{CC}. The special diode raises the threshold level for switching the transistor and hence separates the voltage regions for a 1 and 0 by a large margin, say, 10 V. This logic family is useful to prevent electrical noise, which might leak into the circuit, from affecting the operation. If, for example, your circuit must operate adjacent to a large dc motor or an arc welding machine, you would use HTL circuits.

3. Transistor–transistor logic (TTL) circuits use special-purpose transistors in place of the diodes. These circuits are widely used because they switch rapidly, require

---

[11]One reviewer suggested that these days you might order a special IC with the entire circuit on one chip.

modest power to operate, and are inexpensive. The circuits used in this logic family are considerably more complicated than the DTL circuits.

4. Complementary metal-oxide semiconductor (CMOS) logic circuits use field-effect transistors (FETs). These circuits require very little power to operate and are used where low power consumption is an important requirement, as in battery-operated calculators.

5. Emitter-coupled logic (ECL) gates switch very fast and are used in high-speed circuits such as high-frequency counters.

The design of logic circuits is highly sophisticated, and specialists in this area must become intimately familiar with all the possible products and logic families that are available at a given time.

Realization of Logic Functions

NOT function. Often when a NOT circuit is required, the designer will make one out of a NOR or a NAND circuit. Figure 3.27 shows the two ways to make a NOT (or inverter) out of a NAND circuit. These realizations are based on the first and third rows under AND in Fig. 3.18. The input to be fixed at a digital 1 in the lower realization would be attached to the V_{CC} power supply through a resistor. Similarly, if we required an input to be fixed at 0, this input would be grounded. Grounding an input can be used to realize the NOT function with a NOR gate, which we leave as a problem at the end of the chapter.

Realizing the elevator-door function. In Sec. 3.1, we derived a logical expression for closing an elevator door. The logical expression, recast into the notation we developed, is given in Fig. 3.28. We first realize the function with NAND and/or NOR gates. The realization in Fig. 3.28 utilizes six gates: one NOR, one NAND, and four NOTs, which were accomplished with NANDs. This realization is based on direct translation of the logical expression into logic-gate symbols.

Using De Morgan's theorem. We may accomplish a simpler realization by manipulating the expression for D into a more convenient form. Equation (3.6) shows such a manipulation.

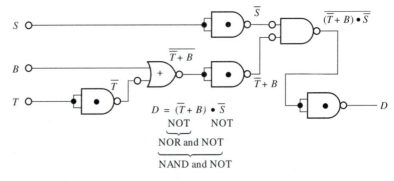

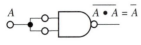

Figure 3.27 NOT from a NAND.

Figure 3.28 Straightforward realization of the elevator-door function using NAND and NOR gates.

$$\overline{\overline{(\overline{T} + B) \bullet \overline{S}}} = \overline{\overline{\overline{T} + B} + \overline{\overline{S}}} = \overline{\overline{\overline{T} + B} + S} \tag{3.6}$$

In Eq. (3.6), the first form is the same expression for D as in Fig. 3.28, except that we NOTed it twice. We do this because we want the final result for D to be the NOT of something, to end with a NOR or NAND gate. The second form results from using De Morgan's theorem to distribute one of the NOTs to the individual terms. The third form is the same as the second, except that we have removed the double complement from S. This final form proves convenient for realization with NOR gates. Figure 3.29 shows the realization. We made a NOT out of NOR similar to the way we realized a NOT with a NAND earlier. De Morgan's theorem thus leads to a simpler realization.

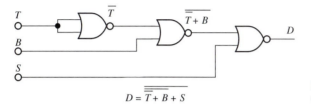

$$D = \overline{\overline{\overline{T} + B} + S}$$

Figure 3.29 Simple realization of the elevator-door function using NOR gates.

EXAMPLE 3.7 | **Minimum gate realization**

What are the fewest NAND or NOR gates to realize $A + BC$?

SOLUTION:
Apply the double NOT and use De Morgan's theorem on the inner NOT:

$$\overline{\overline{A + BC}} = \overline{\overline{A} \bullet \overline{BC}} \tag{3.7}$$

The second expression requires two NANDs and a NOT, which can be made from a NAND. Thus, the answer is three NAND gates.

WHAT IF? What if you try $A(B + C)$?[12]

Binary Arithmetic

binary numbers

Base-2 numbers. High-speed arithmetic is one of the spectacular achievements of digital electronics. Computers perform arithmetic with logic circuits through the representation of numbers as *binary numbers*, that is, in base-2 form. In this section, we investigate how binary arithmetic can be accomplished by digital circuits.

The development of efficient arithmetic methods was retarded for centuries by the lack of a convenient system and notation for the representation of numbers. The break-

---

[12] Two NORs and a NOT, or three NORs.

through came when the 10 Arabic[13] numerals were used to write numbers in base 10, that is, allowing the repetition of numerals, with the position representing powers of 10. For example, the number 806.1 means

$$806.1_{10} = 8 \times 10^2 + 0 \times 10^1 + 6 \times 10^0 + 1 \times 10^{-1} \tag{3.8}$$

The advantage of such a system is that the addition and multiplication tables assume manageable size; and the rules for such operations follow simple patterns.

The triumph of the base-10 number system was so successful that, until recently, few could write and perform arithmetic in some other base. This is no longer true because "modern math" in the secondary school system includes arithmetic in nondecimal number bases. You presumably required little explanation of the binary counting already used in Fig. 3.5. In that table, the first row represents ZERO, the second ONE, and the seventh, for example, represents 6 in binary form:

$$6_{10} = 110_2 = 1 \times 2^2 + 1 \times 2^1 + 0 \times 2^0 \tag{3.9}$$

bit, word (binary)

Thus, it takes three binary digits, or bits, to represent 6 in base 2, or binary, form. In general, n bits can represent numbers 0 to $2^n - 1$ and, if we wish to consider plus or minus numbers, we require another bit to represent the sign. An ordered grouping of binary information, a group of bits, is called a *word*. Thus, a digital computer would represent a number as a word of, say, 32 bits, and hence be limited to numbers smaller than about 4 billion. We are speaking here of an integer format or straight binary for the numbers; we can, of course, use a floating or exponential form to represent a wider range of numbers.

EXAMPLE 3.8 | **How many bits?**

How many bits of information are required to represent 10^6 in binary form?

SOLUTION:
With n bits, we can represent numbers up to $2^n - 1$. Thus, 2^n must exceed 10^6. Take the log

$$n \log 2 > \log 10^6 = 6 \Rightarrow n > 19.9 \tag{3.10}$$

Thus, 20 bits are required.

WHAT IF? | What if you want to represent the English alphabet, including capitals, and the 10 numerals. How many bits are required?[14]

BCD, binary-coded decimal

Binary-coded decimal (BCD). Thus far, we have presented pure decimal and pure binary representation of numbers. A hybrid system, binary-coded decimal (BCD), is frequently used in calculators and digital instrumentation. With BCD, the decimal format

---

[13] Actually, these symbols are thought to have been first used in ancient India and introduced into Western society through Islamic culture.
[14] Six bits.

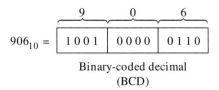

$$906_{10} = \boxed{1\,0\,0\,1\ \ |\ \ 0\,0\,0\,0\ \ |\ \ 0\,1\,1\,0}$$

Binary-coded decimal
(BCD)

Figure 3.30 BCD representation of decimal 906.

of the numbers is preserved, but each digit is represented in binary form. Because we must represent 10 numerals (0, 1, 2, ..., 9), we require 4 bits of digital information for each numeral to be represented. Figure 3.30 offers an example, representing 906_{10} as a 12-bit BCD word. From Fig. 3.30, we deduce that n bits, where n is a multiple of 4, can represent numbers up to $10^{n/4} - 1$ in BCD form. This represents a reduction from what we can represent with pure binary, but facilitates input and output interactions between system and operator. Because we think in decimal, for example, we want 10 keys on our calculators for entering numbers and we also require outputs in decimal. The internal manipulation of the binary information is complicated somewhat by the BCD form, but this is a problem for calculator designers, who obviously are up to the challenge.

hexadecimal number

Hexadecimal system. With the emergence of microcomputers, which work with words of 8 and 16 bits, the hexadecimal system has gained importance. *Hexadecimal* is base 16 and requires 6 "new" number symbols in addition to the 10 Arabic numerals. For convenience in using standard printers, the letters A through F are used for 10 through 15, respectively. Thus, in hexadecimal the number A8F represents

$$\text{A8F}_{16} = 10 \times 16^2 + 8 \times 16^1 + 15 \times 16^0 = 2703_{10} \tag{3.11}$$

The numbers 0 through 16 are represented in decimal, hexadecimal, and binary in Table 3.2.

TABLE 3.2 Numbers in Decimal, Hexadecimal, and Binary

Decimal	Hexadecimal	Binary
0	0	0000
1	1	0001
2	2	0010
3	3	0011
4	4	0100
5	5	0101
6	6	0110
7	7	0111
8	8	1000
9	9	1001
10	A	1010
11	B	1011
12	C	1100
13	D	1101
14	E	1110
15	F	1111
16	10	10000

Digital Arithmetic Circuits

BCD adder. To illustrate digital computation, we design a logic circuit to add two decimal digits represented in BCD form. The problem is symbolized by

$$A + B = S \implies A_4 A_3 A_2 A_1 + B_4 B_3 B_2 B_1 = S_4 S_3 S_2 S_1 \qquad (3.12)$$

where in the second form the A's and B's with the subscripts are binary variables representing the 4 bits required for the BCD representation and the $+$ represents addition.

First stage of the adder. Addition in binary is illustrated in Eq. (3.13):

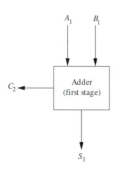

Figure 3.31 The first stage of the adder has two inputs and two outputs.

$$\begin{array}{r} \overset{1}{} \\ 1001_2 \\ 0101_2 \\ \hline 1110_2 \end{array} \qquad (3.13)$$

We have shown the carry for the binary addition in the traditional manner to emphasize that we must consider carries in the design of our circuit. That is, when we add in the lowest-order bits, 1 and 1, we obtain 10_2, which is written as a 0 with a carry of 1. Thus, the binary circuit that adds the lowest-order bits must produce two binary outputs: the lowest-order bit of the sum and the carry to be added with the next-higher-order bits. Figure 3.31 indicates what the first stage of addition must accomplish. It must take two binary inputs, A_1 and B_1, and produce two outputs, the lowest-order bit of the sum, S_1, and the carry, C_2, to the next higher stage of addition. Examination of the truth table in Fig. 3.32 reveals that the carry bit is the AND of the inputs and the sum bit is the EXCLUSIVE OR of the two inputs:

$$C_2 = A_1 B_1 \quad \text{and} \quad S_1 = A_1 \oplus B_1 \qquad (3.14)$$

We thus need a realization for the XOR.

Making an XOR. We may express the EXCLUSIVE OR in terms of OR and AND functions through either of the following equivalent forms:

$$A_1 \oplus B_1 = A_1 \overline{B}_1 + \overline{A}_1 B_1 = (A_1 + B_1)\overline{A_1 B_1} \qquad (3.15)$$

The first form expresses the two ways the function can be 1: either (A_1 is 1 AND B_1 is 0) OR (A_1 is 0 AND B_1 is 1). The second form uses the ordinary OR and then removes the case where both A_1 and B_1 are 1 by ANDing with $\overline{A_1 B_1}$. We can manipulate either expression into a form suitable for realization with NAND and NOR gates, but in this case, the second expression should be chosen because it involves the AND of the two inputs, which we need for the carry bit. Using the double NOT and De Morgan's theorems, we express the EXCLUSIVE OR in the form

$$\overline{\overline{(A_1 + B_1)\overline{A_1 B_1}}} = \overline{\overline{(A_1 + B_1)} + \overline{\overline{A_1 B_1}}} \qquad (3.16)$$

We use only NAND and NOR gates in our realization, so there is no benefit in cancelling the double NOT on the $A_1 B_1$ term. Figure 3.33 shows the realization of the lowest bit adder and represents in detail what is indicated by Fig. 3.31.

Higher stages of the adder. The second and higher stages of the adder must consider the carry from the next-lower stage. Thus, these stages of addition must function

Figure 3.32
Truth-table representation of the single-bit adder inputs and outputs.

Inputs		Outputs	
A_1	B_1	C_2	S_1
0	0	0	0
0	1	0	1
1	0	0	1
1	1	1	0

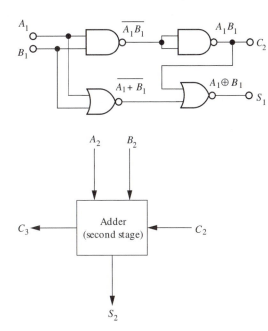

Figure 3.33 Realization for the first stage of the adder.

Figure 3.34 The second stage of the adder has three inputs and two outputs.

Inputs			Outputs	
A_2	B_2	C_2	C_3	S_2
0	0	0	0	0
0	0	1	0	1
0	1	0	0	1
0	1	1	1	0
1	0	0	0	1
1	0	1	1	0
1	1	0	1	0
1	1	1	1	1

Figure 3.35 Truth-table representation of the second-stage inputs and outputs.

with three inputs and two outputs, as indicated by Fig. 3.34 for the second stage. Figure 3.35 shows the truth table relating the two outputs to the three inputs. We will not complete the design of the higher-order stages of the BCD adder beyond developing Boolean expressions for the outputs.

sum of products   The truth table shows there to be four ways to achieve a 1 in the output bit, S_2, corresponding to the four 1's in the S_2 column. This last way is easiest to see because $A_2 B_2 C_2 = 1$ when all three variables are 1. The first way requires 1 when $A_2 = 0$, $B_2 = 0$, and $C_2 = 1$; hence, $\overline{A_2}\,\overline{B_2}C_2$ gives this 1. We then OR the group of ANDs. This form is called a *sum of products*. In this form, the second output bit of the sum would be

$$S_2 = \overline{A_2}\,\overline{B_2}C_2 + \overline{A_2}B_2\overline{C_2} + A_2\overline{B_2}\,\overline{C_2} + A_2 B_2 C_2 \tag{3.17}$$

Similarly, the carry bit from the second stage can be expressed as

$$C_3 = \overline{A_2}B_2 C_2 + A_2\overline{B_2}C_2 + A_2 B_2\overline{C_2} + A_2 B_2 C_2 \tag{3.18}$$

The theorems of Boolean algebra permit simplification of Eq. (3.18). The last two terms can be combined as shown in Eq. (3.19). Thus, we can replace two terms by a simpler term.

$$A_2 B_2\overline{C_2} + A_2 B_2 C_2 = A_2 B_2(\overline{C_2} + C_2) = A_2 B_2(1) = A_2 B_2 \tag{3.19}$$

We could have used the same trick by combining the last term with either of the first two terms. Fortunately, the theorems of Boolean algebra allow us to insert[15] two addi-

---

[15]Notice the third property of the OR in Fig. 3.18: $A = A + A + A + ...$, where we have expanded the expression to as many A's as we require.

tional $A_2B_2C_2$ terms in Eq. (3.18), and then combine them with each of the first two terms in the manner shown in Eq. (3.19). Thus, we reduce Eq. (3.18) to

$$C_3 = B_2C_2 + A_2C_2 + A_2B_2 \tag{3.20}$$

Given this success in simplifying Eq. (3.18), we might try to simplify Eq. (3.17) in a similar way, but our effort would be unsuccessful because Eq. (3.17) is already in its simplest form. Clearly, we could complete the realization of the BCD adder with NAND or NOR gates by manipulating these expressions for the sum and carry bits of the higher-order stages of the adder into suitable forms, as we did in Eq. (3.16).

Karnaugh Maps

Karnaugh map

In the previous section, we used the theorems of Boolean algebra to reduce Eq. (3.18) to Eq. (3.20). It is unclear, however, when such simplifications can be achieved. The Karnaugh map furnishes a technique for simplifying a Boolean expression. By arranging the truth table into a geometric representation, one can identify algebraic relationships through the patterns of ONEs and ZEROs.

Making a map.   The procedure for making a Karnaugh map is simple: Make a rectangular grid having 4, 8, 16, and so on, bins. Figure 3.36 shows such a map for C_3 as a function of A_2, B_2, and C_2. This map has eight bins because there are eight states to be represented. Each bin corresponds to one row of the truth table and is marked at top and side with its coordinates. For example, the bin in the lower-right corner of the map corresponds to $A_2 = 1$, $B_2 = 0$, and $C_2 = 1$, as marked at the top and side.

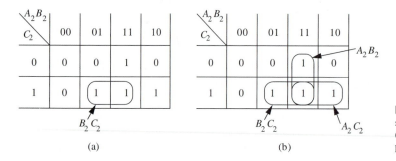

Figure 3.36   (a) Karnaugh map showing C_3 as a function of A_2, B_2, and C_2; (b) the same Karnaugh map with all three patterns identified.

The ordering of the bins is arbitrary, provided (1) all combinations are represented only once and (2) each bin differs from adjoining bins by a change of only 1 bit. For this reason, we cannot count across the top in the usual way (00, 01, 10, 11) because 2 bits change between the second and third states. This second rule for arranging the coordinates ensures that spatial proximity is linked to algebraic closeness.

Grouping the 1's for a sum-of-products expression.   The values of C_3 from the truth table go into the bins; hence, the four 1's under C_3 in Fig. 3.35 correspond to the four 1's in the eight bins in Fig. 3.36, each in the bin representing its row in the truth table. We now look for square and rectangular patterns of 1's. We have circled one such pattern in Fig. 3.36(a) and marked it with B_2C_2 because $B_2 = 1$ AND $C_2 = 1$ uniquely identifies those adjacent bins. In other words, A_2 drops out because it is both 0 and 1 in that rectangle. The reason why we do not circle the three adjacent 1's is that this property of variables dropping out occurs only in rectangular patterns of two, four, eight, and so on, adjacent bins.

Figure 3.36(b) shows the same Karnaugh map with three patterns circled and identified. We now see that the 1's in the map may come from $B_2 C_2$ OR $A_2 B_2$ OR $A_2 C_2$, which leads directly to Eq. (3.20).

product of sums

Grouping the 0's for a product-of-sums expression.
We can derive an alternate expression for C_3 based on the zeros in the Karnaugh map. Figure 3.37(a) shows the Karnaugh map of Fig. 3.36 with a rectangle of zeros marked. We identify this group with $C_2 + A_2$ because both C_2 and A_2 must be ZERO to get ZERO in this rectangle. Figure 3.37(b) shows the Karnaugh map with all rectangles of ZEROs identified and marked with the variables that are ZERO in those rectangles. We build the function by ANDing these components, because the function is ZERO when one of these components is ZERO. This produces a *product-of-sums* [16] form for C_3:

$$C_3 = (C_2 + A_2)(C_2 + B_2)(A_2 + B_2) \tag{3.21}$$

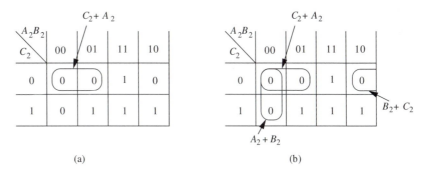

Figure 3.37 (a) Karnaugh map of C_3 with a rectangle of ZEROs identified; (b) all rectangles identified. This leads to the product-of-sums form.

The product-of-sums form for C_3 is equivalent to the sum-of-products form, Eq. (3.20), as may be confirmed by expanding Eq. (3.21) and using the theorems in Fig. 3.20. However direct realization of Eq. (3.21) leads to a different circuit than Eq. (3.20) and may be simpler in some cases, though not in this case.

EXAMPLE 3.9 — DON'T CARE

DON'T CARE

A precision electronic oscillator uses a crystal that must be kept in an oven for stable operation. The oscillator has three states: OFF, STANDBY, and ON. The oven is kept operating in STANDBY and ON. We represent the states with a digital code: $S_1 S_2 = 00$ (OFF), 01(STANDBY), and 10(ON), and design logic to operate the oven ($OV = 1$).

SOLUTION:
Figure 3.38(a) shows the truth table and Fig. 3.38(b) the Karnaugh map. In both, the X represents a DON'T CARE state, meaning a state that cannot occur. In our design, we may assign

---

[16]"Sums" and "products" are not defined in Boolean algebra. The names are based on appearance.

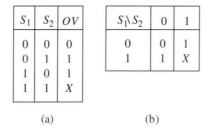

S_1	S_2	OV
0	0	0
0	1	1
1	0	1
1	1	X

$S_1\backslash S_2$	0	1
0	0	1
1	1	X

(a) (b)

Figure 3.38 (a) Truth table for the oven operation. The X represents a DON'T CARE, which cannot happen in practice; (b) Karnaugh map for the oven function. The DON'T CARE can be considered a ZERO or ONE, whichever leads to the simpler expression.

a 1 or 0 to the DON'T CARE, whichever leads to the simplest realization, because that state can never occur.

Our realization function is based on the Karnaugh map. If we call the DON'T CARE a 0, we would have no simplifying rectangles formed, but if we call it a 1, we would have a simple product of sums form based on the one ZERO. Because $X = 1$, the one ZERO is described by $OV = S_1 + S_2$. Thus, our realization is an OR gate.

WHAT IF? What if we insist on calling the X a 0? What are the product-of-sums and sum-of-products functions in that case?[17]

Further properties of Karnaugh maps. Other properties of the Karnaugh map are as follows:

- There is no "edge" to the map. Bits on the far left are adjacent to bits on the far right. Figure 3.39(a) shows the same information as Fig. 3.36(b) rearranged in a different order. The pattern for A_2C_2 now rolls over to the left.

- Patterns corresponding to independent variables that are 0 are identified as complements. For example, the four 1's in Fig. 3.39(b) correspond to $D = 0$; hence, we may identify them as $\overline{D} = 1$. The 1's in Fig. 3.39(b) thus may be identified as $\overline{D}$ OR $\overline{E}F$.

- DON'T CAREs are marked with X's in the Karnaugh map and may be considered either as 0's or 1's, whichever gives the largest patterns. Because these states never occur in practice, we may safely assign them either value to give the simplest algebraic expression.

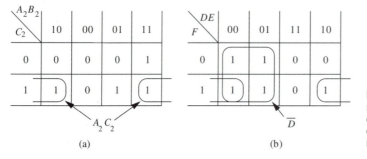

(a) (b)

Figure 3.39 (a) The same Karnaugh map as Fig. 3.36, with a different ordering of the columns. The A_2C_2 pattern now rolls over the edge of the map. (b) The identified pattern corresponds to $\overline{D} = 1$.

[17] Sum of products: $OV = S_1\overline{S_2} + \overline{S_1}S_2 = S_1 \oplus S_2$; product of sums: $OV = (S_1 + S_2)(\overline{S_1} + \overline{S_2}) = S_1 \oplus S_2$.

We leave for a homework problem the Karnaugh map for S_2, in Eq. (3.17), which has no rectangular groupings of 1's. Thus, no simplification of the expression for S_2 is possible.

Check Your Understanding

1. What is 1011_2 in hexadecimal (base 16)?

2. What is C_{16} in decimal?

3. How many bits are required to represent a hexadecimal number?

4. Why would it be incorrect to label a Karnaugh map with 01, 11, 00, and 10 across the top?

5. If the state $A_2B_2C_2 = 010$ were impossible (a DON'T CARE) in the Karnaugh map in Fig. 3.36, what would be the simplest expression for S_2?

Answers. **(1)** B_{16}; **(2)** 12_{10}; **(3)** 4 bits; **(4)** because more than 1 bit changes between 11 and 00 and also between 10 and 01; **(5)** $S_2 = B_2 + A_2C_2$.

3.5 SEQUENTIAL DIGITAL SYSTEMS

combinational logic, sequential logic

The logic circuits in Sec. 3.4 are called *combinational* logic circuits because the output responds immediately to the inputs and there is no memory. A *sequential* logic circuit has memory; its output depends on the inputs *plus* its history. In this section, we show how memory is developed in logic circuits and how memory elements increase greatly the applications of logic circuits.

> **LEARNING OBJECTIVE 6.**
>
> To understand how flip flops are used to store and process digital information

Bistable Circuit

stage of amplification

Two-stage amplifier. The basic memory circuit is the bistable circuit. Figure 3.40 shows two amplifier-switch circuits in cascade, the output of the first, T_1, providing the input to the second, T_2. This is a two-stage amplifier, for amplification takes place in two distinct *stages*.

Input–output characteristic of the amplifier. Each of these stages is identical to the original amplifier switch analyzed in Sec. 2.3, the input–output characteristic shown in Fig. 2.47. Cascading the two stages of amplification requires that we consider the output of the first stage as the input of the second stage. Figure 3.41 shows the overall input–output characteristic of the amplifier in Fig. 3.40. We derived this characteristic by increasing the input voltage starting with zero volts. With zero volts input, the first transistor, T_1, is cut off and T_2 is saturated. As the input voltage rises, T_1 leaves cutoff as the input voltage rises above 0.7 V, but T_2 remains saturated until the output of the first stage drops to about 4 V. This output of 4 V requires an input of about 3 V (see Fig. 2.47); hence, the input to T_1 must increase to approximately 3 V before T_2 comes out of saturation and the output voltage of the entire two-stage amplifier begins to rise. In this region, both transistors are in the active region and the output rises rapidly. The second transistor reaches cutoff when the output of T_1 falls below 0.7 V, which occurs when its input exceeds 4 V. Thus, both transistors are in the active region for input voltages between 3 and 4 V as shown in Fig. 3.41.

The latch circuit. What will happen if we connect the output of the two-stage amplifier to its input? Figure 3.42(a) shows the circuit redrawn with this connection and

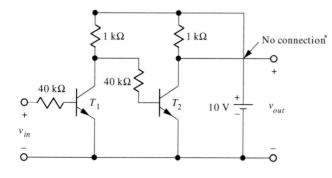

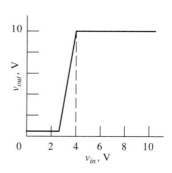

*When wires cross in a circuit diagram, no electrical connection is implied unless a dot is placed at the intersection.

Figure 3.40 Two-stage amplifier.

Figure 3.41 Input–output characteristic of two-stage amplifier.

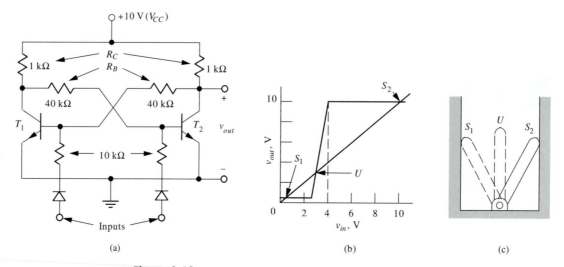

(a) (b) (c)

Figure 3.42 (a) Two-stage amplifier with output connected to input; (b) the circuit has two stable and one unstable operating point; (c) mechanical analog.

with T_1 turned around to emphasize the symmetry of the resulting circuit. We also have added inputs, which we discuss presently. This connection requires $v_{in} = v_{out}$, which defines a straight line passing through the origin and having a slope of unity. Figure 3.42(b) adds this line to the amplifier characteristic, which also has to be satisfied. This straight line is not a load line, but the same reasoning that we followed in thinking about load lines applies here: To satisfy both characteristics, the solution must lie at their intersection(s). The two intersections labeled S_1 and S_2 are stable solutions, but the intersection labeled U is unstable.

Mechanical analog. Figure 3.42(c) suggests a mechanical analog: The lever will have stable equilibria when resting against either wall, but with a frictionless pivot the balanced position will be unstable and will not occur in practice.

Latch action. The stable position marked S_1 occurs with transistor T_2 saturated and transistor T_1 cut off, and the stable position marked S_2 has transistor T_2 cut off and transistor T_1 saturated. The circuit will remain in one of these stable states forever unless an external signal forces it to the other stable state, just as the lever in Fig. 3.42(c) will lean against one wall unless an external force moves it to the other wall. By applying sufficient positive voltage to the input of the transistor that is cut off, we can switch the state of the circuit. The diodes are placed in the inputs to isolate the input drivers from the state of the circuit.

latch circuit

Latch function. The circuit shown in Fig. 3.42(a) is called a *latch circuit* and it provides electronic memory. When you depress a button on your calculator, the signal sets latch circuits in the calculator to retain the keyed information after you release the button. The information thus retained is then available for processing after all numerical information is entered.

Latches and Flip Flops

S–R latch. We can realize the latch function with standard logic gates. Figure 3.43 shows a latch constructed from two NOR gates. The output of each NOR provides one of the inputs for the other NOR. The other inputs are labeled S (for SET) and R (for RESET). The outputs are labeled Q and $\overline{Q}$ because the latch provides complementary outputs. This circuit, called an *S-R* latch, is similar in its operation to the circuit in Fig. 3.42(a).

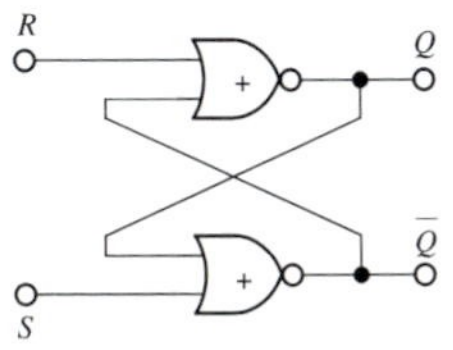

Figure 3.43 Latch from NOR gates.

Latch states. We can make a brute-force analysis of the circuit in Fig. 3.43 by pretending that the inputs are independent of the outputs, as shown in Fig. 3.44(a). Thus, the inputs to the NOR gates are R, S, Q_{in}, and $\overline{Q}_{in}$. With these "inputs," we can determine the outputs, Q_{out} and $\overline{Q}_{out}$, from the characteristics of the NOR gates. Only the states that are self-consistent can actually exist in the latch in Fig. 3.43. The truth table in Fig. 3.44(b) gives the results. We require 16 rows for four input variables, with two output variables, Q_{out} and $\overline{Q}_{out}$. We fill in the 0's and 1's in the first four columns to cover all possible input states by the usual method of binary counting. The columns under Q_{out} and $\overline{Q}_{out}$ are filled in by using the outputs of the NOR gates, Fig. 3.14(b). For example, in the second row $R = 0$ and $\overline{Q}_{in} = 1$, so $Q_{out} = 0$; and $S = 0$ and $Q_{in} = 0$, so $\overline{Q}_{out} = 1$. After filling in all 0's and 1's, we compare the input Q's with the output Q's, and where they are different we know that this state is impossible. Impossible states are labeled "inconsistent."

Interpretation of the stable states. We identify five consistent states, which therefore are stable states for the circuit. Two of these stable states we have identified as the MEMORY states; One state we have labeled SET, one RESET, and one FORBIDDEN. Thus, we reduce the truth table in Fig. 3.44(b) to that in Fig. 3.45(a).

The interpretations in Figs. 3.44(b) and 3.45(a) make sense in the following context: Information comes to the latch in pulses of 1's that come to S or R. If we get a pulse at S, the latch output, Q, is set to 1. If we get a pulse at R, the output is reset to 0. If we get no pulse at either input, the output state remains in, or remembers, its present state. If it gets simultaneous pulses at S and R, both outputs go to 0, which does not hurt anything, but then go to an indeterminate state when the inputs return to 0. This is undesirable because it leads to an unpredictable result; hence, this state is called FORBIDDEN.

(a)

R	S	Q_{in}	$\overline{Q_{in}}$	Q_{out}	$\overline{Q_{out}}$	Interpretation
0	0	0	0	1	1	Inconsistent
0	0	0	1	0	1	Consistent, MEMORY with $Q_{out} = 0$
0	0	1	0	1	0	Consistent, MEMORY with $Q_{out} = 1$
0	0	1	1	0	0	Inconsistent
0	1	0	0	1	0	Inconsistent
0	1	0	1	0	0	Inconsistent
0	1	1	0	1	0	Consistent, SET
0	1	1	1	0	0	Inconsistent
1	0	0	0	0	1	Inconsistent
1	0	0	1	0	1	Consistent, RESET
1	0	1	0	0	0	Inconsistent
1	0	1	1	0	0	Inconsistent
1	1	0	0	0	0	Consistent, but forbidden
1	1	0	1	0	0	Inconsistent
1	1	1	0	0	0	Inconsistent
1	1	1	1	0	0	Inconsistent

(b)

Figure 3.44 (a) S–R latch with inputs independent of outputs; (b) truth table for S–R latch with identification and interpretation of the possible states.

R	S	Q	$\overline{Q}$	Interpretation
0	0	0	1	MEMORY with $Q = 0$
0	0	1	0	MEMORY with $Q = 1$
0	1	1	0	SETS $Q \rightarrow 1$
1	0	0	1	RESETS $Q \rightarrow 0$
1	1	0	0	FORBIDDEN $Q = \overline{Q} = 0$

(a)

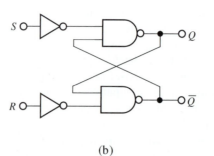

(b)

Figure 3.45 (a) Truth table for an S–R latch made from NOR gates; (b) an S–R latch made with NAND gates. The properties of this latch are the same as shown in (a) except the outputs go to $Q = 1$ and $\overline{Q} = 1$ in the FORBIDDEN state.

The *S–R* latch made from NOT and NAND gates shown in Fig. 3.45(b) has the same properties as the *S–R* latch we analyzed except that both its outputs go to 1's when 1's appear at both inputs. This type of *S-R* latch forms the basis for the gated latch.

EXAMPLE 3.10 **Stable states**

Find the stable states of the logic circuit in Fig. 3.46(a).

SOLUTION:
Figure 3.46(b) treats A and B as independent inputs and then from them deduces the output from the NAND gate truth table in Fig. 3.46(b). We find the only stable state to be $A = 0$ with $B = 1$.

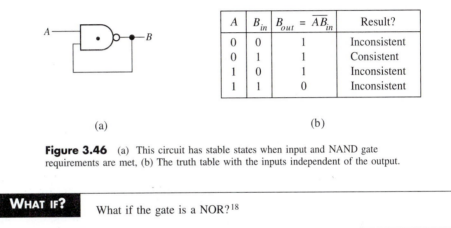

A	B_{in}	$B_{out} = \overline{AB_{in}}$	Result?
0	0	1	Inconsistent
0	1	1	Consistent
1	0	1	Inconsistent
1	1	0	Inconsistent

(a)   (b)

Figure 3.46 (a) This circuit has stable states when input and NAND gate requirements are met, (b) The truth table with the inputs independent of the output.

WHAT IF?   What if the gate is a NOR?[18]

---

glitch

flip flop

Gated flip flops. The *S–R* latch requires a number of refinements to achieve its full potential for memory and digital signal processing. Such latches are often called *flip flops*. One problem is that the *S–R* latch responds to its input signals at *S* and *R* immediately and at all times. Timing problems can occur when logic signals that are supposed to arrive at the same time actually arrive at slightly different times due to separate delays. Such timing problems can create short, unwanted pulses called *glitches*.

The gated latch in Fig. 3.47(a) responds to the *R* or *S* inputs only when a gating signal arrives at the *G* (gate) input. Here we have built the latch out of NAND gates. In this form, the forbidden state at the inputs to the cross-coupled NANDs is 00, which corresponds to 11 at the *R* and *S* inputs, as before. This latch also has Preset (*Pr*) and Clear (*Cr*) inputs that set the latch independent of the input gates. These are active when in the 0 state, as indicated by the circle at their inputs on the logic symbol in

---

[18] One stable state with $A = 1$ and $B = 0$.

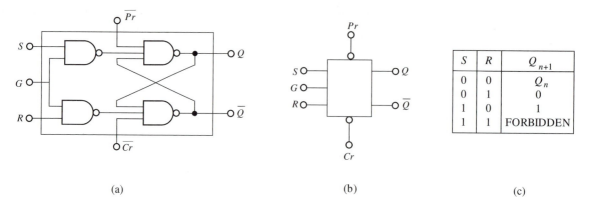

	(a)		(b)			(c)

S	R	Q_{n+1}
0	0	Q_n
0	1	0
1	0	1
1	1	FORBIDDEN

Figure 3.47   (a) Gated latch; (b) logic symbol; (c) truth table.

Fig. 3.47(b). The truth table in Fig. 3.47(c) now lists the output state *after* the gating pulse, Q_{n+1}, as a function of the R and S inputs and the state, Q_n, *prior* to the gating pulse, Q_n. For example, with $RS = 00$, the gating signal produces no change in the output state, $Q_{n+1} = Q_n$. By using an inverter on the gate input, we could have the gating occur at $G = 0$. This would be indicated by a circle at the gating input (G) of the flip-flop symbol in Fig. 3.47(b).

The R and S inputs are thus active when the signal at the gate input is 1. Normally, such timing, or synchronizing, signals are distributed throughout a digital system by clock pulses, as shown in Fig. 3.48. The symmetrical clock signal provides two periods during each cycle when switching may be accomplished, that is, when $Ck \Rightarrow 1$ and when $\overline{Ck} \Rightarrow 1$.

edge triggering

Edge-triggered flip flops.   The gating time of the inputs can be further reduced by making the clock input sensitive to transitions in the clock signal, which is known as *edge triggering*. The circuit can be designed to trigger at the leading or trailing edge of the clock. The symbol for an edge-triggered flip flop is shown in Fig. 3.49 for both (a) leading and (b) trailing edge triggering. The distinguishing mark for edge triggering is a triangle at the clock input. Triggering at the edges of the waveform limits the time during which the inputs are active and thus serves to eliminate glitches. By using circuits that trigger at either the leading or trailing edges, the designer can pass signals in a circuit at two times in each clock cycle.

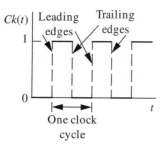

Figure 3.48   Clock signal.

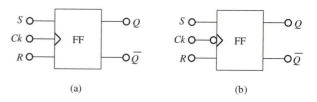

Figure 3.49   Logic symbols for edge-triggered flip flops: (a) leading edge triggering; (b) trailing edge triggering.

toggle

J–K flip flops. Another problem with the basic *S–R* latch is the forbidden state at the input. This can be eliminated by ANDing the inputs with the output of the flip flop, thus blocking one of the inputs, as shown in Fig. 3.50. The added gates here have the effect of inhibiting the 1 input to the gate whose output is 1. Therefore, with $J = 1$ and $K = 1$, the input that is passed will always change the state of the output. The truth table for the *J–K* flip flop, Fig. 3.51, is the same as the truth table in Fig. 3.47(c), except that we indicate a change of output state, $Q_{n+1} = \overline{Q}_n$, for the hitherto forbidden input state. The *J–K* flip flop thus gives us, in addition to a latched memory of the input, the capacity to *toggle*[19] at each clock pulse when both inputs are 1. This toggle feature reveals why we must use edge triggering for this flip flop: If the clock pulse were extended in time, the state would oscillate back and forth and the eventual output would be indeterminate. The toggle mode of the *J–K* flip flop is useful in counters and frequency dividers.

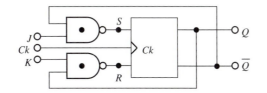

Figure 3.50 *J–K* flip flop.

J	K	Q_{n+1}	MEANING
0	0	Q_n	MEMORY
0	1	0	RESET
1	0	1	SET
1	1	$\overline{Q}_n$	TOGGLE

Figure 3.51 Truth table for the *J–K* flip flop. Q_n represents the state before the clock pulse and Q_{n+1} the state after the clock pulse.

D-type flip flops. The *J–K* flip flop can be converted to a *D*-type flip flop[20] by connecting an inverter between the inputs, as shown by Fig. 3.52. This has the effect of shifting the input to the output at the active clock edge, as shown in Fig. 3.53.

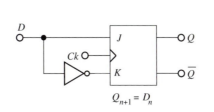

$$Q_{n+1} = D_n$$

Figure 3.52 *D*-type flip flop.

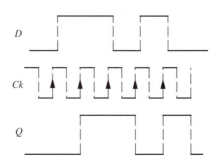

Figure 3.53 The output is delayed until the next active clock edge.

T-type flip flops. Tying the *J* and *K* inputs together produces a *T*-type flip flop.[21] The *T*-type flip flop toggles with the clock pulse when $T = 1$ and does not toggle when

---

[19] A lever-actuated switch, like the ordinary light on–off switch, is called a toggle switch. Thus, to toggle means to switch from one state to another.

[20] *D* for delay.

[21] *T* for toggle.

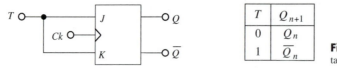

T	Q_{n+1}
0	Q_n
1	$\overline{Q}_n$

Figure 3.54 *T*-type flip flop with truth table.

$T = 0$. This is useful for counters and divide-by-2 applications. The logic symbol and truth table for the *T*-type flip flop are shown in Fig. 3.54.

Summary. In this section, we presented the *S–R* latch as the basic memory element in logic circuits. A variety of refinements were given, leading to *J–K, D,* and *T* flip flops. In all cases, we have shown only one way to realize the characteristic of the different flip flops. In the various logic families, these circuits could be realized through many variations, depending on the properties of the specific family. In the next section, we indicate how flip flops can be used in digital systems.

Flip-Flop Applications

Frequency dividers. The clock frequency can be halved with a *T* flip flop by setting the *T* input to 1 and letting the clock toggle the output. This process can be continued to divide by 4, 8, etc.

Counters. Figure 3.55 illustrates the counting process. We begin with a cleared counter, that is, with all output *Q*'s at ZERO. As shown, every trailing edge of the clock

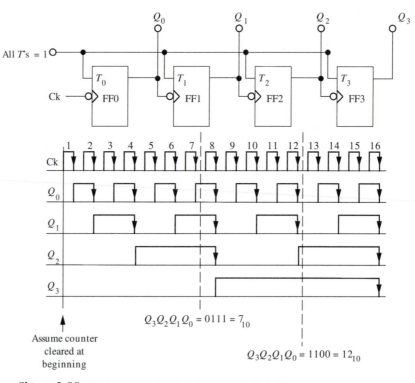

Figure 3.55 Basic counter operation. The states count in binary, with Q_0 interpreted as the lowest-order bit.

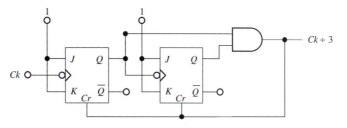

Figure 3.56 Divide-by-3 circuit.

pulse changes the state of FF0; hence, the output frequency of FF0 is half the clock frequency. Because this output is used as a clock for FF1, the frequency is again divided by 2, and so on down the counter. We have drawn two dashed lines to verify that the repeated divide-by-2 operation counts the input clock pulses. For example, after the end of the seventh input pulse, the state of the Q's is 0111 ($= 7_{10}$), considering Q_0 as the lowest-order bit. The reader can verify the count after 12 pulses. Note that the chain counts input pulses in binary even if the clock pulses are unevenly spaced.

Division by a factor that is not an exact power of 2 can be accomplished by using an AND gate to detect the appropriate state. Figure 3.56 shows a divide-by-3 counter. If the two flip flops start off cleared (both Q's $= 0$), then at the end of the third clock cycle, both Q's would be 1, which would produce a 1 at the AND output. This 1 clears the flip flops and acts as output. Hence, we get one output pulse for every three input pulses.

ripple counter, synchronous counter

A decade counter. Figure 3.57 shows how to convert a binary counter to a decade counter. The AND gate detects a count of 10, $Q_3\bar{Q}_2Q_1\bar{Q}_0 = 1$, clears the counter, and provides an input to the next stage. The $Q_3Q_2Q_1Q_0$ output from the stage could be transferred to a BCD-to-decimal display before the counter begins counting another sample of the input.

The types of counters that we have presented are called *ripple counters* because the flip-flop transitions move in sequence from left to right through the counter. By contrast, a *synchronous counter* uses a simultaneous clock pulse at all flip flops and controls the counting operation with external gates. Also possible are up–down counters that increase or decrease the count depending on an input command.

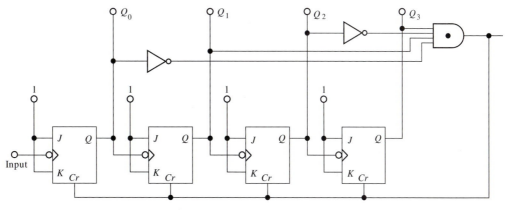

Figure 3.57 One stage of a decade counter.

The digital designer has available frequency dividers, counters, and many other useful system functions on LSI (large-scale integration) chips. Such LSI circuits manage the internal connections and provide the external inputs and outputs.

register, byte, word

Registers. A *register* is a series of flip flops arranged for organized storage or processing of binary information. Before describing several types of registers, we must introduce some concepts that relate to the use of registers in computers.

Information is represented in a computer by groups of 1's and 0's called *words*. Thus, a word in an 8-bit microprocessor might be 01101101, or 6D in hexadecimal. An 8-bit word is also called a *byte* and is a convenient unit for digital information. One byte can represent two BCD digits or an ASCII (American Standard Code for Information Interchange) alphanumeric symbol. Current microprocessors work with words of 4, 8, 16, or 32 bits, whereas larger computers work with words of 32 or more bits. A register in a computer with 8-bit words would require eight flip flops to store or process simultaneously the 8 bits of information.

bus

Words of information are moved around in a computer or other digital system on a *bus*. As on a city bus line where passengers can enter or leave at a variety of points along the way, so on a computer bus the words can originate at any of the several registers or arrive at any of several destination registers. The bus itself consists of the required number of wires[22] connecting all potential source registers with all potential destination registers.

Figure 3.58 shows a bus of four wires connected to a destination register of four *D*-type flip flops. At the leading edge of the LOAD signal, the information on the bus is stored in the register. The information does not come down the bus and get off at the register; rather, the information appears simultaneously all along the bus and may be loaded simultaneously into several registers.

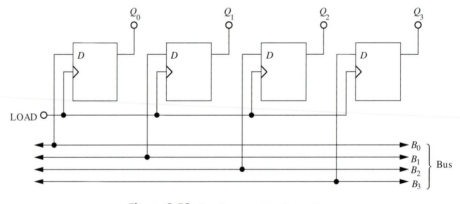

Figure 3.58 Loading a register from a bus.

Three-state gates. Whereas more than one register can be loaded simultaneously from the bus, only one register can put information on the bus at one time. We need a way to connect the outputs of all source registers to the bus such that only one register

---

[22] Plus the common, which would be the signal ground. Actually, "conducting paths" is more appropriate, for no wires are used for internal information transfer on a microprocessor chip.

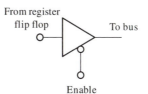

Figure 3.59 Three-stage gate. When ENABLE is 1, the output is disconnected from the bus.

can transfer its output word to the bus at any time. An ordinary gate will not accomplish this, for its output must be either 1 or 0 and hence connecting the outputs of all source registers would result in a tug of war. A three-state gate, shown in Fig. 3.59, is required. In the absence of an ENABLE signal, the output of the gate approximates an open circuit and the gate is disconnected from the bus. When the ENABLE signal is active, the gate is connected to the bus and the output is 1 or 0 according to the input. Thus, we can connect all source registers to the bus with three-state gates and ENABLE one register at a time to transfer data to the bus.

parallel information, channel, serial information

Shift register. The data-storage register described in the preceding section transfers information with *parallel* input and parallel output. Sometimes digital information must be sent over one *channel*, as when a telephone circuit is used. In this case, bits are sent in time sequence, or *serial* form. When digital information must be received in serial form, a shift register may be used to accept the serial information and convert it to parallel form.

The D-type flip flops in Fig. 3.60 will act as a shift register. Recall that the input to the D flip flop is shifted to the output by the clock pulse. Thus, the input to D_3 will appear at Q_3 after one clock pulse, at Q_2 after two, at Q_1 after three, and at Q_0 after four clock pulses. Hence, the 4-bit sequence (word) input at D_3 will fill the register $Q_3Q_2Q_1Q_0$ after four clock pulses. The word can then be transferred to a parallel bus with three-state gates.

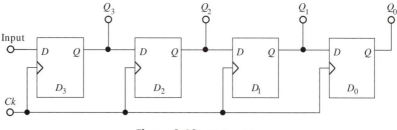

Figure 3.60 Shift register.

EXAMPLE 3.11 | **Shift register**

The shift register in Fig. 3.60 receives a hexadecimal F (= 1111) in serial form at its input. Assuming that the register is initially cleared, what are the states after two clock pulses?

SOLUTION:

We assume the lowest-order bit arrives first, so after one clock pulse, this bit occupies Q_3 and Q_2, Q_1, and Q_0 are still cleared. Thus, the pattern is $Q_3Q_2Q_1Q_0 = 1000$. After two clock pulses, the state of the register is 1100.

The shift register takes a serial input and produces parallel output when operated in the manner of Fig. 3.60. The same register can accept a parallel word at the preset inputs to the flip flops. The word can then be driven out Q_0 by the clock to produce serial output.

Binary multiplication. The shift register in Fig. 3.60 shifts its bit pattern one unit to the right with each clock pulse. A more versatile shift register results when the flip-flop inputs are connected with gates to their neighbors on left or right. Such a register can shift its bit pattern to the right or left, depending on which set of gates is ENABLED. This facility is useful in arithmetic operations. Multiplication by 10 in the decimal number system can be accomplished by shifting the decimal point one digit to the right. Similarly, multiplication by 2 in binary can be accomplished by moving the binary point one place to the right. This can be effected in hardware by shifting all bits one place to the left in a shift register, and clearly multiplication by 2^n requires n left shifts. A combination of shifting and adding intermediate results is required for multiplication by numbers that are not exact powers of 2. In like manner, division can be accomplished by shifting bits to the right in a shift register.

Summary. In this section, we have shown how flip flops are used to store digital information. Groups of flip flops called registers can receive or deliver words of information in parallel with a bus, or information can be stored and delivered in serial form. The shift register is also useful in binary arithmetic operations. These are the principal components of computers, to which we now turn.

Check Your Understanding

1. In the S–R latch made with NOR gates in Fig. 3.43, the forbidden state is $R \bullet S = 1$ or 0?

2. Tying J and K together in a J–K flip flop makes what type of flip flop?

3. If $Q = 1$ and $T = 0$ in the T-type flip flop of Fig. 3.54, what will $\overline{Q}$ be after the clock pulse (Ck)?

4. How many latches are required to store a byte of information?

5. Which of the following have a forbidden state: S–R latch, J–K FF, D-type FF, T-type FF?

Answers. **(1)** $R \bullet S = 1$; **(2)** T, or toggle FF; **(3)** $\overline{Q} = 0$ after the clock pulse; **(4)** eight; **(5)** S–R latch.

---

[23] $\overline{Q_3 Q_2 Q_1 Q_0} = 0010$.

Introduction

Electrical engineers have produced some passing fads, but computers are here to stay. These versatile devices increasingly influence modern business and pleasure; seers predict an even broader place for computers in the future. Computers have changed drastically since their development five decades ago. The original computers were big and expensive; only large institutions could justify their purchase for demanding computational tasks. Costs were in hundreds of thousands, if not millions, of dollars. Large (mainframe) computers still command an important place in the computer market. Then came the minicomputer, about the size of a suitcase. The electronics of minicomputers utilize LSI digital circuits. These computers made possible the automation of process control and the processing of data in real time; costs were in tens of thousands of dollars.

Now we have microcomputers. The low cost of the basic computer chip is revealed by its use in toys that retail for less than $25. Small personal computers, including memory, keyboard, and elementary software, sell for hundreds of dollars; you provide a CRT for display. The computer on a chip finds more and more applications: smart typewriters, cash registers, appliances, and sewing machines; in the automobile, in electronic instruments, at the video arcade, in the nursery. Perhaps the desktop computer itself best demonstrates the potential of microprocessor technology.

Computers are complicated systems, not easily explained. Aside from the fundamental principles of the computer itself, a myriad of related topics could be discussed: interfacing the computer to peripheral devices such as keyboards, CRTs, and printers; software development, including programming languages, editors, assemblers, and compilers; information representation matters such as fixed- and floating-point representation of numbers, data structures, codes for representing text; and potential applications such as numerical calculation, word processing, accounting systems, real-time process control, and time sharing. To add to the complexity, the world of computers has developed its own esoteric and colorful language.

We have listed topics that, for the most part, we are *not* going to discuss in this section. Our brief introduction to this subject addresses two basic questions: What is a computer, and how does it work?

Computer Architecture

The four basic elements of a computer are memory, an arithmetic-logic circuit, a control circuit, and input–output. These elements are present as well in a hand-held calculator. In that case, you supply the program by pushing the various buttons that sequence the calculations. When the calculator has the capability of storing a "program" of keystrokes, it is a computer, albeit a limited, slow, and highly specialized computer.

Central processing unit (CPU). Figure 3.61 shows the system configuration of a typical computer. The central processing unit (CPU) contains an arithmetic-logic unit (ALU), control circuits, and several registers for storage and general manipulation of words of data. The CPU is the brains of the computer and, at our level of understanding, its most complex and mysterious part. Although we have touched on hardware implementation of binary arithmetic and data storage, the complexity of the CPU places it far beyond the level of this introduction.

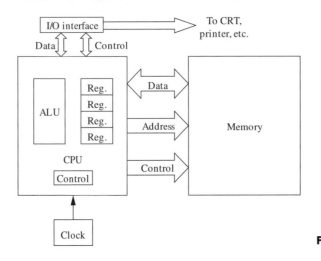

Figure 3.61 Elements of a computer.

The CPU has an instruction register that is sequentially loaded with words from the program in memory. These instructions, mere strings of 1's and 0's, are decoded and executed by the CPU: Words are brought from memory, placed in designated registers, processed according to the instruction, and then stored in memory at a specified location or perhaps transferred to an output device. Within the CPU are a number of registers to hold data and addresses of special places in memory relating to the program or to blocks of data. A program-counter register is incremented after each instruction is "fetched" from memory into the instruction register so that the CPU knows where to get the next instruction. The CPU communicates with memory by means of a two-way data bus, an address bus,[24] that controls the location in memory to furnish or receive data or program words, and a control bus that coordinates timing and order.

random-access memory (RAM)

Memory. Computer memory may be of several types. A *random-access memory* (RAM) is an array of memory registers with which data may be exchanged. The memory access is random because any memory location is equally accessible for reading or writing of data or program instructions. Normally, the program would not write into the portion of the memory containing program instructions, but the entire program must initially be loaded into memory. Most semiconductor memories are "volatile" because they lose the stored information when the computer is turned off. For this reason, and generally to store large amounts of information, magnetic disks or tapes are used for storage of digital information.

read-only memory (ROM)

Also important are *read-only memories* (ROMs), which contain information (usually programs) that can be read but not modified by the computer. Programmable ROMs (PROMs) allow the user to store information by burning microscopic fuses in the ROM. An erasable PROM (EPROM) can be reprogrammed after its information has been erased by ultraviolet light.

Input-output (I/O). Although self-contained for their internal calculations and data manipulations, computers must interact with the outside world. Often computer–person

---

[24] 16 bits for a 64K memory. In computer talk, K means 2^{10}, or 1024. Thus, a 16-bit address can specify one location out of 2^{16}, or 64 K, or 65,536 words.

interaction is provided through terminals with keyboard and display and through printers and plotters. Such devices translate between people-oriented symbols, such as alphanumeric text or graphical representation of information, and computer-oriented representation, bits, 1's and 0's of information.

Computers interact with external systems such as robots, electronic instruments, and manufacturing processes. Such interactions often involve digital-to-analog (D/A) and analog-to-digital (A/D) conversion. The computer communicates with external devices over an external data bus, which is separate from the computer's internal data bus. Several protocols exist for announcing which component has information to transfer to the computer, for keeping two devices from "talking" at once, and for assuring that the target component "heard what was said."

Programming languages. *Programming* is the art of translating a problem into words of 1's and 0's that the computer CPU executes to solve the problem. Figure 3.62 summarizes the communication problem: We think in terms of language, mathematical notation, accounting conventions, and so on, and the computer "thinks" in 1's and 0's. Programming languages provide the bridge between human thought processes and the binary words that control computer operations. Computer languages that are deliberately close to human thought processes or notations are called *high-level languages*: FORTRAN, BASIC, LISP, ALGOL, C, FORTH, and PASCAL are examples of high-level languages. Low-level languages are called *assembly languages*, and these are oriented toward the instructions that the specific computer can execute. The binary words are loaded into computer memory and executed as machine or object code. One line of code in a high-level language can produce many lines of machine code, whereas one line of assembly code, being close to the computer operations, will produce one or at most a few lines of machine code. For the microprocessor, the assembly language is also called the *op code* for the machine.

high-level languages, assembly languages, op code

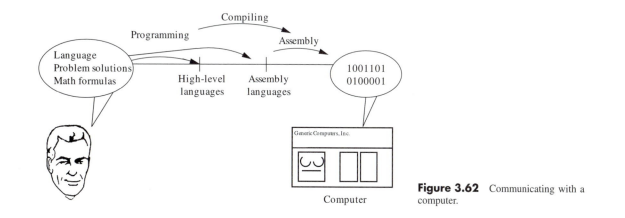

Figure 3.62 Communicating with a computer.

Compilers, assemblers, loaders, and interpreters. A *compiler* is a program that accepts as input a program in a high-level language, usually as a string of ASCII characters, and produces as an output either an assembly-language program or a machine-code program. The compiler is thus a computer program that performs language

compiler, assembler

translation from one computer language to another language closer to what the computer uses. An *assembler* performs the same function, using the assembly (source) code as input and producing output in object code.

loader, bootstrap

Compilers and assemblers are large programs that function on a specific computer. Such programs thus would be developed and furnished by the computer manufacturer, often on a ROM chip, or perhaps by an institution providing computer services. The object code that is produced usually is stored on disk or magnetic tape or in memory at a different location from where it will eventually reside. After the compiler or assembler does its work, it often is erased from memory to make room for the program machine code and whatever data the program requires. A small program called a *loader* loads the machine code into memory for execution. When the program consists of several separate parts (main program, subroutines, functions), the loader puts these into consecutive memory and, in effect, tells each where all the others are so that they can communicate. When one begins with a completely empty memory, a small program called a *bootstrap* can be loaded into memory by means of a keyboard or front panel switches. The bootstrap program can load the loader, which loads the program and signals the beginning of execution.

Some high-level languages, such as BASIC, are interpreted instead of compiled. This means that the program is converted into machine language and executed line by line. This is inefficient in computer time but efficient for finding certain program errors because mistakes can be corrected as they occur without having to reexecute the program up to the point where the error has occurred. This is especially appropriate when the programmer is working on a small computer.

In the past, programs were usually punched on cards or perhaps paper tapes and read into the computer with a card (or paper-tape) reader. Modern systems use editors that allow typing of alphanumeric text directly into the computer and offer the programmer easy modification through a variety of editing commands. Indeed, the storage, editing, and manipulation of text have opened new computer applications for the production of printed matter.

CHAPTER SUMMARY

Information may be represented in a variety of digital forms. The fundamental digital operations of AND, OR, and NOT are introduced. We show how electronic circuits can perform these operations once ONEs and ZEROs are represented as regions of voltage. Boolean algebra is introduced and logic expressions are placed in appropriate forms for realization by combinational logic circuits. Circuits with memory are developed based on the simple latch.

Objective 1: To understand how information is coded in digital form. Representing information in a series of "yes/no" questions might seem limiting, but the technique is quite powerful. Common alphanumeric and ad hoc codes are introduced.

Objective 2: To understand how to perform NOT, OR, NOR, AND, NAND, and XOR operations on binary variables. The basic operations between binary variables are defined and illustrated in the context of an example.

Objective 3: To understand how digital information is represented and manipulated with electronic circuits. A binary state is represented with an electrical signal that has one region for a ONE separated by a forbidden region from a region for a ZERO. By using such regions, simple circuits of diodes and transistors can perform the NOT, NAND, and NOR operations.

Objective 4: To understand how to use Boolean algebra to simplify and manipulate logic expressions. We use Boolean algebra to simplify and manipulate binary expressions into forms that permit minimal realization with logic circuits. De Morgan's theorems produce alternate forms of binary expressions for realization with NAND or NOR gates.

Objective 5: To understand how to efficiently implement logic expressions using standard logic gates. Logic expressions may be placed in forms that suggest direct realization with standard gates. Karnaugh maps allow the designer to reduce a truth table to a sum-of-products or product-of-sums form for realization with logic circuits.

Objective 6: To understand how flip flops are used to store and process digital information. The basic latch has two stable states and can store of 1 bit of information. From this basis, a family of flip flops is derived to store and process digital information.

Objective 7: To understand how a computer uses digital circuits to process information. The basic parts of a computer are described.

Chapter 4 presents the techniques of analog electronics, and Chap. 5 combines the techniques of digital and analog electronics for instrumentation systems.

GLOSSARY

AND function, p. 139, a two-variable logic function which takes a value of ONE if both its variables have a value of ONE.

ASCII, p. 170, an acronym for American Standard Code for Information Interchange. A common 8-bit code for representing text information.

Assembler, p. 176, a program that translates assembly (source) code into the object code used by a specific computer.

Assembly languages, p. 175, a low-level computer language that is oriented toward the instructions that the specific computer can execute.

Binary numbers, p. 153, numbers represented by digital variables, usually involving base-2 representation in some form.

Binary, logic or logical, digital, p. 136, synonyms for systems or devices that use two-valued variables.

Binary-coded decimal (BCD), p. 154, a method of representing numbers in digital form in which the decimal format of the numbers is preserved but each digit is represented in binary form.

Binary variables, p. 136, mathematical variables that have two symbolic values.

Bit, p. 154, the information represented by the state of a binary variable, the smallest unit of information.

Boolean algebra, p. 147, the algebra of two-valued variables, after George Boole, an English mathematician who first investigated this type of mathematics.

Bus, p. 170, a group of parallel paths for moving words of information around in a computer or other digital system.

Byte, p. 170, a convenient unit for digital information. One byte can represent eight bits, two BCD digits, or an ASCII alphanumeric symbol.

CMOS logic family, p. 146, a logic family that uses p-channel and n-channel, or complementary, transistors.

Combinational logic circuits, p. 150, logic circuits in which the output responds immediately to the inputs and there is no memory.

Compiler, p. 175, a program that accepts as input a program in a high-level language, usually as a string of ASCII characters, and produces as an output either an assembly-language program or a machine-code program.

De Morgan's theorems, p. 149, rules for distributing a NOT over variables that are ANDed or ORed. Generally when the NOT is distributed, ANDs are changed to ORs and vice versa.

DON'T CARE, p. 159, a state that cannot occur, which may be assigned a ONE or ZERO, whichever leads to the simplest logic function.

Edge triggering, p. 166, triggering with a state change of a digital signal.

EXCLUSIVE OR (XOR) function, p. 149, a logic function that takes the value ONE if either variable is ONE but takes the value ZERO if both variables are either ONE or ZERO.

Gates, p. 143, circuits that pass or block signals moving through a logic circuit.

Hexadecimal numbers, p. 155, numbers represented in base 16.

High-level languages, p. 175, computer languages that are deliberately close to human thought processes or notations.

Karnaugh map, p. 158, a technique for simplifying a Boolean expression by arranging the truth table into a geometric presentation where algebraic relationships are exhibited through the patterns of ONEs and ZEROs.

Latch circuit, p. 163, the basic memory circuit; stores one bit.

Logic levels, p. 136, regions of voltage or current that represent the two values of a binary variable.

NAND gate, p. 144, a circuit that combines the AND function with the NOT function.

NMOS logic families, p. 146, logic circuits are constructed entirely of n-channel MOS transistors.

NOR gate, p. 143, a circuit that combines the OR function with the NOT function.

NOT function, p. 137, a one-variable logic function that has a value that is the complement, or opposite, to the value of its variable.

OR function, p. 139, a two-variable logic function which takes a value of ONE if one or both of its variables have a value of ONE.

Parallel transmission, p. 171, when all bits in a word are transferred simultaneously.

Product of sums, p. 159, a misnomer for the ANDing of a series of ORs, like $(A + B)(C + B)$.

Random-access memory (RAM), p. 174, an array of memory registers in which any memory location is equally accessible for reading or writing of data or program instructions.

Read-only memory (ROM), p. 174, Memory that contains information (usually programs) that can be read but not modified by the computer.

Realize a logic circuit function, p. 146, to assemble the required logical operations with digital circuits to produce the desired output.

Register, p. 170, a series of flip flops arranged for organized storage or processing of binary information.

Ripple counter, p. 169, a type of counter in which the flip-flop transitions move in sequence through the counter.

Sequential logic circuits, p. 161, logic circuit with memory, whose output depends on the inputs plus its history.

Serial transmission, p. 171, when the bits in a word are transferred sequentially.

Sum of products, p. 157, a misnomer for the ORing of two or more ANDs, such as $AB + CD$.

Synchronous counter, p. 169, a type of counter that uses a simultaneous clock pulse at all flip flops and controls the counting operation with external gates.

Toggle, p. 167, to switch from one binary state to another.

Truth table, p. 137, a systematic enumeration of all possible states of a binary function.

Word, p. 154, an ordered grouping of binary information.

PROBLEMS

Section 3.1: Digital Infomation

3.1. A room has a three-way switch system, meaning that changing either switch changes the state of the light. Let S_A represent the switch at door A and S_B the switch at door B. Let $S_A = 0$ and $S_B = 0$ be a state with the light OFF ($L = 0$).
  (a) Develop a truth table relating the switch states to the digital variable L representing the light.
  (b) Give a Boolean expression for L as a function of S_A and S_B.

3.2. An outside floodlight has an automatic device turning it ON at night ($D = 0$) and OFF during daylight hours ($D = 1$). It also has a switch ($S = 1$ for the switch ON, controlling power to the automatic device and bulb), and $B = 1$ means the floodlight bulb is functional (not burned out).
  (a) Let $F = 1$ indicate that there is light from the floodlight. Give a truth table for F as a function of D, S, and B.
  (b) Write a Boolean expression for F.

3.3. A student is allowed to take a course ($C = 1$) if he or she pays the registration fee ($R = 1$) and either has the prerequisites ($P = 1$) or has the instructor's approval ($A = 1$).
  (a) Give a truth table for C as a function of R, P, and A.
  (b) Write a Boolean expression for C.

3.4. A burglary alarm system sounds an alarm ($A = 1$) if the detectors detect activity ($D = 1$). The alarm system has a key to disable it during working hours ($K = 1$ disables the alarm) and also has a test button ($T = 1$ sounds the alarm if the system is not disabled).
  (a) Give a truth table relating the output A to the inputs D, K, and T. Use X for DON'T CARE states, that is, states that will never happen in practice.
  (b) Give a Boolean expression for A. Count DON'T CAREs as 1.
  Comment: In this situation, some states are impossible. We have excluded the case where $T = 1$ and $K = 1$. Such states are called DON'T CARE states and are indicated by an X under A in the truth table rather than a 1 or 0. The logical function is

still valid; certain combinations of the independent variables never occur in practice. The designer therefore does not care about what the system does in these states. See page 159 for an example.

3.5. A student is sure to accept an invitation ($I = 1$) provided that the student likes the invitor ($L = 1$) and has no test the next day ($T = 0$). But if it's A Certain Friend ($F = 1$), the student will accept even if there is a test the next day.

    (a) Make a truth table relating the dependent variable, I, to the independent variables: L, T, and F.

    (b) Express the invitation function, $I(L, T, F)$, in terms of ANDs and ORs. Count DON'T CAREs as 1. See the comment on DON'T CARE states in Problem 3.4 or the example on page 159.

3.6. In the 1980 NBA playoffs, the Philadelphia 76ers basketball team led the Boston Celtics three games to one; the 76ers could have won the best-of-seven series by winning any of the fifth, sixth, or seventh games. Let G_5, G_6, and G_7 be digital variables to describe the outcome of those games, with $G_5 = 1$ if the 76ers win the fifth game and $G_5 = 0$ if Boston wins, and so forth for G_6 and G_7. For the sake of completing the truth table, we will award the remaining games to the 76ers should they win the series before the seventh game. The dependent variable is $P = 1$ if the 76ers win the series and $P = 0$ if the Celtics win. Make a truth table showing the relationship between G_5, G_6, and G_7 as input (independent) variables and P as the output (dependent) variable. This situation has some DON'T CARE states; see the comment in Problem 3.4.

Section 3.2: The Electronics of Digital Signals

3.7. The circuit shown in Fig. P3.7 has digital input and output, connected by some switches. The input is a digital 1, but the output is 1 or 0, depending on the switches. Let $S_A = 1$ if switch A is closed and $S_A = 0$ if switch A is open, and the same for all switches. Write a Boolean expression for the output, D, as a function of A, B, and C.

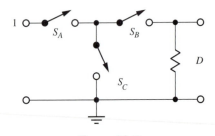

Figure P3.7

3.8. Figure P3.8 shows the input–output characteristic for a transistor circuit. The β of the transistor is 125.

    (a) Draw a circuit with this characteristic. Let one of your resistors be 1 kΩ.

    (b) Define logic levels for 0 and 1 such that the circuit exhibits the NOT function.

    (c) Modify the circuit such that it exhibits NOR behavior.

3.9. The circuit in Fig. P3.9 operates as a digital inverter. Assume a voltage of 0.7 V for a *pn* junction that is ON, and a saturation voltage of 0.3 V for the transistor. The β of the transistor is 50.

    (a) Find the minimum value of v_{in} to saturate the

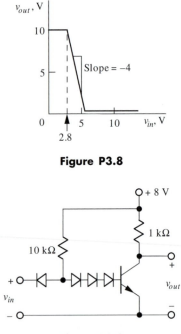

Figure P3.8

Figure P3.9

transistor. What is the collector current in saturation? What current must the input draw to saturate?

    (b) Find the maximum of v_{in} to leave the transistor in cutoff. What is the output (collector) voltage when the transistor is cut off? What current must the input draw to keep in cutoff?

3.10. The circuit in Fig. P3.10 uses switches to create logic inputs to a circuit. $S_A = 1$ indicates closure, and so on. Give the truth table for the circuit

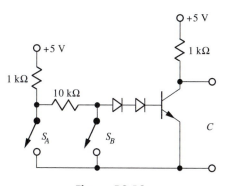

Figure P3.10

Section 3.3: The Mathematics of Digital Electronics

3.12. Make a truth table with two inputs, A and B, and the outputs OR, NOR, AND, NAND, EXCLUSIVE OR, and the equality function.

3.13. With a truth table, verify the second of the absorption rules in Fig. 3.20.

3.14. Simplify the following Boolean expressions

   (a) $\overline{\overline{A} + B(\overline{A} + B)}$.

   (b) $A(B + \overline{A}) + B(B + A)$.

   (c) $\overline{\overline{AB}} + A + B$.

3.15. Use the theorems of Section 3.3 to simplify the following digital functions:

Section 3.4: Combinational Digital Systems

3.17. Show two ways for realizing a NOT function with a NOR gate.

3.18. Give a realization of the Invitation function of Problem 3.5 with NOR and NAND functions.

   (a) Do this directly with the function as you derived it.

   (b) Use De Morgan's theorem to produce a simpler realization.

3.19. Show a way to make a three-input OR gate out of two-input OR gates.

3.20. The radio in a car should be ON ($R = 1$) if the ignition switch is either ON ($I = 1$) or in the accessory position ($A = 1$) and the radio on–off switch is also ON ($S = 1$). Watch for DON'T CAREs.

operation, using the usual convention for the output C. For the transistor, $\beta = 100$, $V_{CE(sat)} = 0.3$ V, and 0.7 V to turn ON a pn junction.

3.11. For the NOR gate in Fig. 3.14 to operate properly, the transistor must be saturated when the input is a digital 1. Assume that the transistor characteristics are those in Fig. 2.39 and that the minimum voltage input for a 1 is 4.0 V. Assume that the pn junctions have a voltage of 0.7 V when ON.

   (a) What is the largest value of R_B (10 kΩ in Fig. 3.14, but now a variable) that allows the transistor to remain saturated for an input 1?

   (b) Keeping R_B at 10 kΩ, what is the smallest value of the transistor β that will keep the transistor saturated? (Now the transistor characteristics are different from those in Fig. 2.39.) Assume that $V_{CE(sat)} = 0.5$ V.

   (a) $\overline{\overline{A} + (B + A)\overline{B}}$.

   (b) $A + \overline{B(1 + \overline{A})AA}$.

   (c) $\overline{ABC} + \overline{A}\,\overline{B}(C + A)$.

   (d) $A\overline{B}C + A + \overline{B}C$.

3.16. Simplify each of the following Boolean expressions by applying the theorems in Sec. 3.3.

   (a) $\overline{ABC} + \overline{AB\overline{C}}$.

   (b) $A + \overline{B}C + \overline{D}(A + \overline{B}C)$.

   (c) $A\overline{B}(C + D) + \overline{C + D}$.

   (d) $(A\overline{B} + \overline{C} + D\overline{E})(A\overline{B} + \overline{C})$.

   (a) Write a Boolean expression for R as a function of I, A, and S.

   (b) Give a realization with NOT, NOR, and NAND gates.

3.21. The safety system in Fig. P3.21 has three inputs. If two or more of these are 1 at the same time, an alarm should sound ($A = 1$).

   (a) Give a truth table for A.

   (b) Give a Boolean function for A.

   (c) Simplify the results of part (b) if possible.

   (d) Use De Morgan's theorem to put the Boolean expression in a form for NAND-gate synthesis.

   (e) Give a logic circuit to perform the alarm function using only two-input and three-input NAND gates.

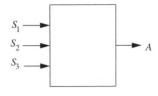

Figure P3.21

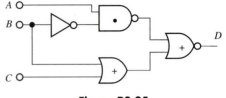

Figure P3.25

3.22. A student will pass ($P = 1$) if he has a good average ($A = 1$) and gets in all the required work ($W = 1$). But if he fails the final exam ($F = 1$), he will not pass.
  (a) Make a truth table for P as a function of A, W, and F.
  (b) Give a Boolean expression for P.
  (c) Give a realization using only two-input NOR gates.

3.23. A telephone-answering device has two incoming lines, A and B. (Consider A and B as digital signals with $A = 1$ for a call on A, etc.) If a call comes in on either line, the "ring" signal is given ($R = 1$) and stays on until the call is completed. If a second call comes in, the "second call" signal is given ($S = 1$) to a light.
  (a) Give a truth table with A and B as inputs and R and S as outputs.
  (b) Develop a combinational logic circuit using NAND and/or NOR gates that perform the necessary logic to activate the output signals.

3.24. Give a truth table for the logic circuit shown in Fig. P3.24. *Hint:* Consider the second input to the NOR gate to be independent of B, and then eliminate those states in which this input is not $\bar{B}$.

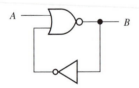

Figure P3.24

3.25. The digital circuit shown in Fig. P3.25 has three inputs and one output. Show the relationship between output and inputs with a truth table.

3.26. A Boolean expression is
$$D = A + (A + \bar{B})C$$
  (a) Give a truth table for D.
  (b) Simplify D if possible. You may use a Karnaugh map if you wish.

(c) Give a realization using only NAND gates. The inputs are A, B, and C, and the output is D.

3.27. Make a Karnaugh map for S_2 in Fig. 3.35 and show that no rectangular patterns of 1's are present. This shows that no simplification of Eq. (3.17) is possible.

3.28. On the Karnaugh map in Fig. P3.28, mark the regions corresponding to the following:
  (a) $A = 1$.
  (b) $\bar{C} = 1$.
  (c) $\overline{AB} = 1$.

C \ AB	00	01	11	10
0				
1				

Figure P3.28

3.29. The digital circuit shown in Fig. P3.29 has three binary inputs, Q_2, Q_1, and Q_0, and one output, L, which activates a light. The input is interpreted as a binary number between 0 and 7, and the output is supposed to activate the light ($L = 1$) if the input is divisible by either 2 or 3. Set $L = 1$ for $Q_0 + Q_1 + Q_2 = 0$ since zero is divisible by everything but itself.
  (a) Give a truth table for $L(Q_2, Q_1, Q_0)$.
  (b) Derive a sum-of-products expression for L using a Karnaugh map.

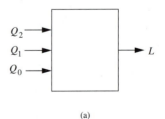

(a)

Figure P3.29

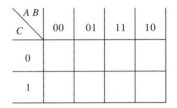

3.30. The digital circuit shown in Fig. P3.30 has three binary inputs, Q_2, Q_1, and Q_0, and one output, L, which activates a light. The input is interpreted as an English letter A through H with 000 representing A, 001 representing B, etc., through 111 representing H. The output is supposed to activate the light ($L = 1$) if the input letter appears in the expression "HARD TEST."

    (a) Give a truth table for $L(Q_2, Q_1, Q_0)$.

    (b) Derive a product-of-sums expression for L using a Karnaugh map.

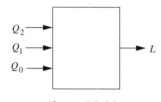

Figure P3.30

3.31. Make a Karnaugh map of the elevator-door function described in Fig. 3.5, and from the map, determine the logic function for D.

3.32. The seven-segment numerical display shown in Fig. P3.32 can display the integers 0 through 9, depending on which segments are illuminated. The driver accepts a BCD input and gives output 1's to the segments that should be lit and 0's to the segments that should not be lit.

    (a) Construct the truth table for segments a through g. Note that inputs corresponding to 10_{10}

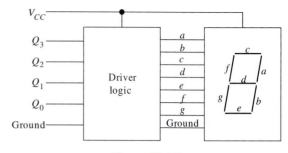

Figure P3.32

through 15_{10} do not occur and result in DON'T CAREs in the truth table.

    (b) Develop a Karnaugh map for the b segment.

    (c) From the map, determine the logic function for b.

    (d) Give a sum-of-products realization using two- and three-input NANDs.

3.33. Give the truth table for the logic circuit in Fig. P3.33.

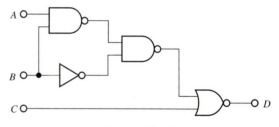

Figure P3.33

Section 3.5: Sequential Digital Systems

3.34. For the bistable circuit in Fig. 3.42(a) to operate as described, sufficient current must flow into the base of the ON transistor to permit saturation (that is, $\beta i_B \geq i_{C(\text{sat})}$). For the resistor values shown, this imposes a minimum value of β.

    (a) What is the minimum value of β for the circuit in Fig. 3.42(a) to function as a bistable? Assume 0.7 V for the ON base–emitter voltage and 0.3 V for $V_{CE(\text{sat})}$.

    (b) Find the formula for the minimum value of β in terms of the power-supply voltage (V_{CC}), base resistor (R_B), and collector resistor (R_C).

    (c) For the circuit in Fig. 3.42(a) and for a β of 60, what voltage at the input to the OFF transistor is required to switch the circuit? Assume 0.7 V to turn ON the diode and transistor pn junctions.

    Note: The bistable will switch if the current into the

ON transistor is dropped below the value required to saturate the transistor. This implies a certain voltage at the collector of the OFF (but coming ON) transistor. This implies in turn a certain collector current, and so on.

3.35. Design an S–R latch circuit with NAND gates rather than the NOR gates shown in Fig. 3.43. Determine the inputs (1 or 0) required to SET and RESET the flip flop.

3.36. The S–R latch in Fig. 3.43 has one input, $v_i(t)$, that goes from logical 0 to logical 1 at $t = 0$. The circuit has an external resistor and capacitor added, as shown in Fig. P3.36. Describe the output that occurs if $Q = 0$ for $t < 0$.

3.37. In a digital circuit, the edge-triggered flip flops could use an input circuit similar to that shown in

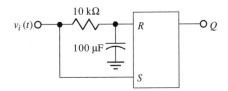

Figure P3.36

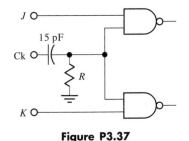

Figure P3.37

Fig. P3.37. Assume the gates treat a logical zero as any voltage in the range from 0 to 0.8 V and a logical one as any voltage in the range 2.3 and 5.0 V. In this case, the AND gates need to be open for at least 1 μs to pass the J and K signals. The capacitor is part of the integrated circuit and has a value of 15×10^{-12} farad (15 pF). A 4- to 5-V pulse is used for the clock. Find the minimum value of R to enable the edge-triggering feature to operate correctly.

3.38. Construct a truth table for the J–K flip flop in Fig. 3.50 with J, K, Q_n, and $\overline{Q}_n$ as inputs, R and S as intermediate outputs, and Q_{n+1} and $\overline{Q}_{n+1}$ as outputs. Use the truth table to confirm the truth table in Fig. 3.51. You must eliminate all rows that contradict the requirement that the outputs be complements.

3.39. The logic circuit in Fig. P3.39 uses a J-K flip flop. There is one input, J, and the output is Q. Find the output, Q_{n+1}, after the clock pulse in terms of the input, J, and the output before the clock pulse, Q_n.

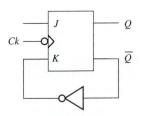

Figure P3.39

3.40. The circuit shown in Fig. P3.40 is not the usual latch because it uses one NAND and one NOR gate.
(a) If A and B are both ZERO, determine which, if any, stable output states exist. That is, which combinations of Q and $\overline{Q}$ can be present?
(b) If A and B are both ONE, determine which, if any, stable output states exist.

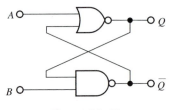

Figure P3.40

3.41. Using T-type flip flops and an AND gate, design a circuit to divide the clock frequency by 5.

3.42. The two T-type flip flops in Fig. P3.42(a) are connected and clocked as shown in Fig. P3.42(b). Give the values of Q_0 and Q_1 at times T_2, T_3, T_4, and T_5 if the states are both zero at T_1.

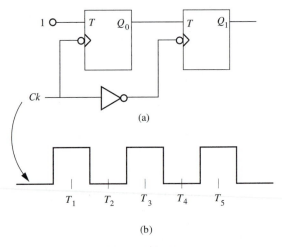

Figure P3.42

3.43. Repeat Problem 3.41, except use a NOR gate rather than an AND gate.

3.44. The S–R latch shown in Fig. P3.44(a) is made with NOR gates, as in Fig. 3.43. The inputs are given in Fig. 3.44(b). At $t = 0$, $Q = 0$. Give Q for $t > 0$. If Q is indeterminate, mark ×'s on the time axis.

3.45. Figure P3.45(a) shows the inputs and the clock of the T-type flip flop in Fig. P3.45(b). Give the output function $Q(t)$ if $Q = 0$ at $t = 0$.

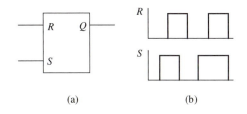

(a)　　　　　　(b)

Figure P3.44

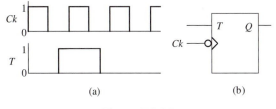

(a)　　　　　　(b)

Figure P3.45

3.46. In the circuit shown in Fig. P3.46, Q_0 and Q_1 are both 1 at $t = 0$. Give Q_0 and Q_1 as functions of time along with a sketch of the clock pulses.

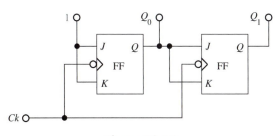

Figure P3.46

3.47. Two $J–K$ flip flops are connected as shown in Fig. P3.47. Give Q_0 and Q_1 under a sketch of the clock pulses. $Q_0 = 0$ and $Q_1 = 0$ at $t = 0$.

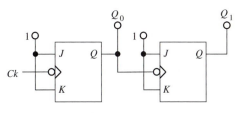

Figure P3.47

3.48. The J-K flip flop shown in Fig. P3.48 has its inputs tied together and tied to the output Q. Give the output as the clock pulses begin, assuming $Q = 1$ at $t = 0$.

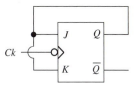

Figure P3.48

3.49 A T-type flip flop has the clock and output shown in Fig. P3.49. The flip flop triggers on the lagging edge of the clock.
  (a) Draw the circuit showing flip-flop type, inputs, and output.
  (b) Give a T input signal that will produce this output.

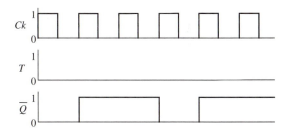

Figure P3.49

3.50. The output of a T-type flip flop can be expressed as $Q_{n+1} = T\overline{Q_n} + \overline{T}Q_n$, where Q_{n+1} is the output after the clock pulse, Q_n is the output before the clock pulse, and T is the input.
  (a) Show by means of a truth table that this gives the same information as the truth table in Fig. 3.54.
  (b) Give the corresponding expression for the output of the D-type flip flop.

3.51. Figure P3.51 shows a chain of T-type flip flops. The initial state is $Q_0 = 1$, $Q_1 = 1$, and $Q_2 = 0$.
  (a) What will be the state of the outputs after one clock pulse?
  (b) What will the circuit do generally?

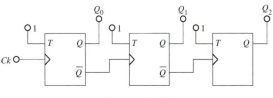

Figure P3.51

3.52. The *J-K* flip flop in Fig. P3.52 has its inputs tied together and tied to the output $\overline{Q}$. Give the output as the clock pulses begin, assuming $Q = 0$ at $t = 0$.

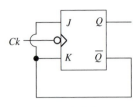

Figure P3.52

3.53. The *J-K* flip flop in Fig. P3.53(a) has *K* set permanently at a digital 1, but *J* varies with time, as shown in Fig. P3.53(b). Show the output *Q* on the same time scale as the clock and *J* if $Q(0^-) = 0$.

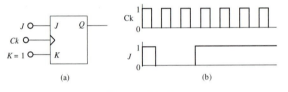

Figure P3.53

3.54. Figure P3.54 shows a *J–K* flip flop with some external logic connected.
  (a) Is the flip flop leading- or lagging-edge triggered?
  (b) Make a truth table with *J* and Q_n as inputs and Q_{n+1} as output.
  (c) Give a Boolean expression for $Q_{n+1}(J, Q_n)$.

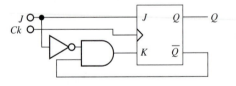

Figure P3.54

3.55. The *J–K* flip flop is connected as shown in Fig. P3.55. Determine the output Q_{n+1} as a function of the input *J* and the output from the previous state, Q_n.

3.56. A chain of *T*-type flip flops are connected as shown in Fig. P3.56, with the clock signal. Give the outputs at Q_0, Q_1, and Q_2 along with the clock signal. Assume $Q_0 = Q_1 = Q_2 = 0$ at $t = 0$.

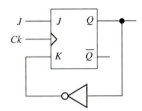

Figure P3.55

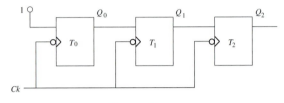

Figure P3.56

3.57. Figure P3.57 shows a clocked circuit with two binary inputs and one binary output. What should go in the box such that the output after the clock pulse is equal to the product of the two inputs, with inputs and output considered base-2 numbers? That is, $C_{n+1} = A_n \times B_n$. You may use standard gates and flip flops.

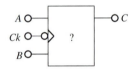

Figure P3.57

3.58. Figure P3.58 shows an ordinary *S–R* latch, except that an *RC* circuit is used in one of the connections. The time constant of the *RC* circuit is 0.5 second.
  (a) If $R = 0$ and $S = 0$ for a long time, what is *Q*?
  (b) Now *R* receives a 1-μs pulse. What happens?
  (c) Then 1 s later, *S* receives a 1-μs pulse. Explain what happens.

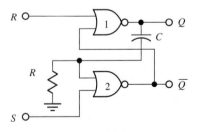

Figure P3.58

3.59. The circuits that perform multiplication in a calculator must handle the signs of the component numbers according to the usual rules of algebra. However, if the product is exactly zero, the sign should be set to positive so that a negative zero is not displayed. In Fig. P3.59, we display the portion of the circuit that determines the sign to be displayed. Consider the multiplication of N_1 and N_2. Let $S_1 = 1$ if the sign of N_1 is $+$, $S_2 = 1$ if the sign of N_2 is $+$, $Z = 1$ if the product is exactly zero, and $S_R = 1$ if a $+$ is to be displayed (including a positive zero).

(a) Make a truth table for the sign of the result, S_R, as a function of S_1, S_2, and Z.

(b) Give a Boolean expression for $S_R(S_1, S_2, Z)$. You may use $\oplus$ for the EXCLUSIVE OR if you wish. You may use a Karnaugh map.

(c) Give a realization using two- and three-input gates.

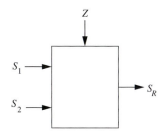

Figure P3.59

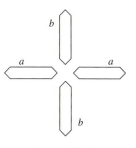

Figure P3.60

3.60. A calculator has a sign formed by two segments, as shown in Fig. P3.60. For plus, both a and b are lighted; for minus, only a is lighted. The calculator must compute the sign of the product of two signed numbers. Let S_1 be the sign of one of the numbers ($S = 1$ means plus) and S_2 be the sign of the other number.

(a) Give a truth table for a and b as functions of S_1 and S_2.

(b) Design a logic circuit for b using NAND gates.

3.61. Figure P3.61 shows four T-type flip flops clocked for synchronous counting. Using AND gates, design logic such that at the clock leading edge, FF_1 changes states when $Q_0 = 1$, FF_2 changes states when $Q_0 Q_1 = 1$, and FF_3 changes states when $Q_0 Q_1 Q_2 = 1$. What is the function of this circuit if $T_0 = 1$? What if $T_0 = 0$?

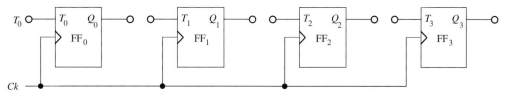

Figure P3.61

General Problems

3.62. The *rise time* of a signal pulse is defined as the time it takes to go from 10% to 90% of the difference between its initial and final values. The circuit in Fig. P3.62 shows a digital inverter with a 15-pF (15×10^{-12}F) capacitor representing the output capacitance of the transistor and the stray capacitance of the output circuit. Assume a voltage of 0.7 V for an ON pn junction and a voltage of 0.2 V for the saturation value of the collector–emitter voltage. Calculate the rise time of the output of this circuit.

3.63. A remote door lock is operated by an electrical relay that requires 10 mA to operate, as shown in Fig. P3.63. The relay can be opened by either switch A or switch B or both. The information from the switches is combined in a logic circuit. The logic levels are 0 to 0.5 V for a ZERO and 4.5 to 5 V for a ONE. The logic circuit can put out only 0.5 mA into a short circuit and hence cannot operate the relay directly. Hence, a transistor is used to switch the relay, and the logic circuit controls the transistor.

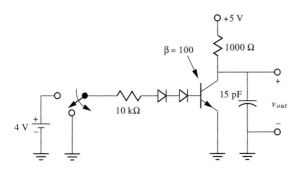

Figure P3.62

The incoming ring signal is a burst of 24-volt, 20-hertz ac voltage, as shown in Fig. P3.65(b). The ring signal is ON for 2 seconds and OFF for 4 seconds.

(a) Design a clock-shaper circuit that will convert the ring signal to a digital signal with nominal +5-and 0-volt logic levels. That is, find suitable values of R and C in the rectifier circuit shown. Assume a high input-impedance level for the logic circuit.

(b) Design a logic circuit, Fig. P3.65(a), that will "count" the rings and furnish signals to the triple-pole, single-throw switch, as shown, for input to the message response. Assume the counter starts in the "clear" state, with 0 at all flip flop Qs. [Assume you have suitable clock signals even if you draw a blank on part (a).]

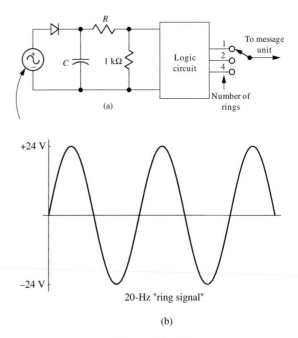

(a)

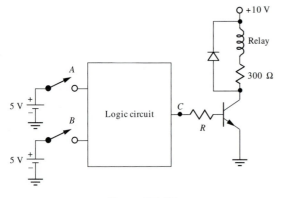

Figure P3.63

(a) Design the required logic circuit using only NAND and/or NOR circuit.

(b) Determine the resistor R. The beta of the transistor is 200, and it requires 0.7 V to turn ON the base–emitter junction.

3.64. Design a digital circuit that will compare a digital signal with its value 1 second earlier and give an output of 1 if the values are the same but zero if different. You may use any kind of latch or flip flop, NAND, NOR, AND, OR, or XOR gates.

3.65. A telephone-answering device has a switch on it that activates the response message ("After the tone, leave a message, etc.") after one, two, or four rings.

(b)

Figure P3.65

Answers to Odd-Numbered Problems

3.1. (a)

S_A	S_B	L
0	0	0
0	1	1
1	0	1
1	1	0

(b) $L = (S_A$ AND NOT $S_B)$ OR (NOT S_A AND $S_B)$.

3.3. (a)

R	P	A	C
0	0	0	0
0	0	1	0
0	1	0	0
0	1	1	0
1	0	0	0
1	0	1	1
1	1	0	1
1	1	1	1

(b) $C = R$ AND (P OR A).

3.5. (a)

L	T	F	I
0	0	0	0
0	0	1	1 or X
0	1	0	0
0	1	1	1 or X
1	0	0	1
1	0	1	1
1	1	0	0
1	1	1	1

(b) $I = F$ OR (L AND NOT T).

3.7. $D = S_A$ AND S_B AND NOT S_C.

3.9. (a) 2.1$^+$ V, 2.7 mA, 154 μA; **(b)** 2.1$^-$ V, 8 V, 520 μA.

3.11. (a) 38 kΩ; **(b)** 24.

3.13.

A	B	$A + B$	$A(A + B)$
0	0	0	0
0	1	1	0
1	0	1	1
1	1	1	1

3.15. (a) $\overline{\overline{A} + (B + A)\overline{B}} = \overline{\overline{A} + A\overline{B}} = \overline{\overline{A}}\,\overline{B} + \overline{\overline{A}}\,\overline{B} + A\overline{B}$

$= \overline{\overline{A} + \overline{B}} = AB$;

(b) $A + \overline{B(1 + \overline{A})}AA = A + A\overline{B} = A$;

(c) $\overline{ABC} + \overline{A}\overline{B}(C + A) = \overline{A} + \overline{B} + \overline{C} + \overline{A}\overline{B}C = \overline{A} + \overline{B} + \overline{C} = \overline{ABC}$; **(d)** $A\overline{B}C + A + \overline{B}C = A + \overline{B}C$.

3.17.

and

3.19.

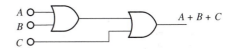

3.21. (a)

S_1	S_2	S_3	A
0	0	0	0
0	0	1	0
0	1	0	0
0	1	1	1
1	0	0	0
1	0	1	1
1	1	0	1
1	1	1	1

(b, c, and d)

$A = S_1 S_2 \overline{S_3} + \overline{S_1} S_2 S_3 + S_1 \overline{S_2} S_3$

$+ \; S_1 S_2 S_3 = S_1 S_2 + S_1 S_3 + S_2 S_3$

$= \overline{(\overline{S_1 S_2})(\overline{S_1 S_3})(\overline{S_2 S_3})}$;

(e)

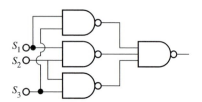

3.23. (a)

A	B	R	S
0	0	0	0
0	1	1	0
1	0	1	0
1	1	1	1

$R = A$ OR B, $S = A$ AND B

(b)

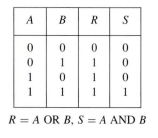

3.25. $D = \overline{\overline{A\overline{B}} + B + C} = \overline{\overline{A} + B + C}$

A	B	C	$\overline{A} + B + C$	D
0	0	0	1	0
0	0	1	1	0
0	1	0	1	0
0	1	1	1	0
1	0	0	0	1
1	0	1	1	0
1	1	0	1	0
1	1	1	1	0

3.27.

A_2 \ B_2C_1	00	01	11	10
0	0	1	0	1
1	1	0	1	0

3.29. (a)

	Q_2	Q_1	Q_0	L
0	0	0	0	1
1	0	0	1	0
2	0	1	0	1
3	0	1	1	1
4	1	0	0	1
5	1	0	1	0
6	1	1	0	1
7	1	1	1	0

(b)

Q_2 \ $Q_1 Q_0$	00	01	11	10
0	1	0	1	1
1	1	0	0	1

$$L = \overline{Q_0} + \overline{Q_2}Q_1$$

3.31.

T \ BS	00	01	11	10
0	1	0	0	1
1	0	0	0	1

$$D = B\overline{S} + \overline{T}S$$

3.33. A drops out and $D = \overline{C + B}$, NOR.

B	C	D
0	0	1
0	1	0
1	0	0
1	1	0

3.35.

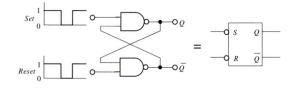

3.37. 120 kΩ.

3.39.

J	K	Q_n	$\overline{Q_n}$	Q_{n+1}
0	0	0	1	0
0	1	1	0	0
1	0	0	1	1
1	1	1	0	0

$$Q_{n+1} = J\overline{Q_n}$$

3.41.

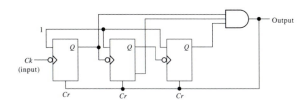

3.43.

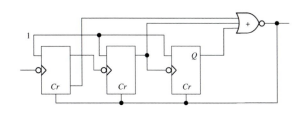

3.45.

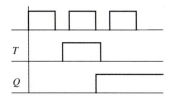

3.47.

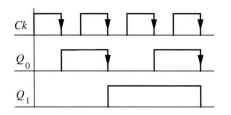

3.49. **(a)** *T*–flip flop, trailing-edge triggering. Input at *T*, output at $\overline{Q}$.

(b)

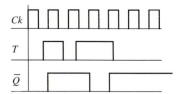

3.51. **(a)**

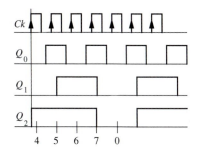

(b) This circuit counts up, modulo 8, starting with 3.

3.53.

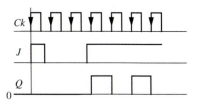

3.55.

J	Q_n	K	Q_{n+1}
0	0	1	0
0	1	0	1
1	0	1	1
1	1	0	1

$$Q_{n+1} = J + Q_n$$

3.57. $C_{n+1} = A_n B_n$, the AND.

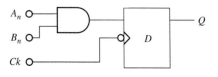

3.59. (a)

S_1	S_2	Z	S_R
0	0	0	1
0	0	1	1
0	1	0	0
0	1	1	1
1	0	0	0
1	0	1	1
1	1	0	1
1	1	1	1

(b) From the Karnaugh map,

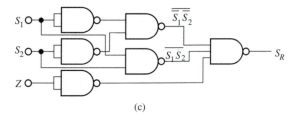

(c)

$$S_R = Z + S_1 S_2 + \bar{S}_1 \bar{S}_2 = Z + S_1 \oplus S_2$$

3.61. The circuit counts to 15 and resets to zero if $T_0 = 1$, and stops counting if $T_0 = 0$.

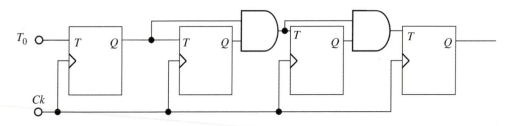

3.63. (a) A NOR gate; **(b)** 13.8 kΩ or less.

3.65. (a) 3800 Ω, 93.2 μF;
   (b)

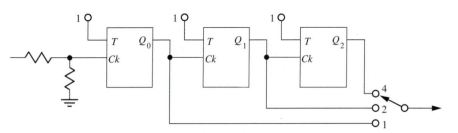

4

Analog Electronics

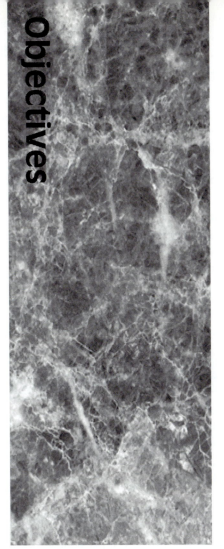

Analog techniques represent information by signals that mimic in some sense the quantity being symbolized, much as the mercury in a thermometer represents temperature. This chapter considers a variety of signal analysis and processing techniques. The operational amplifier is introduced and its basic applications in analog circuits are presented.

4.1 FREQUENCY-DOMAIN REPRESENTATION OF SIGNALS

Introduction

IDEA 4 Analog Information

analog

Contrast between analog and digital electronics. We have already explored how transistors and diodes are used as switches to process information in digital form. In digital electronics, transistors are either ON or OFF. The active region is used only in transition between states.

By contrast, analog electronics depends on the active region of transistors and other types of amplifying devices. In this context, *analog* means that information is encoded into an electrical signal that is proportional to the quantity being represented.

Analog example. In Fig. 4.1, our information originates physically in a musical instrument. The radiated sound is best understood as sound waves, which produce motion in the diaphragm of a microphone, which in turn produces an electrical signal. The variations in the electrical signal are a proportional representation of the sound waves. The electrical signal is amplified electronically, with an increase in signal power derived from the input ac power to the amplifier. The amplifier output drives a recording head that magnetizes a pattern on a moving magnetic tape. If the system is good, every acoustic variation of the air will be recorded on the tape and when the recording is played back through a similar system and the signal reradiated as sound energy by a loudspeaker, the resulting sound should faithfully reproduce the original music.

Systems based on analog principles form an important class of electronic devices. Radio and TV broadcasting are examples of analog systems, as are many electrical instruments used in monitoring deflection (strain gages, for example), motion (tachometers), and temperature (thermocouples). Many electrical instruments—voltmeters, ohmmeters, ammeters, and oscilloscopes—utilize analog techniques, at least in part.

spectrum

Contents of this chapter. Analog techniques employ the frequency-domain viewpoint extensively. We begin by expanding our concept of the frequency domain to include periodic, nonperiodic, and random signals. We will see that most analog signals and processes can be represented in the frequency domain. We introduce the concept of a *spectrum*, which is the representation of a signal as the simultaneous existence of many frequencies. Bandwidth in the frequency domain, the width of a spectrum, is related to information rate in the time domain.

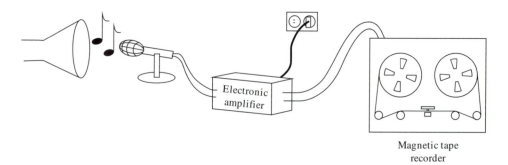

Magnetic tape
recorder

Figure 4.1 Analog system.

This expanded concept of the frequency domain also helps us distinguish the effects of linear and nonlinear analog devices. In this chapter, linear circuits are shown to be capable of "filtering" out unwanted frequency components. Chapter 6 shows how new frequencies, created by nonlinear devices such as diodes and transistors, are used in radios and other communication systems.

Next we study feedback, a technique by which gain in analog systems is exchanged for other desirable qualities such as linearity or wider bandwidth. Without feedback, analog systems such as audio amplifiers would distort signals. Understanding the benefits of feedback provides the foundation for appreciating the many uses of operational amplifiers in analog electronics.

Operational amplifiers provide basic building blocks for analog circuits in the same way that NOR and NAND gates are basic building blocks for digital circuits. We conclude by describing some of the more common applications of op amps.

Frequency-domain concepts. Sinusoidal sources can be represented by complex numbers through

$$v(t) = V_p \cos(\omega t + \theta) = \text{Re}[\underline{V}e^{j\omega t}] \tag{4.1}$$

where the phasor voltage, $\underline{V} = V_p \angle \theta$, is a complex number representing the sinusoidal function in the frequency domain. The $e^{j\omega t}$ term produces sinusoidal behavior; however, when only one frequency is present, the frequency is not stated explicitly in the frequency domain. The time-domain and frequency-domain representations for a sinusoidal signal appear in Fig. 4.2.

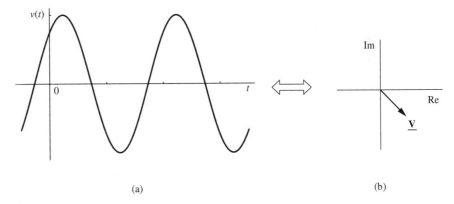

(a)   (b)

Figure 4.2 Sinusoid in (a) the time domain and (b) the frequency domain.

Frequency-Domain Representation of Periodic Functions

The Frequency Domain

A periodic function. Our representation of signals in the frequency domain is enriched when we consider periodic signals that are nonsinusoidal. As an example, we consider the function

$$v(t) = \frac{V}{2} + \frac{2V}{\pi}\left[\cos(\omega_1 t) - \frac{1}{3}\cos(3\omega_1 t)\right]$$

$$= \text{Re}\left[\frac{V}{2} + \frac{2V}{\pi}e^{j\omega_1 t} + \frac{2V}{3\pi}\angle 180°e^{j3\omega_1 t}\right]$$

(4.2)

harmonics

Leaving aside for the moment the significance of this mathematical function, let us consider the time-domain representation in Fig. 4.3. In the time domain, we see nothing unusual except that we now have a more complicated function than a simple sinusoidal function. The dc component of the voltage is easily identified in Eq. (4.2) as $V/2$. The time-varying part of the function consists of two sinusoidal terms. Because the frequency of the second term is an integral multiple of the frequency of the first term, the frequencies are said to be *harmonically related*. The term *harmonic*[1] is used because in music the tones that share harmonics form pleasant-sounding chords. Because the two sinusoids are harmonically related, they form a stable pattern and thus repeat with the period of the lower frequency, $T = 2\pi/\omega_1$. This function in the time domain has three harmonic components: a dc component (zero frequency), a fundamental (or first harmonic), and a third harmonic. The dc component is described by its magnitude, and the two sinusoidal components are described by their amplitudes and phases.

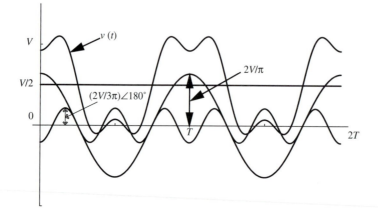

Figure 4.3 The time function.

The spectrum. Because we now have three frequencies involved, we can no longer describe the signal in the frequency domain by a phasor in the complex plane. Frequency becomes important in a new way and cannot be relegated to the memory or the margin of the page. A common way to handle this new complexity is shown in Fig. 4.4. We make frequency, ω, the independent variable and show the three harmonics at their frequencies. The magnitude (for the dc component) and phasors (for the ac components) are placed beside the various frequency components, drawn roughly to scale.

voltage spectrum

Figure 4.4 is a *voltage spectrum*; it shows how much voltage exists at the various frequencies.[2] The voltage spectrum in Fig. 4.4 is the frequency-domain representation of

---

[1] From the Greek, meaning "to fit together."

[2] Phase is expressed notationally but not pictured.

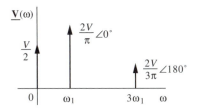

Figure 4.4 Spectrum consists of three frequencies.

the time-domain voltage in Fig. 4.3. Figures 4.3 and 4.4 portray the complementary nature of the two domains: Time and frequency are shown as complementary variables.

EXAMPLE 4.1

What frequencies?

What frequencies exist in the voltage output spectrum of a temperature sensor, with $t =$ hours after midnight?

$$i(t) = 10 + 2.6\ \sin(0.262t) - 9.7\ \cos(0.262t) + 2\ \cos(0.524t + 60°)\ \text{mV} \tag{4.3}$$

SOLUTION:
The constant term is dc or zero frequency. The next two terms have the same frequency of 0.262 radian/hour, 0.0417 cycles/hour, or $T = 24$ h, and the last term has a frequency of 0.524 rad/hour, the second harmonic.

WHAT IF? What if you want the voltage spectrum of the standard power voltage in the United States?[3]

Fourier series. In 1807, a French artillery officer immortalized his name among electrical engineers yet unborn by proposing that any periodic function can be expressed as a series of harmonically related sinusoids. Fourier's theorem expressed mathematically what musicians had long known by ear, that a steady tone consists of many harmonically related frequencies. The mathematical expression of Fourier's theorem, expressed for a periodic voltage, $v(t)$, is

$$v(t) = V_0 + V_1 \cos(\omega_1 t + \theta_1) + V_2 \cos(2\omega_1 t + \theta_2) + \cdots$$
$$+ V_n \cos(n\omega_1 t + \theta_n) + \cdots \tag{4.4}$$

where $\omega_1 = 2\pi/T$, T is the period of $v(t)$, V_n is the amplitude of the nth harmonic, and θ_n its phase. Fourier's theorem shows that any periodic function can be expressed as a spectrum consisting of dc, a fundamental frequency, and all frequencies that are integral multiples of the fundamental.

---

[3] One frequency: $120\sqrt{2}$ V at 60 Hz.

Spectra in music. Fourier's theorem gives a mathematical explanation for many musical phenomena. Musicians know that the overtones of musical instruments are important. The piano sounds different from the harpsichord, for example, because the harpsichord strings are excited in a way that creates more sound energy at the higher harmonics than piano strings produce. In the language of the frequency domain, the various musical instruments sound different because they produce different acoustic spectra.

The musical scale is based on the harmonic relationship between the frequencies of the notes. The octave is a factor of 2 in frequency. Two tones are said to be consonant, to sound good together, when they are harmonically related. For example, in the pure scale the frequency of G would be 3/2 that of C. It follows that the third, sixth, ninth, ... harmonics of C coincide with the second, fourth, sixth, ... of G. This shared harmonic structure defines all major fifth intervals in the pure musical scale, and other consonant intervals share harmonics similarly.

Our digression on music was to show that frequency-domain concepts are commonplace in certain areas of experience. For our purposes, Fourier's theorem, as expressed by Eq. (4.4), shows that all periodic functions can be described by harmonic spectra. We bypass the mathematical procedures for computing the amplitudes and phases of the harmonics of a periodic function;[4] our primary purpose is the exploration of this expanded concept of the frequency domain.

Two spectra. We give two examples of Fourier series. The first is the Fourier series representation of the periodic series of pulses shown in Fig. 4.5(a). The Fourier series for the square wave in Fig. 4.5(a) is

$$v(t) = \frac{V_s}{2} + \frac{2V_s}{\pi}\cos(\omega_1 t) - \frac{2V_s}{3\pi}\cos(3\omega_1 t) + \frac{2V_s}{5\pi}\cos(5\omega_1 t) - \cdots \qquad (4.5)$$

We showed only the first four components of the frequency-domain representation in Fig. 4.5(b). Comparison of Eqs. (4.2) and (4.5) shows that the first three frequency components of the square wave are those plotted in Fig. 4.3. Indeed, you should see some resemblance between the series of pulses and the sum of the dc, the fundamental, and

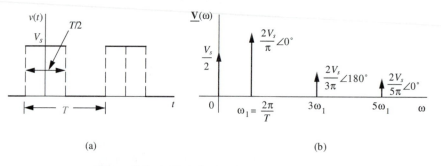

(a)   (b)

Figure 4.5 (a) Periodic time function; (b) voltage spectrum.

---

[4] These procedures are taught in courses in differential equations and signal analysis. Handbooks give the Fourier coefficients for many common periodical signals.

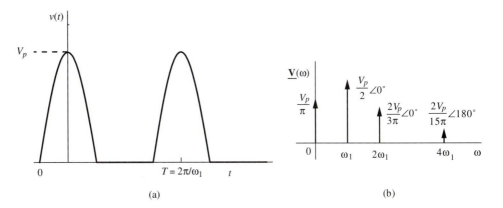

Figure 4.6 (a) Half-wave rectified sinusoid; (b) voltage spectrum.

the third harmonic in Fig. 4.3. If the fifth harmonic were added, it would steepen the slope at the sides and flatten the top and bottom. With the addition of each higher harmonic, the series would better approximate the periodic pulses.

A second example of the Fourier series of a periodic function is shown in Fig. 4.6(a), the half-wave rectified sinusoid. The Fourier series is

$$v(t) = V_p \left[\frac{1}{\pi} + \frac{1}{2} \cos(\omega_1 t) + \frac{2}{3\pi} \cos(2\omega_1 t) - \frac{2}{15\pi} \cos(4\omega_1 t) - \cdots \right] \tag{4.6}$$

The spectrum, Fig. 4.6(b), consists of the fundamental plus all even harmonics, unlike that in Fig. 4.5(b), which contains odd harmonics only. We introduce these spectra for examples to be explored in this and later sections.

Voltage spectra and power spectra. The spectra shown in Figs. 4.5(b) and 4.6(b) are called *voltage spectra* because they correspond to the amplitudes and phases of the harmonics in their respective signals. In many applications, the power in a signal carries great importance; hence, we extend the concept of a spectrum to include power and energy concepts.

Power is defined as energy flow per unit time. Figure 4.7 shows a sinusoidal voltage source connected to a resistive load. The power into the resistor, as given in the figure, depends only on the amplitude and the resistance; frequency and phase do not matter. The factor of 2 in the denominator results from time averaging, and can be absorbed into the voltage term if we use the rms instead of the peak value of the voltage.

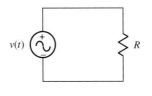

$v(t) = V_p \cos(\omega t + \theta)$ volts

$P_R = V_p^2 / 2R$ watts

Figure 4.7 Average power in a resistor.

Impedance Level

Power in volts². The presence of the resistance in the power relationship in Fig. 4.7 poses a mild dilemma in developing the concept of the power spectrum of a signal. One possibility is to state explicitly the impedance level of an assumed load, say, 50 Ω, and to compute the power that each harmonic would deliver to such a load. With this approach, watts would be the units of the power spectrum. The second possibility leaves the resistance term unstated. In this case, the "power" would be $V_p^2/2$, it being understood that to get actual power, one has to divide by an appropriate resistance. In this second approach, which is more general, the units of "power" are volts². Thus, the *power in volts²* is the power in R multiplied by R.[5] One would be wise to state explicitly these units to stress that the power spectrum is of this second variety. Because of its generality, this second approach is frequently preferred and will be used here.

Conservation Energy

Superposition of power. We stated before that the time-average power in a resistor does not depend on the frequency or phase of the sinusoidal source. It is easily shown, and is verified later, that the total power in a periodic signal is the sum of the power in its harmonics. Thus, the time-average power, P, in the periodic signal in Eq. (4.4) is

$$P = V_0^2 + \frac{V_1^2}{2} + \frac{V_2^2}{2} + \cdots + \frac{V_n^2}{2} + \cdots \tag{4.7}$$

The dc term is not divided by 2 because the effective value is the same as the peak value.

Thus, we can apply superposition to power provided the power contributions of the various sources are at different frequencies. Normally superposition applies only to voltage and current and not to power when we use sources at the same frequency. Thus, we may extend the principle of superposition to power provided (1) we are speaking of time-average power and (2) the sources have different frequencies.

Effective value of a periodic signal. The effective value, V_e, of a periodic signal can be expressed from its spectrum as

$$V_e = \sqrt{V_0^2 + \frac{V_1^2}{2} + \frac{V_2^2}{2} + \cdots + \frac{V_n^2}{2} + \cdots} \tag{4.8}$$

Thus, the effective value of a periodic function is the Pythagorean sum of the effective values of all its component harmonics, including the dc component. In fact, the preceding equation and interpretation are valid for the spectra of nonperiodic functions, provided the time average is taken over a long period of time and provided no two component frequencies are the same.

---

[5] A good compromise is to call these units "watts in a 1-Ω resistor."

| EXAMPLE 4.2 | **Power in a periodic signal** |

Confirm the equivalence of power between the time and frequency domains for the series of pulses in Fig. 4.5(a).

SOLUTION:

Conservation of Energy

The time-average power in volts^2 is

$$P_{avg} = \frac{2}{T} \int_0^{T/2} v^2(t)\, dt = \frac{2}{T} \int_0^{T/4} V_s^2\, dt = \frac{2V_s^2}{T} \times \frac{T}{4} = \frac{V_s^2}{2} \text{ volts}^2 \tag{4.9}$$

The power in the frequency domain would be the sum of the powers in the harmonics in Eq. (4.5)

$$P = \frac{V_s^2}{4} + \frac{2V_s^2}{\pi^2} + \frac{2V_s^2}{9\pi^2} + \frac{2V_s^2}{25\pi^2} + \cdots$$

$$= V_s^2\left[\frac{1}{4} + \frac{2}{\pi^2}\left(1 + \frac{1}{9} + \frac{1}{25} + \cdots\right)\right] \text{ volts}^2 \tag{4.10}$$

The infinite sum in Eq. (4.10) converges rapidly to the value $\pi^2/8$, as you can confirm from math tables or by summing terms with your calculator, and Eqs. (4.9) and (4.10) give the same result.

| **WHAT IF?** | What if you confirm the power equivalence for the half-wave rectified signal in Fig. 4.6(a)?[6] |

Summary. We showed in this section that a periodic signal can be represented by an infinite series of sinusoidal components at frequencies harmonically related to the fundamental frequency of the signal. The voltage spectrum of a periodic signal consists, therefore, of a series of harmonics, each being described by an amplitude and phase. The power spectrum has no phase information and may be described in either watts or volts^2. The power spectrum accounts for the total power as the sum of the powers in the individual harmonics.

Spectra of Nonperiodic Signals

An important question. Granted this success in representing periodic signals in the frequency domain through Fourier series, we are emboldened to ask: How far can we carry this line of development? This is an important question, for in electronic systems, we must deal with both periodic and nonperiodic signals. Examples of periodic

---

[6] $P = \dfrac{V_p^2}{4}$ volts^2 (time domain);

$$P = V_p^2\left\{\frac{1}{\pi^2} + \frac{1}{2}\left[\left(\frac{1}{2}\right)^2 + \left(\frac{2}{3\pi}\right)^2 + \left(\frac{2}{15\pi}\right)^2\right]\right\} = 0.2497 V_p^2 \text{ volts}^2.$$

signals are ac voltages, pulses sent out by radar transmitters, timing signals in a TV signal, clocking signals in a computer, and steady musical tones. Examples of nonperiodic signals are music, speech, information pulses in a computer, and radar pulse returns from a maneuvering target or a diffuse target, such as a thunderstorm. Thus, we must press for the extension of frequency-domain concepts to include a wider class of signals.

A strategy.   How, then, might we explore the possibility of having a spectrum for a nonperiodic signal, such as a pulse that happens only once? One approach, which works in this case, begins with a periodic series of pulses and then increases the period, keeping the width of the individual pulses constant. As we let the period of the repeating pulses become larger and larger, in the limit we would have a single pulse. This limiting process in the time domain affects the spectrum, and the resulting spectrum corresponds to the spectrum of a single pulse.

Spectrum of a series of narrow pulses.   Figure 4.8 suggests such a series of pulses by showing three pulses. The individual pulse widths are τ, which is kept constant, and the period is T, which is allowed to increase so as to isolate a single pulse. The time origin is the center of one of the pulses; hence, the signal has even symmetry about the origin. This means that the Fourier series for these pulses contains harmonics in the form of Eq. (4.4), but every harmonic has a phase angle (θ) of 0 or 180°, because these phases alone possess even symmetry. Because the phase of 180° is equivalent to a negative sign in front of the amplitude, the Fourier series for this signal must be of the form

$$v(t) = V_0 + V_1\cos(\omega_1 t) + V_2\cos(2\omega_1 t) + \cdots + V_n\cos(n\omega_1 t) + \cdots \tag{4.11}$$

where $\omega_1 = 2\pi/T$ and V_n is the amplitude of the nth harmonic, which may be negative. The amplitudes can be computed by the standard methods of Fourier series, with the result

$$V_n = \frac{2V_s\tau}{T}\frac{\sin(n\omega_1\tau/2)}{n\omega_1\tau/2} = \frac{2V_s\tau}{T}\frac{\sin(n\pi\tau/T)}{n\pi\tau/T} \quad \text{V} \quad (n = 1, 2, 3, \cdots) \tag{4.12}$$

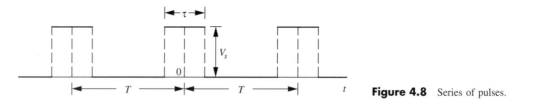

Figure 4.8   Series of pulses.

Recall that the spectrum consists of components at $\omega = 0$, ω_1, $2\omega_1$, ..., $n\omega_1$, ..., with harmonic amplitudes, except for $n = 0$, given by Eq. (4.12). We reserve as a problem the proof that Eq. (4.12) yields the results presented in Eq. (4.5) when $\tau = T/2$.

The harmonics of the spectrum.   As an example, we plot in Fig. 4.9(b) the spectrum for the case where $T = 10\tau$. We see in Fig. 4.9(b) that there are now many

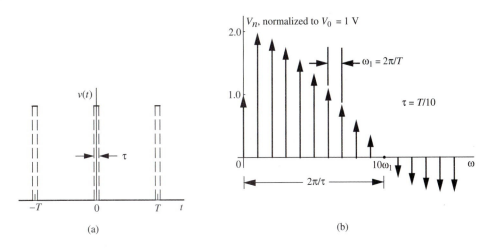

Figure 4.9 (a) Time-domain and (b) frequency-domain representations for a series of pulses with period T and width $\tau = T/10$.

harmonics that are significant, unlike the spectrum in Fig. 4.5(b), where the amplitudes fall off rapidly after the first few harmonics. The negative amplitudes indicate a phase of 180° from the tenth to the twentieth harmonics. The spectrum consists of a large number of harmonics spaced apart $\Delta f = 1/T$, where T is the period of the time function.

The amplitude of the spectrum. Although for graphing purposes we have normalized the harmonic amplitudes to unity relative to the dc component, Eq. (4.12) reveals that the individual harmonics become small as T is made large. There are two reasons for this decrease. As the period is increased, with τ kept constant, the pulses come less and less frequently and hence the power in the signal decreases accordingly. Specifically, the average power must decrease as $1/T$ due to the spreading of the pulses in time. Compounded with this effect, the power in any individual harmonic must decrease even faster because the power in the series of pulses distributes between an increasingly larger number of harmonics. As Eq. (4.12) shows, the power in, say, the first harmonic ($n = 1$) decreases as $(1/T)^2$ because power is proportional to the square of the amplitude.

Isolating a single pulse. With the aid of Fig. 4.9(b) and a measure of imagination, we can now anticipate what will happen to the spectrum as we allow the period, T, to approach infinity. As $T \rightarrow \infty$, the frequencies of the individual harmonics crowd closer and closer together and, in the limit, constitute a continuous spectrum. Individual harmonics must disappear because there is no longer any specific period; hence, frequency becomes a continuous variable. We conclude that a single pulse in the time domain contains all frequencies in the frequency domain.

The voltage spectrum of a single pulse. The shape of the spectrum will not change as T becomes large. The period, T, affects the spacing of the harmonics, but the shape of the spectrum depends only on τ. Thus, the shape of the spectrum retains the form in Eq. (4.12), which approaches

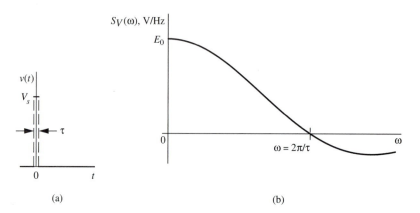

(a)   (b)

Figure 4.10   (a) Time-domain and (b) frequency-domain representations for a single pulse.

$$S_V(\omega) = E_0 \frac{\sin(\omega\tau/2)}{\omega\tau/2} \text{ V/Hz} \tag{4.13}$$

where the continuous variable ω replaces the discrete variable $n\omega_1$, E_0 is a constant, and $S_V(\omega)$ is the voltage spectrum of the pulse. This continuous spectrum is shown in Fig. 4.10(b).

Power or energy spectrum?

We wish also to investigate the power spectrum, but here we must be careful. As implied earlier, the single pulse carries no average power because the finite energy in the pulse is spread over all time, whereas power implies the continuous flow of energy as a time process. Consequently, we may meaningfully discuss the energy in the pulse but not the power. The voltage spectrum in Eq. (4.13), when squared, becomes an energy spectrum, not a power spectrum. The energy in the pulse distributes between the infinite number of frequencies making up the spectrum. Each individual frequency carries an infinitesimal quantity of energy, but taken together the spectrum accounts for the total energy of the pulse.

As in the case of periodic power spectra, we have the option of including resistance explicitly or omitting it to leave the definition of an energy spectrum more general. The units of the spectrum given in Eq. (4.13) are volts per hertz. The energy spectrum has the units joule-ohms per hertz, which requires division by a resistance and multiplication by a bandwidth in hertz to have the units for energy.

Spectrum amplitude.

Conservation of Energy
IDEA 2

Let us apply this interpretation of the square of the voltage spectrum in computing the energy in the pulse in the frequency domain. We can equate this energy to the energy in the time domain to evaluate the constant E_0 in the spectrum in Eq. (4.13). To have a physical picture, and to get the units to work out to the proper units of energy, let us consider that the pulse has an amplitude of V_s and is applied to a resistor of value R, as shown in Fig. 4.11. The total energy in the pulse is the integral of the power, Eq. (1.20). The required calculation is

$$W_R = \int_{-\infty}^{+\infty} vi \, dt = \int_{-\tau/2}^{+\tau/2} \frac{v^2}{R} \, dt = \frac{V_s^2}{R} \int_{-\tau/2}^{+\tau/2} dt = \frac{V_s^2 \tau}{R} \text{ J} \tag{4.14}$$

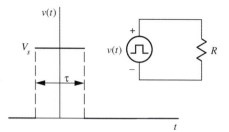

Figure 4.11 Computation of the energy in a single pulse.

The frequency-domain calculation follows the same reasoning used in Eq. (4.10), except that here we replace the sum of the contributions of each individual harmonic with the integral of the continuous spectrum. We use the rms value of each contributing sinusoid; hence the amplitudes must be divided by $\sqrt{2}$. We no longer must treat the dc term separately because it carries an infinitesimal amount of energy. Equation (4.15) summarizes the required computation.

$$W_R = \frac{1}{R}\int_0^\infty \left[\frac{E_0}{\sqrt{2}}\frac{\sin(\omega\tau/2)}{\omega\tau/2}\right]^2 d\left(\frac{\omega}{2\pi}\right) = \frac{E_0^2}{2R}\left(\frac{1}{\pi\tau}\right)\int_0^\infty \frac{\sin^2 x}{x^2}dx$$

$$= \frac{E_0^2}{4R\tau} \quad \text{J} \tag{4.15}$$

You will note that we are integrating with respect to $\omega/2\pi$, which is the frequency in hertz. The second form of the integral results from a change in variables from ω to $x = \omega\tau/2$. The definite integral has the value $\pi/2$, leading to the final result. We now equate the two energy calculations in Eqs. (4.14) and (4.15), with the result

$$\frac{E_0^2}{4R\tau} = \frac{V_s^2\tau}{R} \quad \Rightarrow \quad E_0 = 2V_s\tau \quad \text{volt-seconds} \tag{4.16}$$

Thus the spectrum of the pulse in Fig. 4.10(a) is

$$S_V(\omega) = 2V_s\tau\frac{\sin(\omega\tau/2)}{(\omega\tau/2)} \quad \text{V/Hz} \tag{4.17}$$

EXAMPLE 4.3	**Pulse spectrum**

Find the total energy and spectrum amplitude of a 10-V, 0.2-μs pulse.

SOLUTION:
The total "energy" is given by Eq. (4.14), except that we remove the resistance, R, to make the equation more general. The energy is

$$W_V = V_s^2\tau = (10 \text{ V})^2 \times 2 \times 10^{-7} = 2 \times 10^{-5} \text{ V}^2\text{-s} \tag{4.18}$$

The voltage spectrum of this pulse has the amplitude given by Eq. (4.16)

$$E_0 = 2V_s\tau = 2 \times 10 \times 2 \times 10^{-7} = 4 \times 10^{-6} \text{ V/Hz} \qquad (4.19)$$

The rms value of this spectrum height is $4 \times 10^{-6}/\sqrt{2} = 2.83 \text{ μV/Hz}$.

IDEA 7 The Frequency Domain

LEARNING OBJECTIVE 2.

To understand how to calculate the bandwidth required to pass pulses and other signals

Bandwidth. The spectrum of the single pulse in Eq. (4.17) illustrates an important general relationship between the time and frequency domains. In Fig. 4.12, we show the energy spectrum of the pulse, $S_P(\omega)$,[7] which is the square of the voltage spectrum in Fig. 4.10. Clearly, most of the energy is contained in the frequency bandwidth below the first null at $2\pi/\tau$. Integration of the energy spectrum reveals that 91.2% of the total energy of the single pulse lies in the bandwidth between 0 and the first null in the spectrum. Thus, if we were to pass this spectrum through an ideal low-pass filter that eliminated all frequencies above the first null of the spectrum, most of the energy would go through and the pulse would appear in the time domain without serious change. Similarly, the representation of the periodic pulses, Fig. 4.5(a), by the dc and first two harmonics contains 90.5% of the energy in the entire series of pulses, as you can verify from Eq. (4.5). Figure 4.3 verifies that the shape of the pulses is relatively unchanged by the elimination of the higher harmonics. Later in this chapter, we investigate filtering in the frequency domain; here we merely wish to establish that there is an effective bandwidth in the frequency domain associated with a pulse.

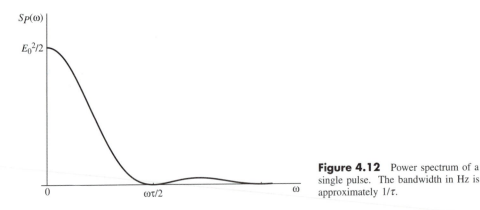

Figure 4.12 Power spectrum of a single pulse. The bandwidth in Hz is approximately $1/\tau$.

We may determine this effective bandwidth by setting $\omega\tau/2$ equal to π, the angle in radians where the spectrum first goes to zero. Solving this equation and converting to frequency in hertz, we obtain

$$\frac{\omega\tau}{2} = \pi \implies \omega = B = \frac{2\pi}{2\pi\tau} = \frac{1}{\tau} \text{ Hz} \qquad (4.20)$$

bandwidth

where B is the significant *bandwidth* in hertz.

---

[7]The subscript "P" indicates a power spectrum. In this context, this means that the voltage spectrum has been squared, not that the units are watts.

Bandwidth and pulse width. Important consequences follow from this approximate relationship between pulse duration in the time domain and bandwidth in the frequency domain. For one, we must provide adequate bandwidth if pulses are to pass through communication or computing systems without serious distortion. If, for example, we have a digital system that utilizes pulses of 1-μs duration, we must provide at least 1×10^6 Hz, or 1 MHz, of bandwidth to send those pulses to another location for recording or processing.

This reciprocal relationship between pulse length and bandwidth applies also to periodic pulses. The spectrum of the periodic pulses shown in Fig. 4.5(b) has the same general shape, and hence the same bandwidth, as that of the single pulse. Furthermore, this relationship between time duration and bandwidth does not depend on the exact shape of the pulses. The pulses may be rounded on top or triangular in shape: The same approximate bandwidth would be needed to preserve the identity of the pulses in passing through a communication system.

EXAMPLE 4.4	**Bandwidth of a pulse**

What is the bandwidth of the 10-V, 0.2-μs pulse in the previous example?

SOLUTION:
Using Eq. (4.20), we find

$$B = \frac{1}{\tau} = \frac{1}{2 \times 10^{-7}} = 5 \times 10^6 \, \text{Hz} \qquad (4.21)$$

WHAT IF?	What if you pass this pulse through an amplifier with a 1-MHz bandwidth? What would happen to the pulse?[8]

Example from mechanics. Let us imagine that we have a large bell and we strike this bell with a metal hammer. We know, of course, that the bell "rings", meaning that the mechanical resonances of the bell are excited to produce a ringing sound. The force between the hammer and the bell is of short duration, and this short pulse of force excites the bell structure. Because the energy spectrum of the short pulse is broad in bandwidth, the higher resonances of the bell are excited.

Now pick up a rubber hammer and again strike the bell. We would expect the bell to sound less harsh, more mellow. We can understand this change in sound in terms of the energy spectrum of the force from the rubber hammer. The rubber hammer remains in contact with the bell much longer because the rubber hammer is more compliant than the metal hammer. The result of this longer pulse of force is a smaller bandwidth of excitation; consequently, the higher-frequency resonances of the bell are not excited.

---

[8] The pulse would lose approximately 80% of its energy, and its shape would be smeared out to a width of 1 μs. Basically it wouldn't be a pulse anymore.

These higher resonances, which give the bell a harsh sound, are eliminated; hence, a mellow sound results with a rubber hammer.

Spectra of Random Signals

Importance of random signals. A *random signal* is any signal that is unpredictable to us, either because it originates from chance events or because its complexity defies routine analysis. Examples of random signals are broadcast signals over radio and TV channels, digital signals passing from computer to computer, and the roar of a jet aircraft engine.

random signal

Information is carried by random signals. We say that a signal contains information when we know something new because we received the signal. This implies that a signal containing the information is unpredictable to us, that is, is random in the sense given earlier. The examples given before illustrate the information-carrying possibility for random signals.

Characterization of random signals. If we are to design electronic circuits to monitor, record, transmit, receive, and process random signals, we must develop appropriate concepts to characterize such signals. Probability and statistics deal with random phenomena. These types of mathematics do not allow us to predict random events, but they can describe the structure of the randomness. Consequently, the application of probability theory to communication systems has yielded many important results.

It is not our aim here to explore communication theory. We limit our characterization of random signals to that required for the discussion of simple communications systems, such as an AM radio. Our major goal is to show that random signals can be characterized by a power spectrum in the frequency domain. First, we discuss the power in a random signal; then we discuss the time structure and show, in consequence, how the power in the signal is distributed in the frequency domain.

DC component of a random signal. Figure 4.13 shows a random signal. This might represent an electrical signal monitoring the temperature at a certain point in a chemical plant. The average temperature would be indicated by the average value of the voltage, v_{dc}. We can denote this time average as

$$v_{dc} = \langle v(t) \rangle \qquad (4.22)$$

where the angle brackets indicate time average. This average could be approximated by

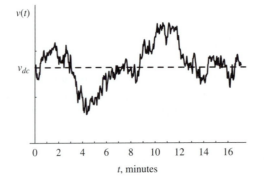

Figure 4.13   Random signal.

averaging over a long period of time:

$$v_{dc} = \langle v(t) \rangle = \frac{1}{T} \int_0^T v(t) \, dt \tag{4.23}$$

Here T would have to be large enough to include many random fluctuations, say, a period of hours for the function in Fig. 4.13. We acknowledge the possibility that at some future time, the plant might shut down and cool off, and hence the average temperature might change. But for normal operation, we assume there to be a meaningful average temperature as represented by the dc voltage in Eq. (4.23).

ac component

AC component. We may define the *ac component* of the random signal as the remainder after the average is subtracted:

$$v_{ac}(t) = v(t) - v_{dc} = v(t) - \langle v(t) \rangle \tag{4.24}$$

By "ac" we mean everything but the dc, the fluctuations in the signal. By definition, the time average of the ac component is zero:

$$\langle v_{ac}(t) \rangle = \langle v(t) - v_{dc} \rangle = \langle v(t) \rangle - v_{dc} = v_{dc} - v_{dc} = 0 \tag{4.25}$$

Because averaging is a linear process, we have distributed the averaging operation to the individual terms in Eq. (4.25).

Power structure. Leaving out the resistance for generality, we may define the power, P, to be

$$\begin{aligned} P &= \langle v^2(t) \rangle = \langle [v_{ac}(t) + v_{dc}]^2 \rangle \\ &= \langle v_{ac}^2(t) + 2v_{ac}(t)v_{dc} + v_{dc}^2 \rangle \text{ volts} \end{aligned} \tag{4.26}$$

Again we may distribute the time average.

$$\begin{aligned} P &= \langle v_{ac}^2(t) \rangle + 2\langle v_{ac}(t) \rangle v_{dc} + \langle v_{dc}^2 \rangle \\ &= \langle v_{ac}^2(t) \rangle + v_{dc}^2 \text{ volts}^2 \end{aligned} \tag{4.27}$$

The middle term drops out because the time average of the ac component vanishes, as shown in Eq. (4.25). Equation (4.27) indicates that the total power of the random signal may be considered as the sum of the powers in the dc and ac components, acting independently.

$$P = P_{ac} + P_{dc} \tag{4.28}$$

These two types of power produce distinct components in the power spectrum of the random signal.

correlation time

The Time Domain

IDEA 8

Time structure of random signals. The fluctuations of a random signal are characterized by a *correlation time*. This is the time, on the average, during which the fluctuations in the signal become independent of the past. In Fig. 4.13, for example, the correlation time is about 1 min. This correlation time suggests how far into the future we might be able to predict the signal. If we know the voltage in Fig. 4.13 at some instant of time, we could predict its value 1 or 2 s later with confidence. But our uncertainty would grow as the prediction moves into the future, and we can say little if anything about the value of the fluctuations, say, some 3 minutes hence. Thus, correlation time characterizes the time structure of a random signal in a crude way.

Statistics gives a precise definition of the correlation time and furnishes numerical methods for estimating it for a random signal. Investigation of such mathematical techniques, however, would lead us far beyond our goal of developing the ideas we need to understand basic communication systems.

Given that we know the correlation time for a signal, we can approximate the signal with a train of pulses, as shown in Fig. 4.14. The approximation is poor for details of the fluctuation of the random signal, but does yield a crude representation of the time structure of the signal.

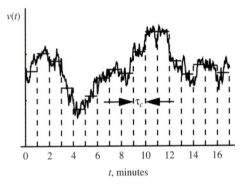

Figure 4.14 Random signal approximated by pulses.

IDEA 7
The Frequency Domain

Spectra of random signals. Let us now consider the spectrum of the random signal. A spectrum represents a time-domain signal as a sum of sinusoids. At first glance, it would appear to be impossible to represent an unpredictable random signal with stable, predictable sinusoids. This is true in part, for a random signal cannot be represented with a stable sum of sinusoids. The random signal contains sinusoids, but the phases of these sinusoids never stabilize, but rather vary randomly with time. An amplitude spectrum exists for random signals, but phase is meaningless. Thus, a power spectrum exists for a random signal but a voltage spectrum cannot be meaningfully defined.

Bandwidth of a random signal. The exact shape of the power spectrum of a random signal depends on the details of the nature of the fluctuations, but the meaningful bandwidth depends on the correlation time. If we were to consider the power spectrum of the pulses approximating the random signal in Fig. 4.14, all of them would have a significant bandwidth of

$$B \approx \frac{1}{\tau_c} \text{ Hz} \tag{4.29}$$

where τ_c denotes the correlation time. Thus, we would require approximately 1/60 Hz of bandwidth to handle the random signal in Fig. 4.13, but if the correlation time were 1 µs, we would require about 1 MHz of bandwidth.

Summary. The power spectrum, $S_P(f)$, of a random signal will have one component due to the dc component and one component due to the ac component, as shown in Fig. 4.15. The dc component is represented by a discrete line at zero frequency, but the ac component is represented by a continuous spectrum. The units of this continuous spectrum are volts²/hertz or volts²-second. As shown, the area of the continuous power spectrum represents the total power in the random signal fluctuations and must give the same power as the time-domain representation,

$$P_{ac} = \langle v_{ac}^2(t) \rangle = \int_0^\infty S_{ac}(f)df \tag{4.30}$$

where $S_{ac}(f)$ represents the power spectrum of the fluctuations in the random signal. Figure 4.15 also shows the relationship between significant bandwidth and correlation time, as asserted in Eq. (4.29).

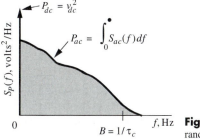

Figure 4.15 Power spectrum of a random signal.

EXAMPLE 4.5

Spectrum of the ac component

An electrical signal has a total power of $100\ V^2$. The time average of the signal is 5.5 V and the correlation time is 0.1 ms. Find the "power" and the bandwidth of the fluctuating component of the signal.

SOLUTION:

From Eq. (4.28), $P_{ac} = P - P_{dc} = 100 - (5.5)^2 = 69.8\ V^2$. The bandwidth is approximately

$$B \approx \frac{1}{\tau_c} = \frac{1}{0.1 \times 10^{-3}} = 10^4\ \text{Hz} \tag{4.31}$$

The average power density in the spectrum of the fluctuations would be

$$S_{P(avg)} = \frac{P_{ac}}{B} = \frac{69.8}{10^4} = 6.98 \times 10^{-3}\ \text{V}^2/\text{Hz} \tag{4.32}$$

WHAT IF? What if the dc voltage were −5.5 V?[9]

---

[9] No difference in the ac spectrum.

Bandwidth and Information Rate

Information.   Information has meaning in both technical and ordinary language. We might describe a book as containing information, and we might understand the gestures of a police officer directing traffic as offering information to motorists. In these instances, we think of information as offering knowledge—matters to be observed and acted upon.

The technical definition of information can cover these possibilities but deals primarily with the machine transmission or storage of symbols. With this theory, we may study the exchange of information between two computers or we may calculate the information stored in the genetic code. This mathematical theory of information is of recent origin, springing from the work of Claude Shannon (1916–). This theory deals with the probability of deciphering a known code that has been transmitted or stored in the presence of random disturbance. Information theory generally supports and enhances the commonsense understanding of information but also allows nonsense to be considered valid information. According to the mathematical theory, for example, the received message "3 = 5" is valid information if that is what the transmitter sent. In other words, the theory measures and describes objective information (messages sent and received), not subjective information (the meaning, validity, and importance of such messages).

IDEA 7
The Frequency Domain

Information rate and bandwidth.   Our purpose here is to show an important relationship between the information rate of a communication system and the bandwidth required by that communication system in the frequency domain. For our purposes, it is unnecessary to distinguish between the common understanding of information and that of the mathematical theory because this relationship is valid in both senses. Our examples are based on the commonsense understanding of information. The conclusion we reach, however, can be supported with the mathematical theory. We consider one analog and one digital example.

IDEA 4
Analog Information

Speech.   What is the band of audio frequencies necessary to communicate, say, over a telephone line? Experiments have established that the essential bandwidth of audio frequencies lies between about 800 and 2400 Hz. That is, if someone spoke to you via a telephone system that passed that band of frequencies, you could certainly understand what they said. Although this would be the minimum bandwidth for communication to occur, the phone company allocates the band of frequencies between about 400 and 3300 Hz. The additional bandwidth is not required for intelligibility but rather for recognition of the speaker. For satisfactory telephone service, we require not only to hear and understand the message, but also to recognize the speaker's voice. This recognition factor represents additional information content to the phone message.

An AM radio signal uses audio frequencies between 100 and 5000 Hz. This additional bandwidth is required for satisfactory transmission of music. Of course, FM radios sound better because their audio bandwidth is from 50 to 15,000 Hz. The large bandwidth enhances the quality of the music, a subjective measure of its information content.

A TV signal contains audio (voice) and video (picture). The audio bandwidth is similar to that of an FM radio, but the video requires much more bandwidth because the information rate is much higher. Of the 6-MHz bandwidth required for transmission of the entire TV signal, over 99% of the bandwidth is required for the video and synchronization signals, the remainder being used for the audio. Thus, we see a clear relationship between bandwidth and information rate in analog systems.

Digital Information

baud

Digital communications. Early communication systems (signal fires, signal flags, Morse-code telegraph) were digital in their coding. The invention of the telephone in the 1870s inaugurated a century of analog communications, but in recent decades, digital methods have made a comeback. Indeed, except for local telephone service, modern communication systems use digital codes.

The unit of digital information rate is the bit/second, or *baud*. The maximum baud rate of a communications system is determined by its bandwidth and its coding sophistication. Let us consider data rates for an ordinary phone line. Originally, data rates were deliberately slow, 110 baud, because the usual recipient of the information, a teletype, was essentially a mechanical typewriter that decoded and typed out the alphanumeric symbols as they arrived. Baud rates were increased for electronic "dumb" terminals to 1200 baud, which is about the limit for a phone line transmitting simple pulses, as shown by the next example. Modern modems utilize sophisticated codes to increase baud rates to about 20 kbaud. Such data rates are adequate for desktop computers, but the higher data rates required for business communications, for example, an airline reservation system, are achieved through increased bandwidth.

Increased bandwidths have been achieved by use of coaxial-cable and microwave systems at 100 Mbits/second, or megabaud, and a variety of optical communication technologies. Current optical systems use optical fibers to carry 3.4 gigabits/second,[10] or Gbaud, and experimental systems have achieved 34 Gbaud.

EXAMPLE 4.6   **Baud rate**

Determine the bandwidth required for a data rate of 1200 baud, using simple pulses to represent the bits of information.

SOLUTION:
Each bit (pulse) will require 1/1200 s to transmit, so the individual pulses will look as shown in Fig. 4.16. The pulse itself would occupy half the time space and hence the pulse width would be about 1/2400 s. Using the relation given in Eq. (4.20), we would expect that a bandwidth of about 2400 Hz would be required to send such a pulse without significant distortion; and therefore we would expect that an ordinary phone line would be adequate for connecting a computer with the remote terminal.

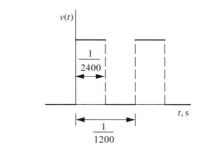

Figure 4.16 Digital pulse train, 1200 pulses/second.

---

[10]"G" = giga = 10^9.

Summary. We have expanded frequency domain concepts to periodic, nonperiodic, and random signals. We have investigated the relationship between time structure and bandwidth. For random signals, bandwidth determines the maximum information rate for a system.

Check Your Understanding

1. If the period of a periodic signal is 12 ms, what is the frequency in hertz of the third harmonic?

2. Which is the most important harmonic output frequency of a power supply: the zeroth, first, second, or third harmonic of the input frequency.

3. If a pulse has a width of 10 ms, what is the bandwidth in hertz required to carry the pulse without significant distortion?

4. If a computer terminal writes 30 characters/second, and each character requires 8 bits, what are the baud rate and the required bandwidth of the channel supplying data to the terminal?

5. Is the amplitude and phase information given by the voltage spectrum or the power spectrum?

6. A random signal has a correlation time of 2 ms. What is the required bandwidth for a communication channel to carry this signal?

7. Which type(s) of signals has (have) a continuous spectrum: sinusoidal, periodic, a single pulse, a random signal?

Answers. **(1)** 250 Hz; **(2)** the dc, the zeroth harmonic, is the *only* desired output of a power supply; **(3)** 100 Hz; **(4)** 240 baud, 480 Hz; **(5)** voltage spectrum; **(6)** 500 Hz; **(7)** single pulse and random signal.

4.2 ELECTRICAL FILTERS

LEARNING OBJECTIVE 3.

To understand how filters modify spectra

Introduction. The frequency-domain representation of signals is important because we can use electrical filters to shape the spectrum of a signal. In this section, we analyze the properties of simple low-pass, high-pass, and band-pass filters. We develop the Bode plot as a convenient means for depicting filter properties.

Filter Concepts

Filtering a rectified signal. In Fig. 4.17, we show an *RC* low-pass filter. Consider the input voltage to be the half-wave rectified sinusoid in Fig. 4.6(a); we wish to deter-

---

[11] The required bandwidth would be 200 MHz.

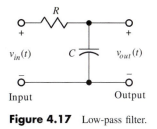

Figure 4.17 Low-pass filter.

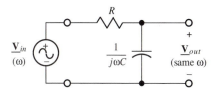

Figure 4.18 Low-pass filter in the frequency domain. We derive the filter function.

mine the output voltage, v_{out}. This input waveform is chosen because we have studied its spectrum; it does not represent the simple rectifier circuits analyzed in Chap. 2 because the output of these rectifiers would be modified by the input impedance of the low-pass filter.

Time-domain analysis. We can write a DE for the output voltage but the solution is a bit tricky. Furthermore, the time-domain approach generates little insight into the behavior of the circuit. We approach the problem in the frequency domain.

IDEA 7 The Frequency Domain

IDEA 8 The Time Domain

filter function

Filter function. Figure 4.18 shows the RC filter in the frequency domain, with R and C represented by their impedances. The input phasor, $\underline{\mathbf{V}}_{in}$, represents a single sinusoid at a frequency ω. The output phasor, $\underline{\mathbf{V}}_{out}$, is readily determined with a voltage divider:

$$\underline{\mathbf{V}}_{out} = \underline{\mathbf{V}}_{in} \times \left(\frac{1/j\omega C}{R + 1/j\omega C} \right) = \underline{\mathbf{V}}_{in} \times \frac{1}{1 + j\omega RC} \tag{4.33}$$

The *filter function* is the ratio between output and input phasors

$$\underline{\mathbf{F}}(\omega) = \frac{\underline{\mathbf{V}}_{out}}{\underline{\mathbf{V}}_{in}} = \frac{1}{1 + j\omega RC} \tag{4.34}$$

How the filter works. The characteristics of the low-pass filter function, $\underline{\mathbf{F}}(\omega)$, can be understood from the effect of frequency on the impedance of the capacitor. Recall that the impedance of a capacitor has a magnitude of $1/\omega C$. At dc, the capacitor acts as an open circuit; Hence, all the input voltage appears at the output, independent of R. At very low ac frequencies, the impedance of the capacitor is still very high compared to R. Specifically, as long as $1/\omega C \gg R$, the output voltage is approximately equal to the input voltage, and thus the filter gain is near unity. Thus, $\underline{\mathbf{F}}(\omega) \to 1$ as $\omega \to 0$, as shown mathematically by Eq. (4.34).

low-pass filter

At very high frequencies, the impedance of the capacitor approaches that of a short circuit and hence little voltage appears across the capacitor. As long as $R \gg 1/\omega C$, the current is approximately $\underline{\mathbf{V}}_{in}/R$ and the magnitude of the voltage across the capacitor, which is the output voltage, is approximately $|\underline{\mathbf{V}}_{in}|/\omega RC$ and hence approaches very small values as ω increases. Thus, $\underline{\mathbf{F}}(\omega) \to 0$ as $\omega \to \infty$. Consequently, the gain of the low-pass filter becomes very small at high frequencies. This means that a high-frequency signal would be greatly reduced, whereas low-frequency components would not be reduced by the filter; that is, it will "pass" only components of the signal at low frequencies. The filter characteristic, $\underline{\mathbf{F}}(\omega)$, is shown in Fig. 4.19, plotted on a log scale to show a large range of frequencies.

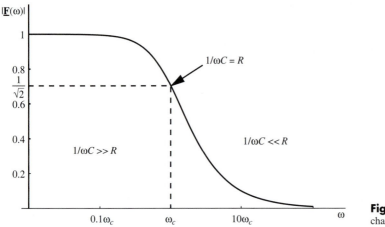

Figure 4.19 Low-pass-filter amplitude characteristic.

Cutoff frequency. The region of transition between these two types of behavior is centered on the frequency where the impedance of the capacitor is equal to that of the resistance:

$$R = \frac{1}{\omega_c C} \Rightarrow \omega_c = \frac{1}{RC} \qquad \text{and} \qquad f_c = \frac{1}{2\pi RC} \tag{4.35}$$

where ω_c is called the *cutoff frequency* and characterizes the region of frequency where filtering begins. The filter function can be expressed in terms of the cutoff frequency:

$$\mathbf{F}(\omega) = \frac{1}{1 + j\omega RC} = \frac{1}{1 + j(\omega/\omega_c)} = \frac{1}{1 + j(f/f_c)} \tag{4.36}$$

At $\omega = \omega_c$, or $f = f_c$, the filter function has the value $1/(1 + j1)$, which has a magnitude of $1/\sqrt{2} = 0.707$.[12] The significance of a voltage gain of 0.707 lies in the relationship between input and output power. Power is proportional to the square of the voltage; hence, at ω_c, the output power is reduced by a factor of 2 from what it would have been in the absence of the filter. For this reason, the cutoff frequency, expressed in either rad/s or hertz, is often called the *half-power frequency*.

Phase effects. Figure 4.19 shows only the magnitude of the filter function, $|\mathbf{F}(\omega)|$. The filter function is a complex function with real and imaginary parts. Thus, the filter affects both the magnitude and the phase of sinusoids passing through it. For example, at the cutoff frequency, the filter function has the value $1/(1 + j1) = 0.707 \angle -45°$. This phase shift is an unavoidable by-product of the filtering process that can cause problems in some applications. In audio systems, few problems occur due to phase shift because the ear is largely insensitive to phase, but in video systems, such phase shifts can degrade the performance.

---

[12] This has nothing to do with the rms value of a sinusoid.

Frequency-domain analysis. We return to the problem of low-pass filtering a half-wave rectified sinusoid. We can express the input voltage as a spectrum, a sum of sinusoids. Each of these represents an ac source and by superposition leads to a straightforward ac circuit problem, which we can solve using phasors and impedance. Specifically, if we have an input sinusoid of frequency ω and phasor magnitude, $\underline{V}_{in}$, the output phasor can be determined from the filter function.

Filtering a spectrum. Let us summarize the approach that we are following. The basic idea is indicated by Fig. 4.20. We use superposition to determine the effect of the filter on each input frequency. We know that the circuit, being a linear circuit, creates no new frequencies; each harmonic in the input produces a harmonic in the output. The filter affects each of these frequencies differently because the impedance of the capacitor varies with frequency. This frequency dependence of the circuit is symbolized by the filter function, $\underline{F}(\omega)$.

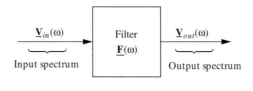

$$\mathbf{V}_{out}(\omega) = \underline{\mathbf{V}}_{in}(\omega) \times \underline{\mathbf{F}}(\omega)$$

Figure 4.20 Filtering a spectrum.

Passing the signal through the filter. We now can determine the effect of the input voltage in Fig. 4.6(a) on the filter in Fig. 4.17 by treating separately each frequency component of the signal. This approach is suggested by Fig. 4.21, where we have indicated that the input can be represented as a series of voltage sources representing the dc component, the fundamental, the second harmonic, the fourth harmonic, and so on. Our analysis is based on superposition, adding up the effect of each input sinusoidal source, considered separately. This exchanges the solution of a DE in the time domain for a host of ac circuit solutions in the frequency domain.

Input harmonics. Fourier analysis allows us to determine the amplitude and phase of the harmonics of the input signal. Equation (4.4) can be combined with Eq. (4.1) to become

$$v_{in}(t) = \text{Re}\{V_{in(dc)} + \underline{V}_{in(1)}e^{j\omega_1 t} + \underline{V}_{in(2)}e^{j2\omega_1 t}$$
$$+ \cdots + \underline{V}_{in(n)}e^{jn\omega_1 t} + \cdots\} \tag{4.37}$$

where $\underline{V}_{in(n)}$ represents the amplitude and phase of the nth harmonic of the input signal, as shown in the voltage spectrum in Fig. 4.6(b).

Output harmonics. We must next evaluate the filter function at each harmonic frequency to see how each is affected in amplitude and phase by the filter. This is symbolized by

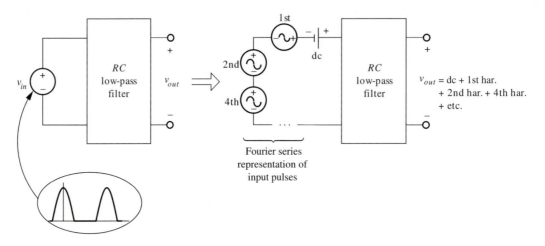

Figure 4.21 Filtering a periodic signal in the frequency domain.

$$V_{out(dc)} = \mathbf{F}(0) \times V_{in(dc)}$$

$$\underline{\mathbf{V}}_{out(1)} = \underline{\mathbf{F}}(\omega_1) \times V_{in(1)} \qquad (4.38)$$

$$\underline{\mathbf{V}}_{out(n)} = \underline{\mathbf{F}}(n\omega_1) \times V_{in(n)}$$

where $\underline{\mathbf{V}}_{out(n)}$ represents the amplitude and phase of the nth output harmonic and $\underline{\mathbf{F}}(n\omega_0)$ is the filter function evaluated at the frequency of the nth harmonic. Finally, we sum the frequency components of the output and transform back to the time domain, as shown in

$$\begin{aligned} v_{out}(t) = \text{Re}\{ V_{out(dc)} &+ \underline{\mathbf{V}}_{out(1)} e^{j\omega_1 t} + \underline{\mathbf{V}}_{out(2)} e^{j2\omega_1 t} \\ &+ \cdots + \underline{\mathbf{V}}_{out(n)} e^{jn\omega_1 t} + \cdots \} \end{aligned} \qquad (4.39)$$

Filtering the half-wave rectified sinusoid. So much for the general theory. We now work through the details for the problem presented in Fig. 4.21. Specifically, we assume a peak amplitude of 18 V and a fundamental frequency of 60 Hz for the half-wave rectified sinusoid. For the filter, we use $R = 500 \ \Omega$ and $C = 20 \ \mu\text{F}$, for a cut-off frequency of

$$\omega_c = \frac{1}{RC} = \frac{1}{500 \times 20 \times 10^{-6}} = 100 \ \text{rad/s} \qquad (4.40)$$

which is about one-fourth the fundamental frequency of 120π rad/s. The first few components in the output are as follows:

DC term. The dc input is $18/\pi = 5.73$ V. The filter function at dc is $\underline{\mathbf{F}}(0) = 1$, so the output at dc is $\underline{\mathbf{V}}_{out}(0) = 5.73$ V.

Fundamental. The phasor for the fundamental is from Eq. (4.6):

$$\underline{\mathbf{V}}_{in}(120\pi) = \frac{V_p}{2} \angle 0° = 9.0 \angle 0° \text{ V} \tag{4.41}$$

The filter function at the frequency of the fundamental is

$$\underline{\mathbf{F}}(120\pi) = \frac{1}{1 + j(120\pi/100)} = 0.256 \angle -75.1° \tag{4.42}$$

Thus, the output phasor for the fundamental is

$$\underline{\mathbf{V}}_{out}(120\pi) = 9.0 \angle 0° \times 0.256 \angle -75.1° = 2.31 \angle -75.1° \text{ V} \tag{4.43}$$

Higher harmonics. Similarly, we can calculate the second and fourth harmonics of the output to be

$$\underline{\mathbf{V}}_{out}(240\pi) = \frac{18 \times 2}{3\pi} \angle 0° \times \frac{1}{1 + j(240\pi/100)} = 0.502 \angle -82.4° \text{ V}$$

$$\underline{\mathbf{V}}_{out}(480\pi) = \frac{18 \times 2}{15\pi} \angle 180° \times \frac{1}{1 + j(480\pi/100)} = 0.0505 \angle +93.8° \text{ V} \tag{4.44}$$

Similarly, the sixth harmonic is $\underline{\mathbf{V}}_{out}(720\pi) = 0.0145 \angle -87.5° \text{ V}$.

Time-domain output. Equation (4.45) gives the time-domain output voltage by converting each harmonic to a sinusoidal time function.

$$v_{out}(t) = 5.73 + 2.31 \cos(120\pi t - 75.1°) + 0.502 \cos(240\pi t - 82.4°)$$
$$+ 0.0505 \cos(480\pi t + 93.8°) + 0.0145 \cos(720\pi t - 87.5°) + \cdots \text{ V} \tag{4.45}$$

From Eq. (4.45) and Fig. 4.22, the results of the filtering are apparent:[13] no loss of dc but substantial reduction of the fundamental and higher harmonics.

Bode Plots

decibels, Bode plot

Decibels of gain. The gain of amplifiers and the loss of filters are frequently specified in decibels, dB. This unit refers to a logarithmic measure of the ratio of output power to input power, as defined in

$$\text{Gain} = 10 \log\left(\frac{P_{out}}{P_{in}}\right) \text{ dB} \tag{4.46}$$

Because the power gain is proportional to the square of the voltage gain, the voltage gain in dB is defined to be

---

[13] Figure 4.22 includes only the five harmonics written in Eq. (4.45). If more harmonics were included, the curve would be smoother.

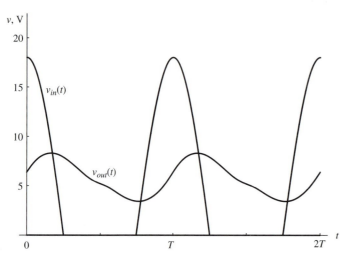

Figure 4.22 The low-pass filter passes the dc but reduces the harmonics.

$$\text{Gain} = 10 \log \left| \frac{\mathbf{V}_{out}}{\mathbf{V}_{in}} \right|^2 = 20 \log \left| \frac{\mathbf{V}_{out}}{\mathbf{V}_{in}} \right| \quad \text{dB} \tag{4.47}$$

We note in comparing Eqs. (4.46) and (4.47) that power ratios require a factor of 10 in the dB calculation and voltage ratios a factor of 20.

EXAMPLE 4.7   **dB of gain**

An audio preamplifier gives an output voltage of 500 mV with an input of 10 mV. Find the dB gain of the amplifier.

SOLUTION:
Using Eq. (4.47), we find

$$G_{dB} = 20 \log \left(\frac{500}{10} \right) = 34.0 \, \text{dB} \tag{4.48}$$

WHAT IF?   What if the input impedance level is 47 kΩ and the output impedance level is 1000 Ω?[14]

Loss in dB.   A loss can also be described by Eq. (4.47), but the "dB gain" is negative for a loss: For example, the gain of a low-pass filter [Eq. (4.36)] at $\omega = \omega_c$ was shown to be $1/\sqrt{2}$; Eq. (4.47) yields

$$G_{dB} = 20 \log \left| \frac{\mathbf{V}_{out}}{\mathbf{V}_{in}} \right|_{\omega = \omega_c} = 20 \log |\mathbf{F}(\omega_c)| = 20 \log \frac{1}{\sqrt{2}} = -3.0 \, \text{dB} \tag{4.49}$$

---

[14] Then the dB of voltage gain is the same but the dB of power gain is 50.7 dB.

Thus, we could say that the gain of the filter at its half-power frequency is negative 3.0 dB or, equivalently, that its loss is 3.0 dB. For this reason, the critical frequency of a filter is often called its 3-*dB frequency* or *half-power frequency*.

The dB scale is also useful for expressing the power in a signal. In this context, the power is expressed as a logarithmic ratio between the signal power and an assumed power level, usually 1 W or 1 mW. Thus dBw, dB relative to 1 W, is defined as

$$\text{dBw} = 10 \log \frac{P}{1 \text{ watt}} \qquad (4.50)$$

where P is a power and dBw is that same power relative to 1 W, expressed in dB. Similarly, dBm normalizes signal power to a milliwatt, expressed in dB.

EXAMPLE 4.8

Satellite transmitter

A satellite transmitter has an output power of 1000 W. Express this power in dBw.

SOLUTION:
From Eq. (4.50), this power would be

$$\text{dBw} = 10 \log \frac{1000\text{W}}{1\text{W}} = 30 \text{ dBw} \qquad (4.51)$$

WHAT IF? What if you want it in dBm?[15]

Why dB is useful. There are three reasons why electrical engineers use the dB scale for describing gains, losses, and signal levels. First, because of the compressive nature of the logarithmic function, the numbers involved in a dB calculation are quite moderate compared to a linear scale. This feature of dB measure is particularly useful when plotting quantities that vary greatly in magnitude.

cascaded systems **Cascaded systems.** The second virtue of dB measure is that the dB gains add when a signal is passed through cascaded systems. Figure 4.23 shows an amplifier *cascaded* with a filter, meaning that the output of the amplifier is the input to the filter. The combined voltage gain of the cascaded system is

$$\left| \frac{\mathbf{V}_3}{\mathbf{V}_1} \right| = \left| \frac{\mathbf{V}_2}{\mathbf{V}_1} \right| \times \left| \frac{\mathbf{V}_3}{\mathbf{V}_2} \right| = A \times \left| \mathbf{F}(\omega) \right| \qquad (4.52)$$

where A is the voltage gain of the amplifier. The dB gain of the cascaded system is

---

[15] 60 dBm. In general, dBw = dBm −30 dB.

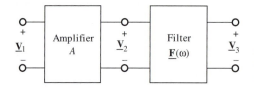

Figure 4.23 Cascaded systems. The output of the amplifier is the input to the filter.

$$G_{dB} = 20 \log \left| \frac{\underline{\mathbf{V}}_3}{\underline{\mathbf{V}}_1} \right| = 20 \log A + 20 \log |\underline{\mathbf{F}}(\omega)| \tag{4.53}$$

$$= A_{dB} + F_{dB}$$

where A_{dB} is the gain of the amplifier in dB and F_{dB} is the gain of the filter in dB. Thus, dB gains add for cascaded systems. This feature results from the familiar technique for multiplying numbers by adding their logarithms. In these days of calculators, this adding of dB gains seems no great benefit, but the custom of using dB became universal among electrical engineers in an earlier day and will no doubt continue in the future.

LEARNING OBJECTIVE 4.

To understand how to describe filters with Bode plots

Bode plots. A third virtue of dB measure involves a special way of plotting the characteristic of a filter, amplifier, or spectrum. The Bode plot (named for H. W. Bode, 1905–1982) is a log power versus log frequency plot. For example, Fig. 4.24 shows the Bode plot of the function:

$$\underline{\mathbf{A}}(f) = \frac{100}{1 + j(f/10)} \tag{4.54}$$

which combines an amplifier (voltage gain of $100 = 40$ dB) with a low-pass filter ($f_c = 10$ Hz). In Fig. 4.24, the vertical axis is proportional to log power expressed in dB. The horizontal axis expresses frequency on a log scale, although normally the frequency (not the logarithm of the frequency) is marked on the graph, as in Fig. 4.24. The log plot gives equal spacing on the graph to equal ratios of frequency—1:10, 10:100, etc.

decade

Advantages of Bode plots. The Bode plot is useful in two ways. As mentioned earlier, the log scales allow the representation of a wide range of both power and frequency. In this case, the vertical scale of 0 to 40 dB represents a range of power gain from 1 to 10^4, or a voltage gain from 1 to 100. Similarly, the horizontal scale represents

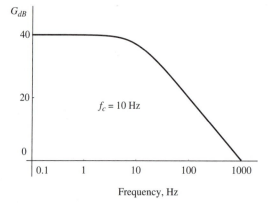

Frequency, Hz

Figure 4.24 Bode plot.

four *decades*, factors of 10, in frequency. These wide ranges are possible because of the compressive nature of the logarithmic function.

The other advantage of the Bode plot is that a filter characteristic can often be well represented by straight lines. The straight lines are evident in Fig. 4.24: The characteristic is quite flat for frequencies well below $f_c = 10$, and slopes downward with constant slope for frequencies well above this value. Only in the vicinity of f_c does the exact Bode plot depart significantly from these straight lines. In the following section we examine in more detail the relationship between the exact Bode plot and the straight-line approximation, which is called the *asymptotic Bode plot*.

Bode plot for a low-pass filter.

As shown in Eq. (4.36), the low-pass filter has the characteristic

$$\underline{F}(f) = \frac{1}{1 + j(f/f_c)} \tag{4.55}$$

The Bode plot represents the magnitude of the voltage gain and hence requires the absolute value

$$F_{dB}(f) = 20 \log \left| \frac{1}{1 + j(f/f_c)} \right| = 20 \log \frac{1}{\sqrt{1 + (f/f_c)^2}}$$
$$= -10 \log \left[1 + \left(\frac{f}{f_c} \right)^2 \right] \tag{4.56}$$

We may generate the exact Bode plot using normalized frequency f/f_c for numerical calculation, Fig. 4.25(a), as shown in Fig. 4.25(b), where we show that the exact Bode plot approaches a straight line above and below the cutoff frequency, $f/f_c = 1$. At the cutoff frequency, the filter gain is −3.0 dB, as noted earlier.

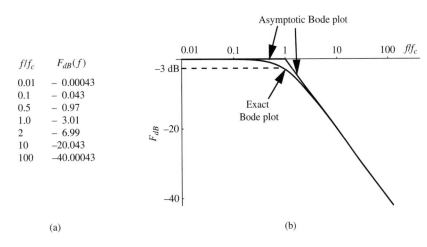

f/f_c	$F_{dB}(f)$
0.01	− 0.00043
0.1	− 0.043
0.5	− 0.97
1.0	− 3.01
2	− 6.99
10	−20.043
100	−40.00043

(a)

(b)

Figure 4.25   (a) Exact calculations; (b) Bode plot for the low-pass filter.

Asymptotic Bode plots. We may derive the asymptotic Bode plot by taking the limiting forms of Eq. (4.56). For frequencies well below the cutoff frequency, $f \ll f_c$, the gain is approximately

$$F_{dB} \approx -10 \log(1 + \text{small}) \approx -10 \log(1) = 0 \qquad (f \ll f_c) \qquad (4.57)$$

This accounts for the flat section. For frequencies well above the cutoff frequency, the gain approaches

$$F_{dB} \to -10 \log\left(\frac{f}{f_c}\right)^2 = -20 \log f + 20 \log f_c \qquad (f \gg f_c) \qquad (4.58)$$

Recalling that our horizontal variable is $\log f$, we see that Eq. (4.58) has the form of a straight line, $y = mx + b$, where $m = -20$, x is $\log f$, and b is $20 \log f_c$. Thus, the high-frequency asymptote will be a straight line when plotted against log frequency. When $\log f = \log f_c$, the y of the straight line has zero value and hence the line passes through the horizontal axis at $f = f_c$. The slope is -20: Usually, we say that the slope is negative 20 dB/decade because dB of gain and decades of frequency are the units for a Bode plot.

The low- and high-frequency asymptotes of the exact Bode plot combine to form the asymptotic Bode plot. As shown in Fig. 4.25(b), the exact Bode plot is well represented by this approximation, the largest error being -3.0 dB at the cutoff frequency. For most applications, the asymptotic Bode plot gives an adequate picture of the filter characteristics.

| **EXAMPLE 4.9** | **Low-pass filter** |

An analog data acquisition system has a low-pass filter with a cutoff (-3 dB) frequency of 50 Hz. From the asymptotic Bode plot, find the frequency where the gain is -6 dB.

SOLUTION:
This is in the part of the characteristic described by Eq. (4.58)

$$-6 = 20 \log f + 20 \log 50$$
$$20 \log f = 39.98 \quad \Rightarrow \quad f = (10)^{1.999} = 99.76 \text{ Hz} \qquad (4.59)$$

| **WHAT IF?** | What if you use the exact Bode plot?[16] |

---

high-pass filter

High-pass filter. Figure 4.26 shows a high-pass filter, which blocks low frequencies and passes high frequencies. The filter characteristic can be determined by considering the circuit as a voltage divider in the frequency domain

---

[16] The exact -6 dB frequency is 86.33 Hz.

$$\mathbf{F}(\omega) = \frac{\mathbf{V}_{out}(\omega)}{\mathbf{V}_{in}(\omega)} = \frac{R}{R + 1/j\omega C} = \frac{j\omega RC}{1 + j\omega RC} = \frac{j(f/f_c)}{1 + j(f/f_c)} \qquad (4.60)$$

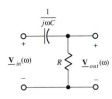

Figure 4.26
High-pass filter.

where $f_c = 1/2\,\pi RC$ is the cutoff frequency.

How the filter works. At low frequencies, $f \ll f_c$, the capacitor has a high imped-ance and allows little current; hence, little voltage develops across the resistor. Indeed, at zero frequency, dc, the capacitor acts as an open circuit and no voltage appears at the output. At high frequencies, $f \gg f_c$, the capacitor impedance is low, and essentially all the voltage appears across the resistor. Consequently, the filter passes high frequencies and blocks low frequencies. The transition between these two regimes occurs when the impedance of the capacitor is comparable to that of the resistor, and the cutoff fre-quency occurs where they are equal, as given again by Eq. (4.35).

Asymptotic Bode plot for high-pass filter. The asymptotic Bode plot for the filter function given in Eq. (4.60) may be derived by the same method used for the low-pass filter. The Bode plot is derived from the absolute value of the filter function,

$$F_{dB} = 20\,\log\left|\frac{j(f/f_c)}{1 + j(f/f_c)}\right| = 20\,\log\frac{f/f_c}{\sqrt{1 + (f/f_c)^2}} \qquad (4.61)$$

At frequencies well below the cutoff frequency, the frequency term in the denominator can be neglected and the asymptote becomes

$$F_{dB} = 20\,\log\frac{f}{f_c} = 20\,\log f - 20\,\log f_c \qquad (4.62)$$

Plotted against $\log f$, this is a straight line with a slope of $+20$ dB/decade, passing through the horizontal axis at $f = f_c$, as shown in Fig. 4.27. For high frequencies, the fil-ter function approaches unity, or 0 dB, also shown in Fig. 4.27. The exact Bode plot for the filter combines these two asymptotes and makes a smooth transition near the cutoff frequency. At the cutoff frequency, the exact Bode plot lies 3 dB below the intersection of the low- and high-frequency asymptotes.

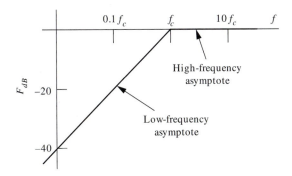

Figure 4.27 Asymptotic Bode plot for a high-pass filter.

Combining filter characteristics. Figure 4.28 shows a combination of a low-pass filter, an amplifier, and a high-pass filter. As derived in Eq. (4.53) and shown in

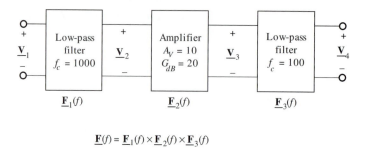

$$\mathbf{F}(f) = \mathbf{F}_1(f) \times \mathbf{F}_2(f) \times \mathbf{F}_3(f)$$

$$G_{dB} = F_{1(dB)} + G_{2(dB)} + F_{3(dB)}$$

Figure 4.28   Two filters and an amplifier in cascade.

Fig. 4.28, the dB gain of cascaded circuits is the sum of the dB gain of the components:

$$G_{dB} = F_{1(dB)}(f) + G_{2(dB)}(f) + F_{3(dB)}(f) \tag{4.63}$$

band-pass filter

where G_{dB} is the dB gain of the system. Thus, the Bode plots add to give the combined Bode plot of the cascaded system. Figure 4.29 shows the effect of adding up the individual Bode plots. The amplifier has a constant gain of 20 dB and raises the sum of the filter Bode plots by that amount. The high-pass filter blocks the low frequencies, and the low-pass filter blocks the high frequencies. The result is a *band-pass filter*; the "band" being passed in this case contains the frequencies between 100 and 1000 Hz. Thus, we combine filter effects by adding the Bode plots.

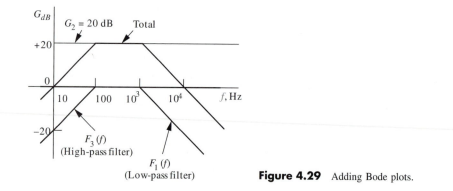

Figure 4.29   Adding Bode plots.

EXAMPLE 4.10   **Band-pass filter**

A band-pass filter for an audio equalizer is created by combining a low-pass filter with a cutoff frequency of 2 kHz with a high-pass filter with a cutoff of 1 kHz. What is the gain in dB at 1.5 kHz?

SOLUTION:

From the asymptotic Bode plots, we must say 0 dB. But the exact value from Eqs. (4.60) and (4.36) is

$$G_{dB} = -20 \log \frac{1.5/1}{\sqrt{1 + (1.5/1)^2}} - 10 \log [1 + (1.5/2)^2]$$

$$= -1.60 - 1.94 = -3.54 \text{ dB}$$

(4.64)

Thus, this filter does not work well when the relative bandwidth is not large.

A narrow-band _RLC_ filter. As shown before, we can produce a filter to pass the frequencies in a certain band by combining high-pass and low-pass _RC_ filters. This approach works well only when the relative passband of frequencies is rather large. However, most communication systems, such as radios, require filters that pass a narrow band of frequencies. This is normally accomplished with _RLC_ filters, such as appears in Fig. 4.30. This particular _RLC_ filter produces a parallel resonance between the inductor and capacitor at about 1125 kHz, which is near the middle of the AM radio band. The resistance shown in parallel with the inductor is not present in the physical circuit but is placed in the circuit model to represent the losses of the inductor. For frequencies far away from the resonant frequency, the impedance of either the inductor or the capacitor becomes small and the filter response drops. At the resonant frequency, the inductance and capacitance resonate and the output is maximum.

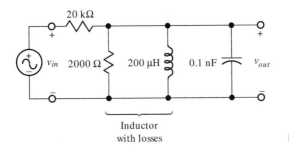

Figure 4.30 _RLC_ band-pass filter.

Figure 4.31 shows the band-pass characteristics of this filter in linear and dB scales. It might be noted that this filter has an inherent loss, even at its maximum output, -20.8 dB in this case. Thus, such a filter must be used in conjunction with an amplifier to compensate for this loss. This is no disadvantage in most communication circuits, for the radio already must provide considerable amplification—this merely requires a bit more. In Chap. 6, after we discuss the properties of nonlinear circuits, we discuss the role of such filters in typical radio circuits.

Check Your Understanding

1. A filter normally changes the phase as well as the amplitude spectrum of the input signal. True or false?

2. An _RC_ high-pass filter passes a dc signal. True or false?

3. If one decade is a factor of 10, what ratio corresponds to one-fourth of a decade?

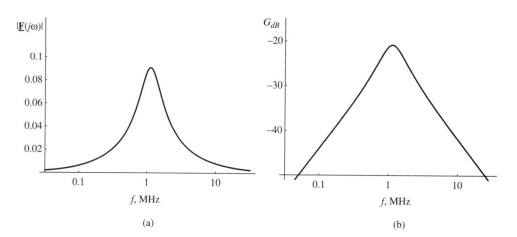

Figure 4.31 Narrow-band *RLC* filter characteristics: (a) linear plot; (b) dB plot.

4. If the input to an amplifier is 5 mW and the output is 0.1 W, what is its gain in dB if input and output impedance levels are the same?

5. If an amplifier reduces the signal voltage by a factor of 4, what is its gain in dB?

Answers. **(1)** True; **(2)** false; **(3)** 1.778; **(4)** 13.0 dB; **(5)** −12.0 dB.

4.3 FEEDBACK CONCEPTS

feedback

Introduction to feedback. *Feedback* brings a sample of the output back to the input of an amplifier, device, or composite system. The feedback signal is combined with the input signal to modify the input/output characteristics of the system. This technique, first proposed by Harold S. Black in 1928, has profoundly affected the development of electronics, for reasons that are explained in this section.

We first examine feedback effects with an amplifier and then generalize to a system model. The many benefits of feedback are discussed. Section 4.4 shows how feedback is used with operational amplifiers to achieve a family of circuits useful for amplification and signal processing generally.

A Feedback Amplifier

**LEARNING
OBJECTIVE 5.**

**To understand the
nature and
benefits of
feedback**

Feedback

⑨

Amplifier with feedback. Figure 4.32 shows an amplifier that has been modified from straight amplification by having a sample of the output brought back to interact with the input as a feedback signal. The main amplifier is represented by the top box: Its input, v_i, and its output, v_{out}, are related by the main amplifier gain A, as indicated in the box. We assume an ideal amplifier with A real and positive, infinite input impedance, and zero output impedance. In the main amplifier, the signal goes from left to right.

A feedback circuit, represented by the bottom box, consists in this example of two resistors arranged as a voltage divider. In the feedback circuit, the signal goes from right to left, from the output of the main amplifier back to its input. The input to the feedback circuit is v_{out} and its output is v_f.

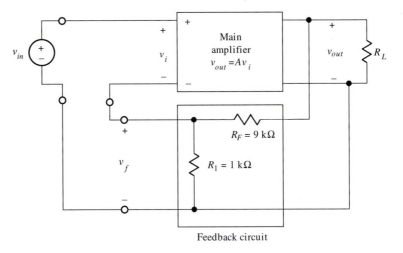

Figure 4.32 Amplifier employing feedback.

A voltage source, v_{in}, is located at the input to the entire "amplifier with feed-back," that is, the entire system, feedback and all. The output of the feedback circuit, v_f, is part of the input loop, which also contains the voltage source and the input to the main amplifier.

Analysis of amplifier gain. We calculate the gain of the amplifier with feedback, A_f, which is the ratio of the output to the source voltage:

$$A_f = \frac{v_{out}}{v_{in}} \tag{4.65}$$

We can write three equations describing the system. They are as follows:

1. Definition of the main amplifier:

$$v_{out} = Av_i \tag{4.66}$$

2. Voltage divider for the feedback circuit:

$$v_f = \frac{R_1}{R_1 + R_F} v_{out} \tag{4.67}$$

3. KVL around the input loop:

$$-v_{in} + v_i + v_f = 0 \tag{4.68}$$

Equations (4.66) through (4.68) give us three equations relating four voltages. We can eliminate two voltages between the equations, reducing to one equation in two voltages. We require the ratio of v_{out} to v_{in} for computing the gain in Eq. (4.65), so we eliminate v_f and v_i. Some straightforward algebra produces

$$A_f = \frac{A(R_1 + R_F)}{R_1 + R_F + R_1 A} = \frac{A}{1 + [R_1/(R_1 + R_F)]A} \tag{4.69}$$

We conclude from Eq. (4.69) that the gain with feedback, A_f, is smaller than the gain without feedback, A. Having lost gain, what have we achieved? Answer: We have improved the reliability, linearity, bandwidth, and impedance characteristics of the amplifier. We verify these benefits later in this section. First, we investigate how feedback works.

Signal levels. We now examine the signal levels for the amplifier with feedback in Fig. 4.32 to get a picture of how feedback works. The gain of the main amplifier is $A = 200$ and the feedback circuit consists of $R_F = 9 \text{ k}\Omega$ in series with $R_1 = 1 \text{ k}\Omega$. According to Eq. (4.69), the gain of the amplifier with feedback is

$$A_f = \frac{200}{1 + [1 \text{ k}\Omega/(1 \text{ k}\Omega + 9 \text{ k}\Omega)](200)} = \frac{200}{1 + 20} = 9.52 \tag{4.70}$$

We assume that the output voltage is 10 V and calculate the signal levels throughout the circuit. With 10 V at its output, the input voltage to the main amplifier, v_i, must be $10/200 = 0.050$ V. The input voltage to the entire amplifier is $v_{in} = 10/9.52 = 1.050$ V. The output of the feedback circuit, v_f, is

$$v_f = \frac{R_1}{R_1 + R_F} v_{out} = \frac{1 \text{ k}\Omega}{1 \text{ k}\Omega + 9 \text{ k}\Omega}(10) = 1.000 \text{ V} \tag{4.71}$$

Hence, we confirm KVL around the input loop: $v_{in} = v_f + v_i$.

The calculation performed in the preceding paragraph moves from the output back to the input. The signal goes the other way, so let us think of the operation of the amplifier in time sequence. We apply 1.050 V at the input at some instant of time. Electronic circuits respond quickly but not instantaneously; hence the output signal at this initial instant is zero. Thus, the feedback voltage, v_f, is at first zero and the entire 1.050-V input appears at the main amplifier input, v_i. This signal is amplified by the main amplifier and the output voltage increases toward 200×1.050; but as the output increases, so does the feedback signal. Because the feedback signal subtracts from the 1.050-V input, the input voltage to the main amplifier diminishes as the output voltage rises. Thus, we have competing effects: The higher the output voltage rises, the more voltage feeds back and the more the input voltage to the main amplifier is reduced. In equilibrium, the feedback signal subtracts 1.000 V from the input voltage of 1.050 V and the remaining 0.050 V is amplified by the main amplifier to yield an output voltage of 10 V.

System Model

System notation. A system representation of the feedback amplifier is shown in Fig. 4.33. This representation is characterized by the various blocks and circles connected with lines representing signal flow. Although our signals are voltages, customarily denoted with v's, we have used a neutral symbol x to denote the signals on the system diagram of Fig. 4.33. Use of a neutral symbol emphasizes that the *signal*, not the physical variable that represents the signal, is of primary importance from the system

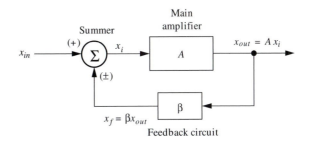

Figure 4.33 System representation of feedback.

viewpoint. At some places in the system, the signal might be represented by a voltage, at others a current, at yet others a pressure or the physical displacement of a mechanical component. The system representation embraces such hybrid systems.

System components. We show four system components in Fig. 4.33. The main
**sampling
connection**
amplifier and the feedback circuit are obvious carryovers from the original formulation of the amplifier except that here we allow A and β to be negative. The direction of signal flow is indicated with arrowheads. At the output, we show a means for *sampling* the output; this represents the parallel connection to the output of the main amplifier. In other feedback arrangements, the feedback network might be placed in series with the load and main amplifier output; with this arrangement, the feedback network would sample the output current. Both possibilities are covered by the dot at the output.

**summer,
comparator**
Summer. At the input, there is a circle containing a summation symbol to represent the *summer*, which represents the interaction of signals at the input of the system. In our previous example, this circle represents KVL in Eq. (4.68), which can be put into the form

$$v_i = + v_{in} - v_f \qquad (4.72)$$

where the $+$ and $-$ signs on the summer in Fig. 4.33 are associated with the input and the feedback signals, respectively. We may think of the summer as having a gain of $+1$ for v_{in} and a gain of -1 for v_f. The summer thus works as a differencer or subtractor in this case, but the symbol covers both possibilities, as indicated by the $+$ and $-$ signs. In other contexts, the summer at the input might be called a *comparator* because it compares the input with a sample of the output, furnishing the difference as input to the main amplifier.

Analysis of the system. The equations of the system are as follows:

1. Main amplifier:

$$x_{out} = Ax_i \qquad (4.73)$$

2. Feedback circuit:

$$x_f = \beta x_{out} \qquad (4.74)$$

3. Summer:

$$x_i = x_{in} \pm x_f \qquad (4.75)$$

In Eq. (4.75), we used a gain of $+1$ for the input signal but allowed for either $+1$ or -1 for the feedback signal. As before, we can eliminate two of the variables and solve for the ratio of the output and input signals to obtain the gain with feedback.

$$A_f = \frac{x_{out}}{x_{in}} = \frac{A}{1 - (\pm 1)(\beta)(A)} \tag{4.76}$$

Equation (4.76) reduces to our earlier result, Eq. (4.69), when the minus sign is used for the summer and $R_1/(R_1 + R_F)$ is used for β.

Loop gain. Another form for Eq. (4.76) is

$$A_f = \frac{A}{1 - L} \tag{4.77}$$

loop gain

where $L = (\pm 1)(\beta)(A)$ is called the loop gain. The *loop gain* is the product of the gains of all the system components around the feedback loop, including signs. The basic idea of the loop gain is suggested in Fig. 4.34. To calculate or measure the loop gain, break the feedback loop at some convenient point, insert a test signal, x_t, and calculate or measure the return signal, x_r. The loop gain is the ratio $L = x_r/x_t$. The loop may be broken, both in analysis and in the laboratory, only at a point where the function of the various system components is unimpaired. Care must be taken to terminate the break with the same impedance as the circuit saw before the break. For example, it would be inconvenient in the circuit in Fig. 4.32 to open the loop between R_F and R_1.

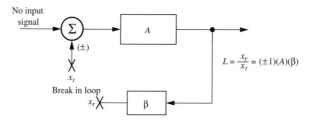

Figure 4.34 Determining the loop gain by breaking the loop at some point, inserting a test signal, and measuring the return signal.

EXAMPLE 4.11 | **Loop gain**

What is the loop gain of the amplifier in Fig. 4.32?

SOLUTION:
The gain of the summer is -1, so the loop gain is

$$L = (-1)(A)(\beta) = -200 \times \frac{1}{10} = -20 \tag{4.78}$$

Comparison of Eqs. (4.70) and (4.77) confirms this value.

WHAT IF? What if the voltage divider were eliminated and the output connected directly back to the input? Find the loop gain and the gain with feedback[17]

Sign of the loop gain. The sign of the loop gain indicates the nature of the feedback. Negative loop gain indicates negative feedback. With *negative feedback*, the feedback signal subtracts from the input signal and the gain is reduced. Positive loop gain indicates *positive feedback* and is rarely used. Positive feedback increases the amplifier gain; indeed, for $L = +1$, the gain becomes infinite. This condition is used in the design of electronic oscillators, which can be considered amplifiers with output but no input.

Magnitude of the loop gain. The magnitude of the loop gain indicates the importance of feedback in influencing the system characteristics. When the magnitude of the loop gain is much larger than unity, the system properties are controlled largely by feedback considerations. We illustrate the importance of loop gain in the next section.

Benefits of Negative Feedback

static stability **Improved static stability.** By *static stability* we mean insensitivity[18] of performance to changes in the system parameters. For instance, in our earlier example, the gain of the main amplifier might decrease due to the deterioration of a circuit component or replacement of a transistor. Assume the gain of the main amplifier decreases from 200 to 100. Without feedback, this change might jeopardize the usefulness of the amplifier. With feedback, however, the new gain, A_f', becomes

$$A_f' = \frac{100}{1 - (-1)(0.1)(100)} = 9.09 \tag{4.79}$$

Thus, a 50% decrease in the gain of the main amplifier results in a 5% decrease in the overall gain of the system. We can see from Eq. (4.77) the cause of this robustness. When the magnitude of the loop gain is much greater than unity, the gain with feedback is approximately

$$A_f = \frac{A}{1 - L} \approx \frac{A}{-L} = \frac{A}{-(-1)(\beta)(A)} = \frac{1}{\beta} \tag{4.80}$$

We see that, as loop gain increases, the gain with feedback approaches a value determined by β. Thus, the reduction of the main amplifier gain has little effect as long as the loop gain remains much larger than unity. The value of β could depend on resistors only, as in Fig. 4.32, and thus be stable over long periods of time. In our example the reciprocal of β is 10, and we see that the gain of the system, both before and after the gain decrease, falls close to this value. Often in a practical system, the main amplifier merely provides sufficient gain to keep the loop gain much larger than unity, for in this case, the β of the feedback circuit determines the overall gain with feedback.

dynamic stability We used the word "static" to distinguish this type of stability from *dynamic stabil-*

---

[17]$L = -200$, $A_f = 0.995$.

[18]Could also be called *robustness* or *immunity to change*.

ity, the tendency of the system to vibrate or oscillate under the influence of external stimuli. With a bridge, for example, the static stability might be good, meaning that the bridge footings are sound and the bridge members are sufficiently stiff to hold the bridge solidly in place. But under the influence of traffic or wind, the bridge might shake and even collapse, as did the Tacoma Narrows bridge in 1940, and hence exhibit poor dynamic stability.

Although negative feedback improves static stability, feedback can cause dynamic instability in a system. Time delays, or the frequency-domain equivalent, phase shift, can convert negative feedback to positive feedback and hence cause dynamic instability or oscillations. This aspect of feedback theory is discussed in Chap. 7.

EXAMPLE 4.12 | **Static stability**

How large does the loop gain have to be such that a 10% decrease in main amplifier gain causes a 1% change in overall gain?

SOLUTION:
Let $A' = 0.9\,A$, which gives $L' = 0.9L$. Then

$$A'_f = 0.99 \times \frac{A}{1 - L} = \frac{A'}{1 - L'} = \frac{0.9A}{1 - 0.9L} \quad \Rightarrow \quad L = -10 \tag{4.81}$$

WHAT IF? | What if the loop gain is –50? What is the change in overall gain with feedback if A changes by 10%?[19]

Improved linearity. Semiconductor devices such as transistors can be fairly linear in their active regions if signal variations are kept small, but large signals are often required. For example, the output amplifier in an audio system must produce a large signal to drive the speakers. Without feedback, such an amplifier would exhibit substantial distortion, but feedback techniques can be used to improve the linearity of the amplifier and thus reduce the distortion.

Equation (4.80) suggests how this is accomplished. The main amplifier introduces distortion into the system. If the loop gain is high, the system gain is determined by the β of the feedback circuit, which depends on two resistors and not directly on the main amplifier gain. The resistors are linear components; hence, the amplifier with feedback produces minimal distortion if the loop gain is high. Thus, a large loop gain is always used when good linearity is required for large-signal amplification.

Disturbance reduction. Feedback can reduce unavoidable disturbances that enter a system. As an example, we consider the thermostatically controlled oven shown in Fig. 4.35. The input voltage to the system is compared with a voltage from a temperature sensor in the oven. The difference, v_d, is amplified and applied to a heater element, v_H.

---
[19] The change is 2.17%.

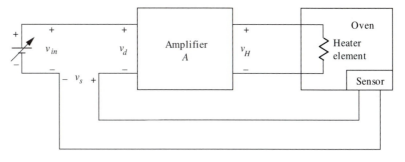

Figure 4.35 Oven controller.

We analyze the static characteristics here and the dynamic characteristics in Chap. 7.

A simplified system diagram is shown in Fig. 4.36. The heater element has been represented as a box having a gain A_H, which is defined as the change in box temperature divided by the change in the heater voltage. Although the relationship between heat (power) and voltage is nonlinear, for simplicity we are treating small changes as linear effects. The oven temperature is influenced both by the internal heater and by the ambient temperature, whose changes represent a disturbance in the system. This effect is represented with a summer. The sensor is also represented by a gain factor, A_S, which would represent changes in sensor output voltage divided by changes in oven temperature.

The loop gain of the system is

$$L = -AA_H A_S \tag{4.82}$$

Thus the relationship between oven temperature and input voltage is

$$T_{oven} = v_{in} \times \frac{AA_H}{1 - L} \tag{4.83}$$

Disturbance signal. Because we modeled the system as a linear system, we can use superposition to study the effect of the ambient temperature. The disturbance signal, T_a, thus represents a second input to the system. We redraw the system diagram in Fig. 4.37 to clarify this role. The gain of the second summer is $+1$, so we can determine

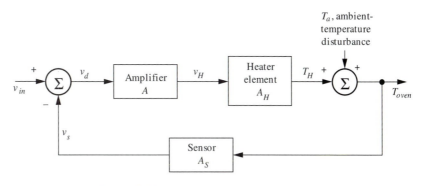

Figure 4.36 System diagram for oven controller.

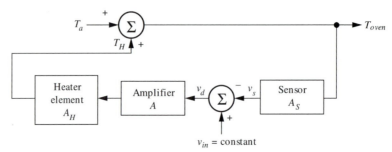

Figure 4.37 System diagram for controller with T_a as input.

a relationship between ambient and oven temperature:

$$T_{oven} = T_a \frac{1}{1 - L} \qquad (4.84)$$

where the loop gain, L, is the same as before. We want a large loop gain to reduce the effect of changes in the ambient temperature.

EXAMPLE 4.13 | **Loop gain**

Find the loop gain in the oven such that a $10°$ change in ambient temperature would cause a $0.1°$ change in oven temperature.

SOLUTION:
Interpreting Eq. (4.84) to apply to differences in temperature, we have

$$0.1° = 10° \times \frac{1}{1-L} \quad \Rightarrow \quad L = -99 \qquad (4.85)$$

The effect of ambient temperature can be reduced to low levels, provided the system can be stabilized dynamically. This system is analyzed in greater detail in Chap. 7.

Improved response time. Negative feedback can improve the response time of a system. Consider, for example, an electrical relay, as shown in Fig. 4.38. A relay is an **relay** electromagnet; when current passes through the coil, a magnetic force is produced at the gap and the lever closes against a spring. Such relays are used typically to activate switches and to operate locks and valves. We now analyze a relay having the following properties:

Current required to activate $= 20$ mA

Maximum current $= 30$ mA

Coil resistance, $R_c = 200\ \Omega$

Coil inductance, $L_c = 2$ H.

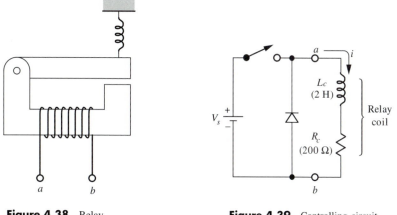

Figure 4.38 Relay.

Figure 4.39 Controlling circuit

Relay circuit. A suitable circuit for energizing the relay is shown in Fig. 4.39. When the relay is to be activated, the switch is closed. Current, initially at $i_0 = 0$, increases exponentially toward a final value of $i_\infty = V_s/R_c$. As current builds up, the relay closes as the current exceeds 20 mA.

EXAMPLE 4.14	**Relay circuit design**

Find the shortest time for relay closure without exceeding relay limitations.

SOLUTION:

To close the relay as quickly as possible, the circuit requires a voltage source to be as large as possible. The voltage source is limited by the maximum current to be $30\,\text{mA} \times 200\,\Omega = 6\,\text{V}$. With that value of voltage, the time required for the relay to close is easily shown to be $\tau \ln 3$, where $\tau = L_c/R_c$ is the time constant, $2\text{H}/200\,\Omega = 10\,\text{ms}$. Hence, in this case, the relay closes about 11.0 ms after the switch is closed.

WHAT IF? What if a resistor were placed in series with the relay?[20]

free-wheeling diode

Why the diode? Throughout the period when the relay current is increasing, the diode remains OFF; but when the switch is opened, the inductor reacts to keep the current going. This causes the diode to conduct and the current flows for a time through coil and diode, even with the switch open. The diode is thus placed across the relay coil to deenergize the inductance.[21] Without the diode, a spark would appear at the switch contacts.

---

[20] That would shorten the time of closure but increases the required voltage and circuit losses.

[21] This diode is called a *free-wheeling diode* because it allows the inductor current to continue after the source is removed.

Using feedback. Let us suppose that the 11-ms response time is too slow. We can use a feedback technique to improve the relay speed. As shown in Fig. 4.40, we add an ideal amplifier, with high input impedance and low output impedance and a resistor to the relay circuit. The current-sampling resistor, R_s, produces a feedback signal that is proportional to the current in the relay coil. After the switch is closed, the equations of the system are, for the amplifier,

$$v_{out} = A\,(V_s - v_f) = A\,(V_s - iR_s) \tag{4.86}$$

and for the coil and feedback resistor,

$$v_{out} = L_c \frac{di}{dt} + (R_c + R_s)i \tag{4.87}$$

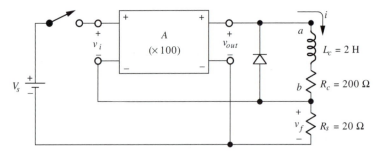

Figure 4.40 Feedback circuit to improve response time. The 20-Ω resistor samples relay current and gives a feedback signal.

When we combine Eqs. (4.86) and (4.87), we obtain a differential equation describing the feedback system:

$$L_c \frac{di}{dt} + [R_c + (1+A)R_s]i = AV_s \tag{4.88}$$

which can be put into standard form of Eq. (1.88)

$$\tau_f \frac{di}{dt} + i = \frac{AV_s}{R_c + (1+A)R_s} \tag{4.89}$$

where

$$\tau_f = \frac{L_c}{R_c + (1+A)R_s} \tag{4.90}$$

This time constant with feedback is shorter than the original time constant. With the numerical values shown in Fig. 4.40, the time constant drops from 10 to 0.91 ms and the response time is reduced accordingly, from 11.0 to 1.0 ms.

EXAMPLE 4.15 Turn-off

What is the turn-off time constant for the circuits shown in Figs. 4.39 and 4.40?

SOLUTION:

For the circuit of Fig. 4.39, the time constant with the switch is the same for turn on and turn off because the battery and ON diode are both low-impedance paths: $\tau = 10$ ms. When the input voltage is removed in the feedback circuit, both v_{out} and v_f drop instantly to zero and the inductor must deenergize through the free-wheeling diode as before. Thus, the turn-off time constant is 10 ms in both cases.

WHAT IF?   What if you want to shorten the turn-off time constant by a factor of 10?[22]

Improvement of frequency response. As might be anticipated from the preceding example, and from the correspondence between the speed of response in the time domain and the bandwidth in the frequency domain, feedback can often improve the frequency response of a system. Let us reconsider the amplifier example in Fig. 4.32, except that we now introduce a bandwidth limitation to the main amplifier. We show the frequency-domain version of the circuit in Fig. 4.41, with the main amplifier now having a low-frequency gain of A_0 and a frequency cutoff of ω_c.

$$A(\omega) = \frac{A_0}{1 + j(\omega/\omega_c)} \tag{4.91}$$

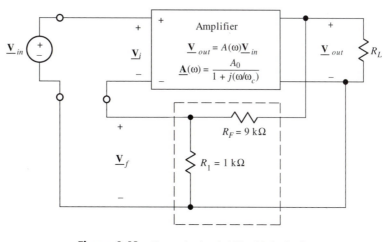

Figure 4.41   Improving bandwidth with feedback.

The main amplifier frequency response has the form of a low-pass filter, the loss of the high frequencies resulting from a limitation within the amplifier. This bandwidth limita-

---

[22] Then put an 1800-Ω resistor in series with the diode.

tion of the amplifier can be alleviated through negative feedback.

The equations of the system, Eqs. (4.66) to (4.68), transform into the frequency domain changed only in notation, and hence the amplifier characteristic with feedback becomes the frequency-domain version of Eq. (4.69):

$$\mathbf{A}_f(\omega) = \frac{A(\omega)}{1 + [R_1/(R_1 + R_F)]A(\omega)} = \frac{A_0/[1 + j(\omega/\omega_c)]}{1 + \beta A_0/[1 + j(\omega/\omega_c)]} \qquad (4.92)$$

where $\beta = R_1/(R_1 + R_F)$. Equation (4.92) can be put into the form

$$\mathbf{A}_f(\omega) = \frac{A_0/(1 + \beta A_0)}{1 + j(\omega/\omega_{cf})} \qquad (4.93)$$

where $\omega_{cf} = (1 + \beta A_0)\omega_c$ is the cutoff frequency with feedback. Recognizing that the loop gain $L = -\beta A_0$, we see from Eq. (4.93) that the amplifier gain at low frequencies is reduced by the factor $1 - L$ and the cutoff frequency is increased by the same factor.

Bode plots for the amplifier with and without feedback are shown in Fig. 4.42. This application furnishes a good example of our claim that negative feedback trades gain for some other useful property. Here gain is exchanged for bandwidth. The inherent bandwidth limitation of the amplifier has been overcome through the use of negative feedback.

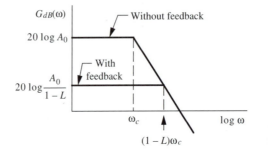

Figure 4.42 Feedback reduces gain but increases bandwidth.

We showed that negative feedback increases the bandwidth of a first-order system. Higher-order systems cannot always be improved by these means.

Impedance control. Feedback techniques can be used to change the impedance level of an electronic circuit. This is useful for reducing loading effects and for effecting maximum power transfer.

common collector, emitter follower

Feedback analysis of emitter-follower circuit. The control of impedance levels through feedback is demonstrated by analyzing the *common collector*, or *emitter-follower*, circuit shown in Fig. 4.43. This amplifier is a variation on the amplifier-switch amplifier in Fig. 2.40 except that (1) the collector resistor is placed between the emitter and ground and renamed R_E, and (2) the output is taken between the emitter and ground instead of the collector and ground.

System analysis. Our analysis considers only the active region, where the relationship between emitter and base currents from Eq. (2.22) is

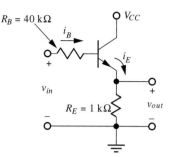

Figure 4.43 For the common-collector or emitter-follower amplifier, the output is taken off the emitter.

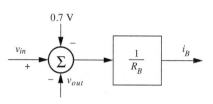

Figure 4.44 Ohm's law in the base-emitter circuit is represented by a summer and a scaler.

$$i_E = (\beta + 1)i_B \qquad (4.94)$$

where β is the current gain of the transistor. In the active region, $v_{BC} = 0.7$, so KVL around the base-emitter loop is

$$-v_{in} + R_B i_B + 0.7 + v_{out} = 0 \quad \Rightarrow \quad i_B = (v_{in} - 0.7 - v_{out}) \times \frac{1}{R_B} \qquad (4.95)$$

where we have written the equation for the base current to suggest a summer followed by a scaler box on a system diagram, Fig. 4.44. The remaining equations describing the circuit operation are Eq. (4.94) and Ohm's law for R_E: $v_{out} = R_E i_E$. All the emitter-follower equations are represented by the system diagram in Fig. 4.45.

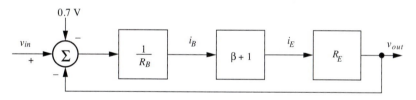

Figure 4.45 System diagram for the emitter follower.

Analysis of the system diagram. We use Eq. (4.77):

$$A = \frac{1}{R_B} \times (\beta + 1) \times R_E$$

$$L = (-1)A \qquad (4.96)$$

$$A_f = \frac{A}{1 - L} = \frac{(\beta + 1)R_E/R_B}{1 + (\beta + 1)R_E/R_B}$$

Because our goal is to determine input and output impedance levels, we will ignore the effect of the 0.7-V term in Eq. (4.95), which merely describes a voltage offset in the output. For this goal, the gain with feedback is best expressed as a relationship between input and output voltages

$$v_{out} = v_{in} \times \frac{(\beta+1)R_E/R_B}{1+(\beta+1)R_E/R_B} \tag{4.97}$$

Input impedance. The input impedance follows from Eq. (4.97) and the system diagram:

$$R_{in} = \frac{v_{in}}{i_B} = \frac{v_{out}/A_f}{v_{out}/(\beta+1)R_E} = R_B(1-L) = R_B + (\beta+1)R_E \tag{4.98}$$

Equivalent Circuits

Output impedance. The output impedance can be derived from the open-circuit voltage and short-circuit current, as expressed in Eq. (1.47). The voltage calculated in Eq. (4.97) is the Thévenin voltage because no load is present at the output terminals. Thus, the open-circuit voltage is given by Eq. (4.97). The effect of shorting the output terminals can be seen from Fig. 4.43. Reducing v_{out} to zero with the short kills the feedback and the Norton current flowing through the short will be the emitter current:

$$i_N = i_E\big|_{v_{out}=0} = \frac{v_{in}}{R_B} \times (\beta+1) \tag{4.99}$$

Thus, the output impedance is

$$R_{eq} = \frac{v_T}{i_N} = \frac{v_{in} \times \dfrac{(\beta+1)R_E/R_B}{1-L}}{v_{in} \times (\beta+1)/R_B} = \frac{R_E}{1-L} \tag{4.100}$$

Conclusions. From the foregoing analysis, we conclude the following:

1. The feedback raises the input impedance from R_B to $R_B \times (1-L)$, Eq. (4.98). This change is desirable because a voltage amplifier with a large input impedance does not load its input source.

2. The final form of Eq. (4.98) shows that the transistor does not isolate the output circuit from the input circuit when the load is connected to the emitter. The input circuit looks through the emitter and sees the emitter resistor, increased by the current gain of the transistor, through a feedback effect.

3. The output impedance is lowered from R_E to $R_E/(1-L)$. This is also a desirable change because a voltage amplifier should have low output impedance to avoid being loaded by its load and to have high available power.

4. The factor $1-L$ appears in the gain, the input impedance, and the output impedance, underscoring the importance of this factor in all feedback calculations.

5. The feedback in this circuit is not introduced by bringing a connection from output to input, but occurs naturally in the circuit operation.

6. The analysis of the input and output impedances of an emitter-follower amplifier furnishes a good example of how circuit equations can be converted to system components such as summers and samplers. Once the conversion is made, the circuit can by analyzed by system techniques.

Check Your Understanding

1. Sometimes the summer in a feedback system represents taking the difference, not the sum, of two signals. True or false?

2. For negative feedback, must the loop gain be positive or negative?

3. The loop gain is calculated or measured with the input source OFF. True or false?

4. If the loop gain in a feedback system is -0.2, feedback effects will not be very important in defining system characteristics. True or false?

5. Negative feedback, compounded with a phase shift in the loop, can cause dynamic instability. True or false?

Answers. **(1)** True; **(2)** negative; **(3)** true; **(4)** true; **(5)** true.

4.4 OPERATIONAL-AMPLIFIER CIRCUITS

Introduction

<table>
<tr><td>

LEARNING OBJECTIVE 7.

To understand the importance and applications of op amps in analog circuits

</td></tr>
</table>

operational amplifier, op amp

Importance of op amps. An *operational amplifier*, *op amp* for short, is a high-gain electronic amplifier, normally controlled by negative feedback, that accomplishes many functions or "operations" in analog circuits. Such amplifiers were originally developed to accomplish operations such as integration and summation for solving differential equations with analog computers. Applications of op amps have increased until, at the present time, most analog electronic circuits are based on op-amp techniques. If, for example, you required an amplifier with a gain of −10, rarely would you design a circuit of the type in Fig. 2.48; convenience, reliability, and cost considerations dictate the use of an op amp. Thus, op amps form the basic building blocks of analog electronic circuits much as NOR and NAND gates provide the basic building blocks of digital circuits.

Op-amp model and typical properties. The typical op amp is a sophisticated transistor amplifier utilizing a dozen or more transistors, several diodes, many resistors, and perhaps a few capacitors. Such amplifiers are mass produced on semiconductor chips and sell for less than $1 each. These parts are reliable and rugged, approaching the ideal in their electronic properties.

Figure 4.46(a) shows the symbol and the op-amp properties that interact with external signals. The two input signal voltages, v_+ and v_-, are subtracted and amplified with a large voltage gain, A, typically 10^5 to 10^6. The input resistance, R_i, is large, typically exceeding 10 MΩ; the output resistance, R_o, is small, 10 to 100 Ω. The amplifier is supplied with dc power from positive $(+V_{CC})$ and negative $(-V_{CC})$ power supplies. For this case, the output voltage lies between the power-supply voltages, as shown by Fig. 4.46(b). Sometimes one power connection is grounded, for example, "$-V_{CC}$" $= 0$, in which case the output lies between 0 V and $+V_{CC}$. The power connections are seldom drawn on circuit diagrams.

Realistic op amps. The model in Fig. 4.46(a) approaches the ideal voltage amplifier: high input impedance, low output impedance, and high gain. Real op amps have these properties but also have undesirable features that limit their performance and influence circuit design. The circuit given in Fig. 4.47 models a number of the limitations of realistic op amps. These are as follows:

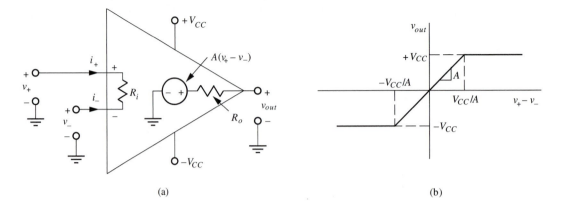

(a)

(b)

Figure 4.46 (a) Op-amp model. The amplifier circuit, whose details are omitted due to its complexity, has two dc power inputs, V_{CC} and $-V_{CC}$, two signal inputs, v_+ and v_-, and one signal output, v_{out}; (b) input–output characteristic of an op amp. The sloped region is the amplifying region. The op amp saturates at the power-supply voltages.

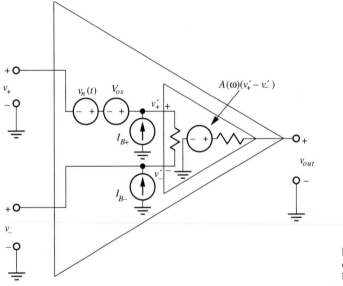

Figure 4.47 An op amp has noise, offset voltage, bias currents, and limited bandwidth.

- A dc offset voltage, V_{os}, is shown to indicate that the output of the op amp is nonzero with zero input voltage, that is, with $v_+ = v_-$.

- Input dc bias currents, I_{B+} and I_{B-}, are shown. Normally, these bias currents are expressed in terms of an offset bias current, $I_{os} = I_{B+} - I_{B-}$, and an average bias current $I_B = (I_{B+} + I_{B-})/2$.[23]

- A noise voltage, $v_n(t)$, represents broadband noise originating in the op amp. The magnitude of this noise is proportional to the square root of the bandwidth of the op amp in the circuit application, to be discussed in Chapter 6.

---

[23] These are the common-mode and difference-mode components in the bias currents, as defined in Chap. 5.

- Gain–bandwidth limitation is indicated by making the op-amp gain $A(\omega)$, a function of frequency. The gain–bandwidth product is limited by inherent capacitance in the circuit and the semiconductor devices in the op amp, and often is deliberately limited by the chip designer to inhibit oscillations at high frequencies.

Reducing the effects of op-amp imperfections. The effect of dc offset voltage and current, V_{os} and I_{os}, can be eliminated by introducing an external dc voltage to null the output voltage with $v_+ = v_-$. Application literature provided by the manufacturer gives information about canceling the effect of offset voltage and current.

The effect of the bias current, I_B, can be eliminated by providing equal dc impedance at the two inputs. In this way, the bias current produces equal voltages that cancel in the subtraction process.

The effect of input noise can be reduced by limiting the bandwidth of the system to that required by the signal. However, noise cannot be eliminated entirely and must be considered in the error analysis of the circuit employing the op amp.

The gain–bandwidth limitation of the op amp limits the product of the gain and bandwidth of the resulting amplifier employing the op amp. Generally, op amps with large gain–bandwidth products are more sophisticated and expensive than basic op amps.

Summary. The high gain of the op amp is converted to other useful features through the use of strong negative feedback. All the benefits of negative feedback are utilized by op-amp circuits. Furthermore, op amps are inexpensive and lead to easy designs that are easy to construct.

Contents of this section. We next analyze two common op-amp applications, inverting and noninverting amplifiers. We derive the gain of these amplifiers by a method that may be applied simply and effectively to any op-amp circuit operating in its linear amplifying region. We then discuss op-amp circuits for adding and subtracting signals, for converting a current signal to a voltage signal, and for integrating signals. We then consider nonlinear applications of op amps: comparators, log, and antilog amplifiers. In the next chapter, we consider additional applications of op amps.

Basic Op-Amp Amplifiers

Inverting amplifier circuit. The inverting amplifier, shown in Fig. 4.48, uses an op amp plus three resistors. The positive $(+)$ input to the op amp is grounded through R_2; the negative $(-)$ input is connected to the input signal via R_1 and to the feedback signal from the output via R_F. Note that $R_2 = R_1 \parallel R_F$ to reduce the effect of bias current. Throughout this section on linear op-amp circuits, we consider that the effect of offset voltage and current has been canceled to produce zero volts at the output for $v_+ = v_-$.

One potential source of confusion in the following discussion is that we must speak of two amplifiers simultaneously. The *op amp* is an amplifier that forms the amplifying element in a *feedback amplifier* that contains the op amp plus the associated resistors. To lessen confusion, we reserve the term "amplifier" to apply only to the overall, feedback amplifier. The op amp will never be called an "amplifier"; it will be called "the op amp." For example, if we refer to the input current to the amplifier, we are referring to the current through R_1, not the current into the op amp.

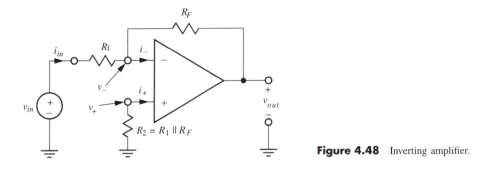

Figure 4.48 Inverting amplifier.

Analysis of the inverting amplifier.   We could solve for the gain of the inverting amplifier in Fig. 4.48 either by solving the basic circuit laws, KVL and KCL, or by attempting to divide the circuit, in the style of Fig. 4.33, into the summer, main amplifier, and feedback system blocks. We, however, present a third approach that assumes that the op-amp gain is very high, effectively infinite. In the following, we give a general assumption, which may be applied to any op-amp circuit; then we apply this assumption specifically to the inverting amplifier. As a result, we establish the gain and input resistance of the inverting amplifier.

1. We assume that the amplifier operates in its linear amplifying region, the sloping region in Fig. 4.46(b). It follows that the output lies between the power-supply voltages. Thus, we assume that the negative feedback stabilizes the amplifier such that moderate input voltages produce moderate output voltages. If the power supplies are $+10$ V and -10 V, the output would have to lie between these limits. This assumption restricts our analysis to linear op-amp circuits.

2. Therefore, the difference between input voltages to the op amp is very small, essentially zero, because this difference is the output voltage divided by the large voltage gain of the op amp:

$$v_+ - v_- = \frac{v_{out}}{A} \approx 0 \;\Rightarrow\; v_+ \approx v_- \tag{4.101}$$

For example, if $|v_{out}| < 10$ V and $A = 10^5$, then $|v_+ - v_-| < 10/10^5 = 100$ μV. Thus, v_+ and v_- will be equal within 100 μV or less. For the inverting amplifier in Fig. 4.48, v_+ is grounded through R_2; therefore, $v_+ \approx 0$ and $v_- \approx 0$. Consequently, the current at the input to the amplifier would be

$$i_{in} = \frac{v_{in} - v_-}{R_1} \approx \frac{v_{in}}{R_1} \tag{4.102}$$

3. Because $v_+ \approx v_-$ and the input impedance of the op amp, R_i, is large, the current into the $+$ and $-$ op-amp inputs is very small, essentially zero:

$$|i_+| = |i_-| = \frac{|v_- - v_+|}{R_i} \approx 0 \tag{4.103}$$

CHAPTER 4 ANALOG ELECTRONICS

For example, for $R_i = 10 \text{ M}\Omega$, $|i_-| < 10^{-4}/10^7 = 10^{-11}$ A.

For the inverting amplifier, Eq. (4.103) implies that the current at the input, i_{in}, flows through R_F, as shown in Fig. 4.49. This allows us to compute the output voltage. The voltage across R_F is $i_{in}R_F$ and, because one end of R_F is connected to $v_- \approx 0$,

$$v_{out} = -i_{in}R_F = -\frac{v_{in}}{R_1} \times R_F \tag{4.104}$$

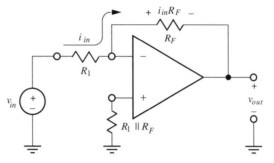

Figure 4.49 The input current flows through the feedback circuit.

Thus, the voltage gain is

$$A_v = \frac{v_{out}}{v_{in}} = -\frac{R_F}{R_1} \tag{4.105}$$

The minus sign in Eq. (4.105) means that the output is inverted relative to the input: a positive signal at the amplifier input produces a negative signal at the output.

Circuit impedance level. Equation (4.105) shows the gain to depend on the ratio of R_F to R_1, which implies that only the ratio matters, not the individual values of R_1 and R_F. This would be true if only the gain of the amplifier were important, but the input impedance to an amplifier is also important. The input resistance to the inverting amplifier follows from Eq. (4.102):

$$R_{in} = \frac{v_{in}}{i_{in}} \approx R_1 \tag{4.106}$$

For a voltage amplifier, the input impedance level is an important factor, for if R_{in} were too low, the signal source of v_{in} could be loaded down by R_{in}. Thus, R_1 must be sufficiently high to avoid this loading problem. Once R_1 is fixed, R_F may be selected to achieve the required gain. Therefore, the values of the individual resistors become important because they affect the input resistance to the amplifier.

IDEA 5 **Impedance Level**

EXAMPLE 4.16 **Amplifier design**

Design an inverting amplifier to have a gain of -8. The input signal comes from a pressure sensor having an output impedance of $100\,\Omega$.

Circuit operation.   Feedback effects dominate the characteristics of the amplifier. When an input voltage is applied, v_- increases, causing v_{out} to increase rapidly in the negative direction. This negative voltage increases to the value where the effect of v_{out} on the $-$ input via R_F cancels the effect of v_{in} through R_1. Put another way, the output will adjust itself to withdraw through R_F any current that v_{in} injects through R_1, because the input current to the op amp is extremely small. In this way, the output depends only on R_F and R_1.

Summary of the method.   The inverting-amplifier voltage gain and input impedance were determined from two principles:

1. The input voltages *at the op amp* are equal: $v_+ = v_-$.
2. The input currents *to the op amp* are negligible: $|i_+| = |i_-| \approx 0$.

These assumptions are valid for any good op amp operating with negative feedback in the linear region. We illustrate their application again in the next section.

Noninverting amplifier.   For the noninverting amplifier shown in Fig. 4.50 the input is connected to the $+$ input through $R_2 = R_1 \parallel R_F$. The feedback from the output connects through a voltage divider to the $-$ op-amp input, as required for negative feedback. We determine the gain with the analysis outlined earlier.

1. Because the input current to the op amp is very small, no signal voltage is lost across R_2 and hence $v_+ = v_{in}$.

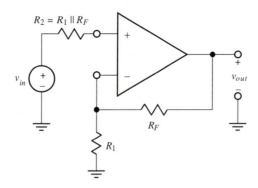

Figure 4.50   Noninverting amplifier.

---

[24]$R_{in} = 10\,\text{k}\Omega, R_F = 80.8\,\text{k}\Omega.$

2. Because $v_+ \approx v_-$, it follows that

$$v_- \approx v_{in} \qquad (4.107)$$

3. Because $i_- \approx 0$, R_F and R_1 carry the same current. Hence, v_{out} is related to v_- through a voltage-divider relationship:

$$v_- \approx v_{out} \times \frac{R_1}{R_1 + R_F} \qquad (4.108)$$

Combining Eqs. (4.107) and (4.108), we establish the voltage gain to be

$$v_{in} = v_{out} \frac{R_1}{R_1 + R_F} \quad \Rightarrow \quad A_v = \frac{v_{out}}{v_{in}} = +\left(1 + \frac{R_F}{R_1}\right) \qquad (4.109)$$

The $+$ sign before the gain expression emphasizes that the output of the amplifier has the same polarity as the input: a positive input signal produces a positive output signal. Again, we see that the ratio of R_F and R_1 determines the gain of the amplifier.

Circuit operation. When a voltage is applied to the amplifier, the output voltage increases rapidly until the voltage across R_1 reaches the input voltage. Thus, negligible input current flows into the amplifier, and the gain depends only on R_1 and R_F. The input impedance to the noninverting amplifier is very high because the input current to the amplifier is also the input current to the op amp, i_+, which is extremely small. Input impedance values exceeding 1000 MΩ are easily achieved with this circuit. This feature of high input impedance is an important virtue of the noninverting amplifier because loading of the input source is eliminated.

EXAMPLE 4.17 | **Output impedance**

Find the output impedance of the noninverting amplifier.

SOLUTION:
Assume $v_{in} = 1$ V and calculate the Thévenin voltage and Norton current at the output. Assuming the loop gain is much greater than 1, the open-circuit output voltage is

$$V_T = 1\text{V} \times \left(1 + \frac{R_F}{R_1}\right) \qquad (4.110)$$

Shorting the output to find the Norton current kills the feedback, $v_- = 0$, and $I_N = 1$ V $\times$ A/R_o, where R_o is the output impedance of the op amp defined in Fig. 4.46 (a). Thus,

$$R_{out} = R_{eq} = \frac{V_T}{I_N} = \frac{1 + R_F/R_1}{A/R_o} = R_o \times \frac{R_1 + R_F}{AR_1} \qquad (4.111)$$

WHAT IF? What if you apply the same analysis to the inverting amplifier?[25]

---

[25] You must get the same answer because the circuits are identical with the input voltage OFF. But seriously, see if you can derive it.

Impedance levels. With the noninverting amplifier, the values of R_1 and R_F may be chosen from a broad range as long as the ratio gives the desired gain. However, if the impedance level is too low, comparable to the output impedance of the op amp, the op amp might be loaded by its own feedback network, which would be undesirable. At the other extreme, very large resistors can act as sources of significant noise in the amplifier, increasing the output noise above that inherent to the op amp. Resistor noise is discussed in Chap. 6.

Linear Op-Amp Circuits

Buffer or voltage follower. An amplifier with a gain of +1 results by eliminating R_1 $(= \infty)$ in the noninverting amplifier, as shown in Fig. 4.51. This *buffer* is used to control impedance levels in the circuit. The input impedance to the buffer is high and its output impedance is low. The output voltage from a source with high output impedance can supply signal to one or more loads that have a low impedance via the buffer. The value of R_F is arbitrary and can be chosen equal to the source impedance, R_s, to eliminate entirely the need for R_2. Often $R_F = 0$ is used.

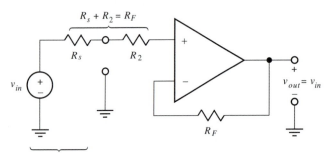

Figure 4.51 The buffer or voltage-follower circuit.

Current-to-voltage converter. Many devices produce current signals, whereas most electronic circuits require voltage signals. Figure 4.52 shows an op-amp circuit for converting a current signal to a voltage signal. This is the inverting amplifier with $R_1 = 0$. It has zero input impedance and hence does not load the input current source. The input current is converted to a voltage by flowing through the feedback resistor, as explained earlier.

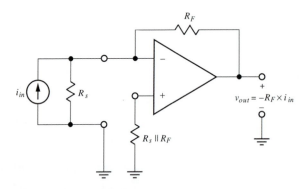

Figure 4.52 A current-to-voltage converter.

CHAPTER 4   ANALOG ELECTRONICS

EXAMPLE 4.18 ## 20-mA current-loop converter

Certain digital devices use a "20-mA current loop" to encode a digital signal. This means that a digital ONE is a current of 20 mA into a short circuit and a digital ZERO is 0 mA. Assuming the output impedance of the loop is 100 kΩ, design an op-amp interface to convert this current signal to a voltage signal with logic levels of 0 and +5 V.

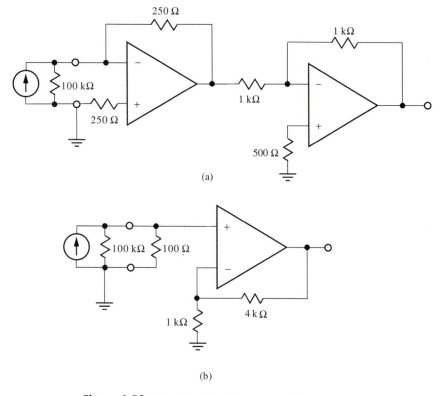

(a)

(b)

Figure 4.53 Two 20-mA ⟹ 5-V current-to-voltage converters.

SOLUTION:
One circuit is shown in Fig. 4.53(a). The feedback resistor must be $R_F = 5\,\text{V}/20\,\text{mA} = 250\,\Omega$. Because we require a positive output voltage, the current-to-voltage converter must be followed by an inverter, as shown in Fig. 4.53(a). Another approach is shown in Fig. 4.53(b). A 100-Ω resistor converts the 20-mA to a 2-V signal and the noninverting amplifier provides the gain to make 5 V. Here no inverter is required.

Summing amplifier. An inverting amplifier can accept two or more inputs and produce a weighted sum. Figure 4.54 shows a summer with two inputs. We may understand the operation of the circuit by applying the same reasoning we used earlier to understand the inverting amplifier. Because $v_- \approx 0$, the sum of the currents through R_1 and R_2 is

$$i_{in} = \frac{v_1}{R_1} + \frac{v_2}{R_2} \tag{4.112}$$

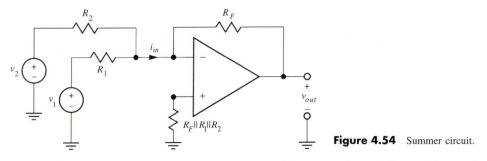

Figure 4.54 Summer circuit.

The output voltage adjusts itself to draw this current through R_F, and hence the output voltage is

$$v_{out} = -i_{in}R_F = -\left(v_1 \times \frac{R_F}{R_1} + v_2 \times \frac{R_F}{R_2}\right) \tag{4.113}$$

The output is thus the sum of v_1 and v_2, weighted by the gain factors, R_F/R_1 and R_F/R_2, respectively. If the inversion produced by the summer is unwanted, the summer can be followed by an inverting amplifier with a gain of -1. Clearly, we could add other inputs in parallel with R_1 and R_2.

EXAMPLE 4.19 **A sinusoid with dc**

An op amp has power-supply voltages of ± 10 V. The available input voltage is $v_{in} = 0.1 \cos(\omega t)$ V. Design a circuit to give an output of $v_{out} = 1 + \cos(\omega t)$ V.

SOLUTION:
We use a summer and an inverter with unity gain. The summer has a gain of 10 for the sinusoid but a gain of 0.1 for the power-supply voltage. Figure 4.55 shows the circuit.

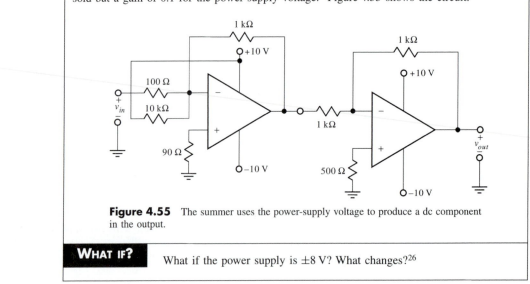

Figure 4.55 The summer uses the power-supply voltage to produce a dc component in the output.

WHAT IF? What if the power supply is ± 8 V? What changes?[26]

---

[26] The 10 kΩ changes to 8 kΩ.

Differencing amplifier. The circuit in Fig. 4.56 produces an output proportional to the difference between the two inputs. The circuit is linear, so we may use superposition. The output due to v_2 is given by Eq. (4.105). The signal to the noninverting input is reduced by the voltage divider of R_1 in series with R_F. The output thus combines a voltage-divider relationship with the gain given by Eq. (4.109):

$$v_{out} = v_1 \times \frac{R_F}{R_1 + R_F} \times \left(1 + \frac{R_F}{R_1}\right) = v_1 \times \frac{R_F}{R_1} \tag{4.114}$$

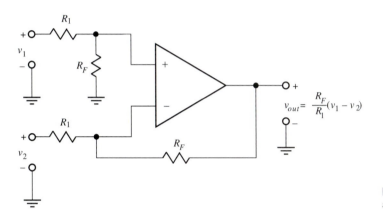

Figure 4.56 This circuit subtracts and amplifies the inputs.

Combining the effects from each input, we determine the output voltage to be

$$v_{out} = \frac{R_F}{R_1}(v_1 - v_2) \tag{4.115}$$

Thus, the amplifier subtracts the inputs and amplifies their difference.

Impedance Level

Two features of this subtractor circuit should be mentioned. Perfect subtraction requires a balance of the resistor values and thus may call for careful adjustment of the resulting circuit. Furthermore, the input impedances to the inputs are different, and therefore the balance could be affected if loading effects are significant. On the other hand, the input impedances at the two inputs can be made equal by reducing the resistors in the voltage divider by the ratio $R_1/(R_1 + R_F)$, but this would forfeit the dc impedance balance seen at the op-amp inputs. In practice, a compromise might be required.

Integrator. Because a capacitor integrates current and an op amp can simulate a controlled current source, an op-amp circuit can integrate its input voltage. The op-amp circuit of Fig. 4.57 uses a capacitor in the feedback path to integrate the input voltage and a voltage source V_1 that is disconnected at $t = 0$ to produce an initial voltage. Let us examine the initial state of the circuit before investigating what happens after the switch is opened. The input current to the amplifier, v_{in}/R, flows through the V_1 voltage source and into the output of the op amp. Because v_+ is approximately zero, so will be v_-, and hence the output voltage is fixed at $-V_1$ with the switch closed.

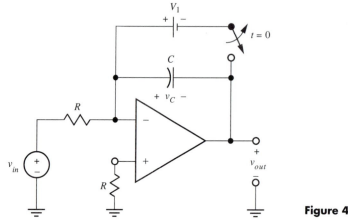

Figure 4.57 Integrator circuit.

After the switch is opened at $t = 0$, the input current flows through the capacitor and hence v_C is

$$v_C(t) = v_C(0) + \frac{1}{C} \int_0^t \frac{v_{in}(t')}{R} dt' \tag{4.116}$$

Thus, the output voltage of the circuit is

$$v_{out}(t) = -v_C(t) = -V_1 - \frac{1}{RC} \int_0^t v_{in}(t') dt' \qquad (t > 0) \tag{4.117}$$

Except for the minus sign, the output is the integral of v_{in} scaled by $1/RC$.

EXAMPLE 4.20 **Integrator design**

Design a noninverting circuit that has an input impedance of 1 kΩ resistive, an initial voltage of +3 V, and an output that is 10 times the integral of the input voltage.

SOLUTION:
To have a noninverted output, we require an inverting amplifier and an integrator, which may be placed in either order. However, if the integrator were placed first, its drift due to unbalance would be amplified, so we use the circuit in Fig. 4.58.

The input impedance requirement fixes R_1 as 1 kΩ, and the initial voltage requirement fixes V as +3 V. Two of the three components (C, R, and R_F) may be chosen arbitrarily and the third adjusted to give the correct overall gain. We choose C as a convenient value of 10 μF and choose R_F as 20 kΩ to give the inverting amplifier a gain of −20. It follows from Eq. (4.117) that the integrator must have an RC value of 2, and hence the value of R must be 200 kΩ.

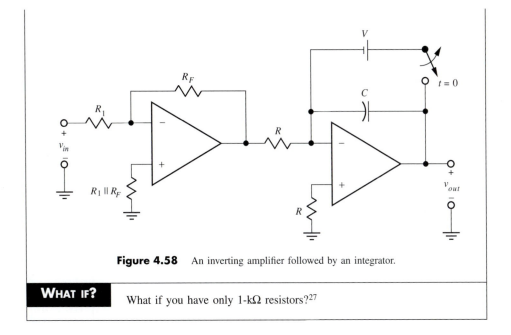

Figure 4.58 An inverting amplifier followed by an integrator.

WHAT IF?   What if you have only 1-kΩ resistors?[27]

Nonlinear Op-Amp Circuits

Many circuits employing op amps are nonlinear, either because nonlinear elements such as diodes are used in the circuit or because the op amp operates outside its linear amplifying region. In this section, we discuss several nonlinear applications of op amps. Many more circuits are detailed in standard handbooks and application literature provided by manufacturers.

Comparators.   Figure 4.59(a) shows a basic comparator, and Fig. 4.59(b) shows its input–output characteristic. No feedback is employed. When the input is less than a threshold voltage, V_{th}, the op amp is saturated at the value of the negative power supply, which is zero in this case. As the input voltage is increased past V_{th}, the op amp passes rapidly through its linear region and saturates at the voltage of the positive power supply. Because there is no feedback, transition between the two saturation voltages requires an input voltage range of a few millivolts at most. This circuit, therefore, gives a digital output for an analog input, depending on the region of the input voltage. Comparators have many applications, including alarm circuits, control circuits, and analog-to-digital converters.

A noisy signal, however, causes problems, as shown in Fig. 4.60(a). The transition region is so narrow that an erratic output is likely to occur in the vicinity of the threshold, as shown in the bottom graph. The circuit shown in Fig. 4.60(b) remedies this problem through the use of positive feedback. Figure 4.60(c) shows the input–output characteristic of the modified circuit.

---

[27]No problem. Use them with $C = 100$ μF.

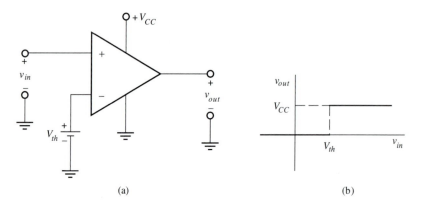

(a)

(b)

Figure 4.59 (a) Basic comparator circuit; (b) the input–output characteristic.

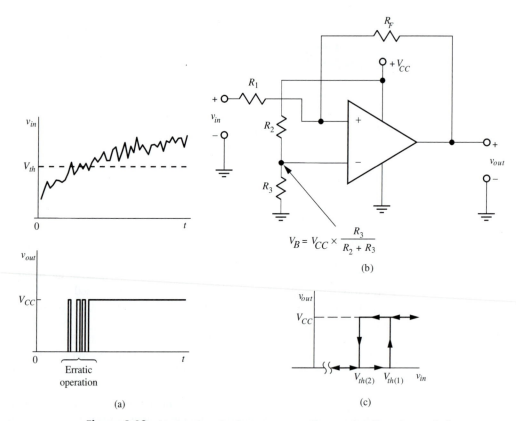

$$V_B = V_{CC} \times \frac{R_3}{R_2 + R_3}$$

(b)

(a)

(c)

Figure 4.60 (a) A noisy signal can cause erratic operation if no hysteresis is provided; (b) a comparator with hysteresis created by positive feedback; (c) the input–output characteristic of the comparator with hysteresis.

Finding the thresholds. As the input voltage increases, the output transition occurs when the input voltage reaches a threshold voltage, $V_{th(1)}$, which in this case is established from the power supply with a voltage divider. Because the input voltage to the op amp is reduced slightly by the feedback resistor, the transition occurs at the value given with the nodal analysis in Eq. (4.118). Because $v_{out} = 0$, KCL at the + input is

$$\frac{V_{th(1)} - V_B}{R_1} = \frac{V_B - 0}{R_F} \implies V_{th(1)} = V_B\left(1 + \frac{R_1}{R_F}\right) \tag{4.118}$$

As the op amp comes out of saturation in the lower state, the positive feedback produces a rapid transition to saturation in the upper state.

As the input voltage is lowered and $v_{out} = V_{CC}$, the input voltage must be reduced to a value, $V_{th(2)}$, given by the nodal analysis in Eq. (4.119).

$$\frac{V_{th(2)} - V_B}{R_1} = \frac{V_B - V_{CC}}{R_F} \implies V_{th(2)} = V_B\left(1 + \frac{R_1}{R_F}\right) - V_{CC} \times \frac{R_1}{R_F}$$

$$= V_{th(1)} - V_{CC} \times \frac{R_1}{R_F} \tag{4.119}$$

Note that the width of the threshold region is $V_{CC} \times R_1/R_F$.

EXAMPLE 4.21 **Comparator design**

Design a comparator with a threshold region 30 mV wide, centered on 5 V. The positive and negative power supplies are 15 and 0 V, respectively. The input impedance to the circuit should exceed 100 kΩ.

SOLUTION:
The specified threshold values are $V_{th(1)} = 5.015$ V and $V_{th(2)} = 4.985$ V. From the width of the threshold region, we find $R_1/R_F = 0.030/15 = 0.002$, and from Eq. (4.118) $V_B = 5.015/(1 + 0.002) = 5.005$ V. The input impedance is the sum of R_1 and R_F because the op amp draws no current. We choose $R_F = 1$ MΩ and hence $R_1 = 2000$ Ω. Here it is unnecessary to balance the input dc impedances at the op-amp inputs because the op amp is always saturated, and therefore bias currents have no effect. The voltage divider of R_2 and R_3 may be chosen at convenient values to give the required value of V_B.

WHAT IF?   What if $V_{th(1)} = 5.5$ V and $V_{th(2)} = 4.5$ V?[28]

Logarithmic amplifier. By placing a diode in the feedback path, as shown in Fig. 4.61, we create an amplifier whose output is proportional to the logarithm of the input. Applying the usual analysis, we see the input current to be v_{in}/R. The output voltage assumes a value to draw this current through the diode. The diode characteristic is well represented by the ideal *pn*-junction equation in Eq. (2.17); hence, the output voltage is

---

[28] $R_F = 1$ MΩ, $R_1 = 66.7$ kΩ, $V_B = 5.16$ V.

$$v_{out} = -\eta V_T \ln\left(\frac{v_{in}}{RI_0} + 1\right) \approx -\eta V_T \ln v_{in} + \eta V_T \ln RI_0 \qquad (4.120)$$

The output thus contains a factor proportional to the logarithm of the input voltage. The circuit in Fig. 4.61 can be followed by another op-amp circuit that subtracts the constant term and adjusts the gain of the log amplifier to a prescribed value. For example, the gain can be adjusted to produce an output in dB relative to 1 V. The circuit in Fig. 4.61 cannot deal with signals that go negative.

Antilog amplifier.   With a diode in the input circuit, as shown in Fig. 4.62, the output voltage can be made proportional to the antilog, or exponential function of the input.

Applications of log and antilog amplifiers.   The existence of log and antilog amplifiers makes possible analog circuits for multiplication and powers of analog signals. For example, if we wished a circuit to produce the product of two analog signals, we could take the log of each, add the logs, and take the antilog. If one or both of the inputs could go both positive and negative, they would have to be put through circuits that produce the absolute value, with the signs handled by a separate logic circuit.

Excellent analog multipliers and other nonlinear circuits are commercially available. Such circuits operate according to the principles outlined earlier but contain more complicated circuits to compensate for temperature effects and generally improve performance. Our purpose here is to show some representative nonlinear applications of op amps.

Summary.   In this section we introduced the op amp as an inexpensive and versatile circuit component for processing analog signals. We described a method for determining circuit operation when the op amp is operating in its linear region. We dealt with representative circuits for amplification, addition, subtraction, and integration of signals. We also described several nonlinear applications of op amps, including comparators, log, and antilog converters. In Chap. 5, we show additional applications of op amps.

Check Your Understanding

1. An ideal op amp has infinite gain, infinite input impedance, and infinite output impedance. True or false?

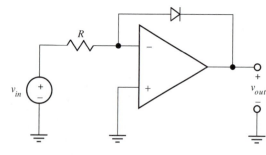

Figure 4.61   Logarithmic amplifier.

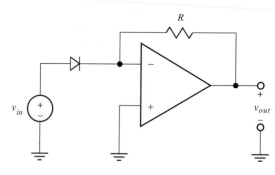

Figure 4.62   Antilog amplifier.

2. In a noninverting op-amp amplifier the feedback resistor is five times the value of the resistor between the inverting input and ground. If the output is $-6\,\mathrm{V}$, what is the voltage at the input?

3. For a voltage gain of $+5$ in an op-amp amplifier, the feedback comes from the output to which input terminal? Which input gets the input signal?

4. What operation is performed in an inverting amplifier if the feedback resistor is replaced by a capacitor? By a diode?

5. If an op amp has $\pm 8\,\mathrm{V}$ for plus-and-minus power-supply values and a gain of 80 dB, at what value of $v_+ - v_-$ does the op amp saturate?

6. The voltage-follower amplifier inverts the signal. True or false?

Answers. (1) False; the output impedance should be zero; (2) $-1.0\,\mathrm{V}$; (3) feedback to inverting $(-)$, input to noninverting $(+)$; (4) becomes an integrator; becomes a log amplifier; (5) $\pm 0.8\,\mathrm{mV}$; (6) false.

CHAPTER SUMMARY

Information may be represented in analog form. Signal spectra for periodic, nonperiodic, and random signals are described. Spectra may be modified with filters to improve signal characteristics. Feedback concepts are introduced generally and illustrated with linear and nonlinear operational-amplifier circuits.

Objective 1: To understand the concept of a spectrum as applied to periodic, nonperiodic, and random signals. Spectra consist of the simultaneous occurrence of sinusoids of different frequencies. Periodic signals have harmonic spectra of discrete frequencies. Nonperiodic and random signals have continuous spectra.

Objective 2: To understand how to calculate the bandwidth required to pass pulses and other signals. The bandwidth of the spectrum of a signal is related to the scale of time structure of the signal. For a random signal, the information rate is proportional to the bandwidth.

Objective 3: To understand how filters modify spectra. The frequency dependence of the impedances of capacitors and inductors permits the design of filter circuits to modify spectra. We discuss low- and high-pass RC filters and two types of band-pass filters.

Objective 4: To understand how to describe filters with Bode plots. A Bode plot is a log-log plot of filter response or signal spectrum. Bode plots accommodate a wide range in frequency and response level, and the Bode plots of simple filters are easily approximated by straight lines.

Objective 5: To understand the nature and benefits of feedback. Feedback combines the input with the output of a system. Negative feedback can improve the robustness, linearity, noise immunity, bandwidth, time response, and impedance characteristics of a system. Positive feedback is used in oscillator circuits.

Objective 6: To understand the importance of loop gain in feedback systems. The sign of the loop gain determines if the feedback is negative or positive.

The magnitude of the loop gain determines if feedback effects in the system are strong or weak.

Objective 7: To understand the importance and applications of op amps in analog circuits. An op amp amplifies the difference between two input signals with high gain. Op amps are the basic building blocks of analog circuits. Using op amps with negative feedback, we can build inverting and noninverting amplifiers, adders and subtracters, current-to-voltage converters and multipliers of analog signals. Using op amps with positive feedback, we can build comparators and oscillators.

Chapter 5 combines the techniques of digital and analog electronics in instrumentation systems.

GLOSSARY

AC component, p. 211, the remainder of a signal after the time average is subtracted, the fluctuations of the signal about its average value.

Analog, p. 196, a means by which information is encoded into an electrical signal that mimics the quantity being represented.

Asymptotic Bode plot, p. 225, the approximation of a Bode plot with straight lines.

Band-pass filter, p. 228, a filter that combines low-pass and high-pass characteristics to pass a band of frequency. Narrow-band band-pass filters are usually realized with a resonant circuit.

Bandwidth, p. 208, the width of the range in frequency in which the significant signal exists, or else the frequencies bounding that range, usually in hertz.

Baud, p. 215, a unit of digital information rate in bits/second.

Bode plot, p. 221, a log power versus log frequency plot of the characteristic of a filter or signal spectrum.

Buffer, p. 252, an amplifier with unity gain, used to isolate one part of a circuit from another in some sense.

Cascaded system, p. 223, a signal processing arrangement in which the signal is passed through two or more subsystems, with the output of one the input of the next.

Correlation time, p. 211, the time during which the fluctuations in a random signal become largely independent of the past.

Cutoff frequency, p. 218, the frequency that characterizes the region between the pass band and the reject band of a low-pass or high-pass filter, generally the half-power frequency.

DC component, p. 211, the time average of a signal.

Decade, p. 224, a factor of 10, especially in frequency.

Decibels, p. 224, a logarithmic measure of the ratio of output power to input power, specifically, $10 \log_{10}(P_{out}/P_{in})$

Dynamic stability, p. 235, the tendency of the system to vibrate or oscillate under the influence of external stimuli or internal noise.

Feedback, p. 230, feedback combines a sample of the output signal with the input signal to modify the input-output characteristics of the system.

Filter function, p. 217, the ratio between output and input signals, expressed in the frequency domain.

Free-wheeling diode, p. 239, a diode placed in parallel with an inductor to give a current path and thus prevent the occurrence of excessive voltage.

Half-power frequency, p. 218, the frequency of a filter where half the input power in volts squared is eliminated, also called the 3-dB frequency.

Harmonics, p. 198, frequencies that are integral multiples of a common frequency.

High-pass filter, p. 226, a filter that eliminates dc and other frequencies below a critical frequency but passes higher frequencies.

Loop gain, p. 234, the product of the gains of all the system components around a feedback loop, including signs.

Low-pass filter, p. 217, a filter that passes dc and other frequencies below a critical frequency but eliminates higher frequencies.

Negative feedback, p. 235, when the loop gain is negative, hence, the feedback signal subtracts from the input signal and the gain is reduced.

Operational amplifier, op amp, p. 245, a high-gain electronic amplifier, normally controlled by negative feedback, that accomplishes many functions or "operations" in analog circuits.

Positive feedback, p. 235, when the loop gain is positive, hence the feedback signal adds from the input signal. This condition is used in the design of electronic oscillators, which can be considered amplifiers with output but no input.

Power in volts squared, p. 202, the square of the rms value of a voltage signal, with units of volts^2 or watts in a 1-Ω resistor.

Power spectrum, p. 201, a spectrum that shows how much power in voltage squared exists at the various frequencies, with units of volts^2/hertz.

Random signal, p. 210, any signal that is unpredictable to us, either because it originates from chance events or because its complexity defies routine analysis.

Spectrum, p. 196, the representation of a signal as the simultaneous existence of many frequencies.

Static stability, p. 235, insensitivity to changes in the system parameters; robustness of performance.

Summer, p. 233, a system symbol representing the sum or difference of two signals, also called a comparator when the difference is indicated.

System notation, p. 232, a means for representing signal flow in complex systems with symbols that represent amplification, filtering, sums and differences, and other linear processes.

Voltage spectrum, p. 198, a spectrum that shows how much voltage exists at the various frequencies, with units of volts/hertz.

PROBLEMS

Section 4.1: Frequency-Domain Representation of Signals

4.1. In Austin, Texas, the maximum average daily temperature occurs in mid-July and is 84.5°F. The minimum daily average temperature is 49.1°F and occurs in mid-January.
  (a) Based on this information, and assuming that the daily average temperature $T(t)$ is well represented by the dc and first harmonic terms of a Fourier series such as Eq. (4.4), find T_0, T_1, θ_1, and ω_1. Let t be the time in months, with $t = 0$ on January 1, 1 on February 1, and so on.
  (b) From your result, estimate the expected average temperature for Christmas Day.

4.2. Consider a square wave similar to the one in Fig. 4.5(a), except that it goes from $+3$ to -1 V and has a pulse width of 2 ms, and has a spectrum similar to that shown in Fig. 4.5 (b).
  (a) What is V_0?
  (b) Find the frequency (in hertz), the amplitude, and the phase (in degrees) of the eleventh harmonic.

4.3. What would be the first three terms of the Fourier series for the half-wave rectified sinusoid in Fig. 4.6(a) if the origin were drawn at the beginning of the pulse? *Hint:* Change variables from $t \Rightarrow t' - T/4$, where t' is the origin in the new time system.

4.4. Show that the Fourier harmonic spectrum of a full-wave rectifier sinusoid is

$$v_{FW}(t) = V_p \left[\frac{2}{\pi} + \frac{4}{3\pi}\cos(2\omega_1 t) - \frac{4}{15\pi}\cos(4\omega_1 t) + \cdots \right]$$

Note that $v_{FW}(t) = v_{HW}(t) + v_{HW}(t + T/2)$.

4.5. A Fourier series is

$$v(t) = -10 + 5\cos(300\pi t) + 3\cos(600\pi t - 90°) + \cdots V$$

  (a) Would this signal have a continuous or a discrete spectrum?
  (b) What is the total "power" in volt² in the three harmonics given?
  (c) If the voltage were increased by 5 V dc, how would the Fourier series change?
  (d) What is the frequency of the third harmonic in hertz?

4.6. A periodic waveform has the form

$$v(t) = 2 + 3\sin(150\pi t) + 0.5\cos(450\pi t)V$$

  (a) What is the time average of $v(t)$?
  (b) Find the period of $v(t)$.
  (c) Find the rms value of $v(t)$.

4.7. Figure P4.7 shows a series of pulses in the time domain. Find the following:
  (a) Frequency of the third harmonic in Hz.
  (b) Power in volt² in the dc component.
  (c) Amplitude of the third harmonic.
  (d) Total power in volt² in all ac harmonics.

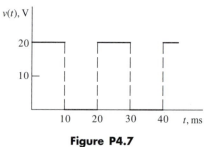

Figure P4.7

4.8. In Washington, DC, the maximum elevation angle of the sun at summer solstice, June 23, is about 73° and at winter solstice, December 23, about 27°. The maximum elevation can be described by a Fourier series of only two terms. What is the minimum length of the shadow of the Washington Monument, 555 feet in height, on the Fourth of July?

4.9. An unfiltered half-wave power supply produces 18.2 V dc from a 60-Hz sinusoid. What is its output power in volt² in the range of frequencies between 100 and 320 Hz?

4.10. A periodic waveform is shown in Fig. P4.10. This voltage can be represented by the following Fourier series:

$$v(t) = V_0 + V_1\cos(\omega_1 t + \theta_1) + V_2\cos(2\omega_1 t + \theta_2) + V_3\cos(3\omega_1 t + \theta_3) + \cdots$$

Find the following:
  (a) V_0.
  (b) ω_1.
  (c) V_1.
  (d) θ_2.
  (e) V_3.

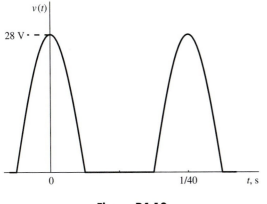

Figure P4.10

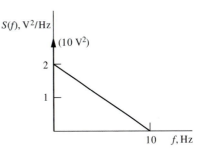

Figure P4.14

4.11. Show that Eq. (4.12) for $\tau = T/2$ gives the same harmonic amplitudes and phases as those in Eq. (4.5).

4.12. The nominal bandwidth of a standard telephone line is 4 kHz. What is the approximate duration of the shortest pulse that can be sent over such a telephone line? Given that the width between pulses should be equal to the pulse width, how many pulses per second can be sent over a phone line using simple pulses?

4.13. For the half-wave rectified sinusoid in Fig. 4.6(a), what fraction of the total power is carried by the dc component?

4.14. The "power" spectrum of a random signal is shown in Fig. P4.14.
  (a) What is the dc value of the signal, assuming that it is negative?
  (b) What power does this signal give to a 5-Ω resistor?
  (c) What is the approximate correlation time for this signal?

4.15. The effective bandwidth of standard AM radio is 5000 Hz. If such a radio were tuned off a station, receiving only static, what would be the approximate correlation time for the static?

4.16. The effective bandwidth of standard FM radio is 15,000 Hz. If such a radio were tuned off a station, receiving only static, what would be the approximate correlation time for the static?

4.17. Figure P4.17 shows the power spectrum of a random signal.
  (a) What is the average value of the signal?
  (b) What is the approximate correlation time of the signal?
  (c) What is the total power from this signal into a 50-Ω resistor?

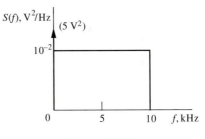

Figure P4.17

4.18. A computer-based communication system has an information rate of 3000 characters/second, where each character is represented by a byte of digital information. What is the approximate bandwidth required by this system? State assumptions.

Section 4.2: Electrical Filters

4.19. A 60-Hz half-wave rectified sinusoid with a peak value of 10 V is filtered by a low-pass filter, as shown in Fig. P4.19. Calculate the dc and the peak value of the fundamental at the output.

4.20. Make asymptotic Bode plots of the magnitude of the following filter functions:

  (a) $\mathbf{F}(\omega) = \dfrac{25}{1 + j(\omega/250)}$.

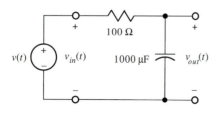

$v(t)$ = half-wave rectified sinusoid
V_p= 10 V
f = 60 Hz

Figure P4.19

(b) $\mathbf{F}(\omega) = 50 \times \dfrac{j\omega}{100 + j4\omega}$.

(c) $\mathbf{F}(f) = 10^{-2} \times \dfrac{1 + j(f/10)}{1 + j(f/1000)}$.

4.21. Show that the loss of the low-pass filter in Fig. 4.17 is approximately 1 dB at a frequency one octave (factor of 2) below the cutoff.

4.22. The gain in dB of a filter is shown in the Bode plot of Fig. P4.22.
  (a) Is this a high-pass, low-pass, or band-pass filter?
  (b) What is the gain in dB at 500 Hz?
  (c) Estimate the cutoff frequency of the filter.
  (d) If the input voltage to the filter were $v_{in}(t) = 5 \cos(2000\,t)$ V, what would be the output in the form $v_{out}(t) = A \cos(2000t + \theta)$? In other words, find A and θ.

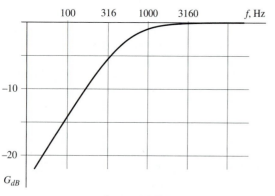

Figure P4.22

4.23. The input to a filter is the time function $v_{in}(t) = 10 + 5\sin(8000t)$. The asymptotic Bode plot of the filter characteristic is shown in Fig. P4.23.
  (a) Is the filter high-pass or low-pass?
  (b) What is the frequency in hertz of the fundamental of the input?
  (c) What is the input "power" in volt²?
  (d) What is the output "power" in volt²?

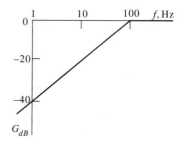

Figure P4.23

4.24. For the square wave shown in Fig. P4.24, determine the following:
  (a) dc component.
  (b) Frequency of the fifth harmonic.
  (c) rms amplitude of the fifth harmonic.
  (d) After passing through a low-pass filter with a cutoff frequency of 100 Hz, what is the output peak amplitude of the fundamental?

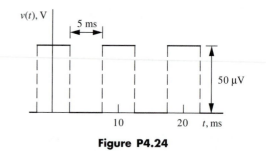

Figure P4.24

4.25. The input signal to an RC low-pass filter is $v(t) = 5 + 5\sin(1000t)$. Draw the circuit diagram of a filter such that the power in the output is 90% dc power. The filter uses a 10-µF capacitor.

4.26. The half-wave rectified sinusoid in Fig. P4.26(a) is passed through an RL filter, as shown in Fig. P4.26(b). What percent of the output "power" in volt² is in its second harmonic?

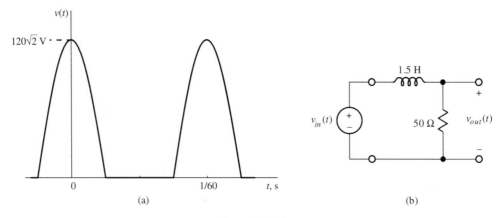

Figure P4.26

Section 4.3: Feedback Concepts

4.27. Work out the signal levels with $v_{out} = 10$ V for the amplifier in Fig. 4.32 with reduced gain, $A = 120$.

4.28. How large does the amplifier gain, A, have to be in the feedback amplifier in Fig. 4.32 before the gain with feedback is within 1% of $1/\beta$ with negative feedback? What is the loop gain for this value of A?

4.29. A voltage amplifier has a gain of -500. Fifteen percent of the output voltage is added to an input voltage to provide the input voltage to the amplifier. The feedback is negative.
(a) What is the gain of the feedback system?
(b) If you wished to double this gain, how could this be accomplished?

4.30. A feedback system employs negative voltage feedback. Assume that for a test, the feedback path is opened at the point shown in Fig. 4.34, and 5 mV is fed into the input to the amplifier. Under these conditions, the output voltage is observed to be 5 V and the return from the feedback loop is 0.5 V.
(a) What is the loop gain?
(b) What is the gain of the feedback amplifier with the loop closed?

4.31. The system diagram in Fig. P4.31 shows a feedback path.
(a) Find the loop gain.
(b) Find the gain of the system with feedback.

4.32. The amplifier in Fig. P4.32 employs feedback derived from a voltage divider. The voltage amplifier has infinite input impedance, zero output impedance, and a gain given as $v_{out} = -50v_i$.
(a) Draw a system diagram of the amplifier, being

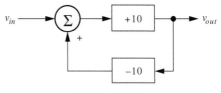

Figure P4.31

sure to put the correct sign on the summer.
(b) Determine the loop gain.
(c) Find the gain with feedback.

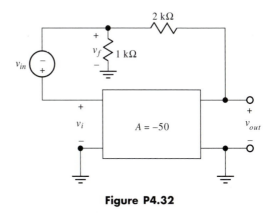

Figure P4.32

4.33. The feedback amplifier shown in Fig. P4.33 uses a current amplifier and has a single resistor, R_F, providing feedback. In your analysis, assume that $R_F \gg R_L$, such that the output voltage is $v_{out} \approx i_{out} R_L$. Assume that the input impedance of the current amplifier is very low and A_i is a positive number.

(a) Derive the gain with feedback, $A_f = v_{out}/i_{in}$, by using KCL, Ohm's law, and the gain equation for the current amplifier.

(b) For $A_i = 500$ and $R_L = 10\,\Omega$, find R_F for an overall gain of $-500\,\Omega$.

(c) Put the results from part (a) in the form of Eqs. (4.73) to (4.75) and draw the system diagram corresponding to this amplifier.

(d) What is the loop gain of the amplifier?

(e) If $A_i \Rightarrow \infty$, what is the limiting form for A_f?

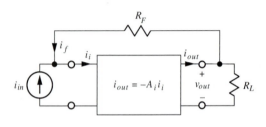

$$A_f = \frac{v_{out}}{i_{in}} \text{ volt/ampere}$$

Figure P4.33

4.34. Assume that the amplifier in Fig. 4.32 has a severe nonlinearity, as shown by the input-output characteristic in Fig. P4.34. Without feedback, this would cause much distortion in the amplifier output. Derive and sketch the overall (with feedback) characteristics of v_{out} vs. v_{in} with feedback to demonstrate the benefits of feedback in reducing the nonlinearity. *Hint*: The final results will consist of straight-line segments. Assume selected values at the

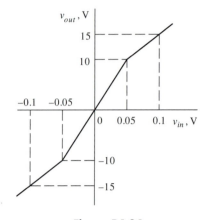

Figure P4.34

output and work back to the input, as was done in Sec. 4.3.

4.35. For the relay circuit with feedback shown in Fig. 4.40, find the following:

(a) What is the maximum input voltage to keep the diode current less than 30 mA.

(b) Assume the relay opens at a current of 12 mA. In the original circuit, how long after the switch opens does it take for the relay to open, assuming that the relay has been closed a long time? Assume an ideal diode and assume that the initial current is 30 mA.

4.36. A signal x_1 has noise x_2 added to it in a system, as shown in Fig. P4.36 (a). To improve the signal-to-noise ratio, a feedback system is developed, as shown in Fig. P4.36 (b). Find the value of the amplifier gain, A, a positive number, such that the signal-to-noise ratio is improved by a factor of 10.

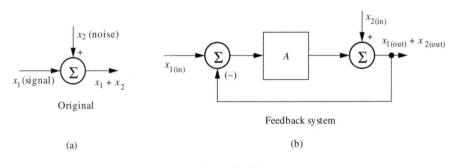

Figure P4.36

Section 4.4: Operational-Amplifier Circuits

4.37. Design an amplifier having a gain of $+15$ dB, a phase shift of $180°$, and an input impedance of $500 \, \Omega$, real.

4.38. Design an op-amp amplifier having a gain of $+10$ dB, a phase shift of $0°$, and an input impedance of $5000 \, \Omega$, real.

4.39. For the amplifier shown in Fig. P4.39, find the following:
   (a) What is the voltage gain in dB?
   (b) What is the input impedance?

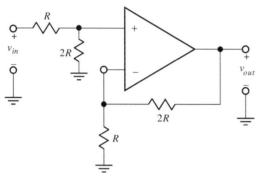

Figure P4.39

4.40. Determine the output of the op-amp circuit shown in Fig. P4.40 as a function of v_{in} and V.

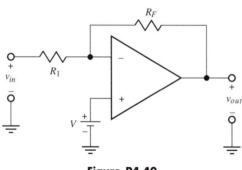

Figure P4.40

4.41. Determine the output of the op-amp circuit shown in Fig. P4.41 as a function of v_{in} and V.

4.42. A noninverting amplifier is designed using op amps for a gain of $+10$, but constructed with $\pm 10\%$ resistors. (The actual value of the resistance can depart from the nominal value by 10%.) What is the range of gain that can result?

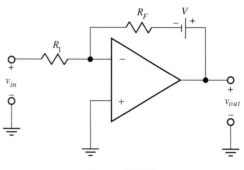

Figure P4.41

4.43. In the op-amp circuit in Fig. P4.43, we assume that the input current to the op amp is zero at both the inverting and noninverting inputs, but we do not assume that $v_+ = v_-$ because we have a finite loop gain.
   (a) Determine the output voltage, v_{out}, as a function of the input voltage, v_{in}, using a nodal analysis; identify the independent node(s); write the nodal equation(s); and solve for the output, v_{out}.
   (b) Compare with standard results of feedback theory using loop-gain analysis.
   (c) Compare with the approximate gain in Eq. (4.105). What inequalities(s) must be satisfied to make the two the same?

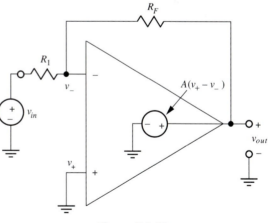

Figure P4.43

4.44. What is the output voltage of the circuit in Fig. P4.44 as a function of v_{in}? Explain your answer.

4.45. An inverting amplifier is designed using op amps for a gain of $+10$, but constructed with $\pm 10\%$ resistors.

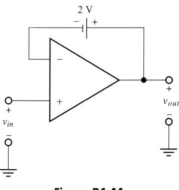

Figure P4.44

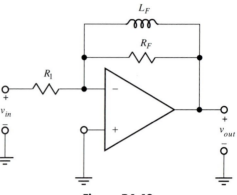

Figure P4.48

(The actual value of the resistance can depart from the nominal value by 10%.) What is the range of gain that can result?

4.46. A transducer puts out a current signal in the range from 5 to 25 mA. For compatibility with a computer analog-to-digital input, this signal must be converted to a voltage signal in the range from −5 to +5 V. Design an op-amp circuit to accomplish this current-to-voltage conversion and add the offset.

4.47. Design an integrator with an input impedance of 10 kΩ that provides an output

$$v_{out}(t) = -10 - 3\int_0^t v_{in}(t')dt'$$

when the switch is opened at $t = 0$.

4.48. The op amp in Fig. P4.48 may be considered ideal.
  (a) Write a DE for the output voltage in terms of the input voltage and the circuit components.
  (b) Solve for the output voltage if the input is a battery of V volts in series with a switch that is closed at $t = 0$. Assume that the op amp behaves in a linear manner.

4.49. For the op-amp circuit shown in Fig. P4.49, what are the voltages at points a and b relative to ground?

4.50. The op-amp circuit shown in Fig. P4.50 has plus-and-minus power supplies at ±10 V. The switch is closed at $t = 0$.
  (a) Find the value that the output would reach if the output transistors in the op amp did not saturate.
  (b) Find the time when the output saturates, assuming that this occurs at 10 V of output voltage.

4.51. For the op-amp circuit shown in Fig. P4.51, find the following:

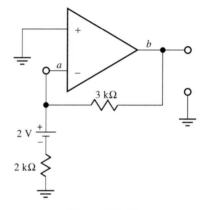

Figure P4.49

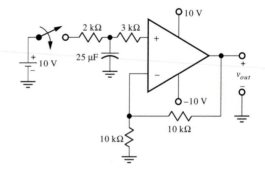

Figure P4.50

  (a) Input resistance seen by v_{in}.
  (b) Voltage gain v_2/v_{in}.
  (c) Voltage gain v_3/v_{in}.
  (d) If $v_{in} = 1$ V and the switch opened at $t = 0$, as shown, how long would it be until the magnitude of the output voltage is 1 V?

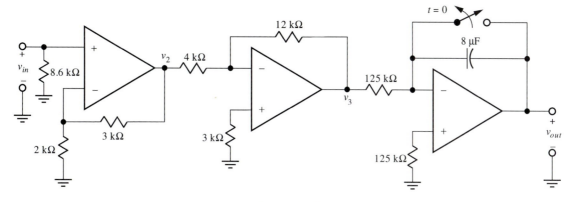

Figure P4.51

4.52. Design a comparator circuit that has output levels of 0 and +5 V (these are the power-supply values) and makes its transition at an input voltage of +3 V for increasing inputs and +2 V for decreasing inputs.

General Problems

4.53. The circuit in Fig. P4.53 is a comparator with hysteresis. The op amp has a voltage gain of 10^5 and its output saturates at ±9 V.
  (a) Is the feedback positive or negative?
  (b) Determine the voltage that will cause the output to change from −9 V to +9 V for increasing inputs.
  (c) Determine the voltage that will cause the output to change from +9 V to −9 V for decreasing inputs.

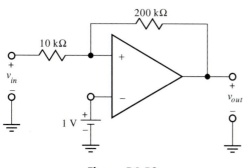

Figure P4.53

4.54. The filter shown in Fig. P4.54 uses a resistor, a capacitor, and an ideal transformer with a turns ratio of 4:1. Make a Bode plot of the gain of this filter. There is no load on the output.

4.55. The frequency characteristic of an *RC* filter circuit is described by the Bode plot shown in Fig. P4.55(a). Find the output voltage of the circuit, $v_{out}(t)$, if a 10-

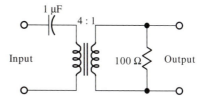

Figure P4.54

V battery is connected for 10 ms, as indicated by Fig. P4.55(b).

4.56. The circuit shown in Fig. P4.56(b) can be considered a filter followed by an amplifier or it can be considered as a filter combining resistors, a capacitor, and an op amp. The input to the filter is the square wave shown in Fig. P4.56(a).
  (a) Taking the second approach, determine the filter function relating the output and input voltages in the frequency domain.
  (b) Determine the frequency in hertz at which the gain of the filter is −5 dB.
  (c) What would be the average value of the output?
  (d) What would be the power in volts² in the first harmonic, or fundamental, of the output voltage?

4.57. The filter shown in Fig. P4.57(b) has no input initially; then, at $t = 0$, a series of pulses begins, as shown in Fig. P4.57(a) and continues for a long time.
  (a) What is the output voltage at the end of the first pulse, at $t = 5$ ms?

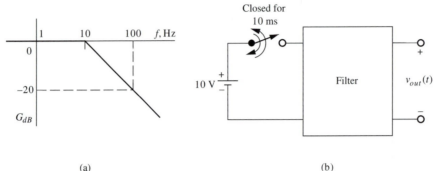

(a)　　　　　　　　　　　　　　　(b)

Figure P4.55

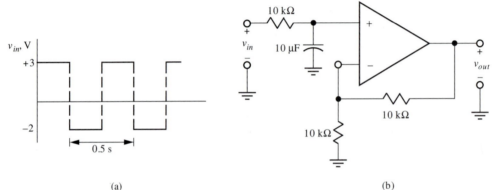

(a)　　　　　　　　　　　　　　　(b)

Figure P4.56

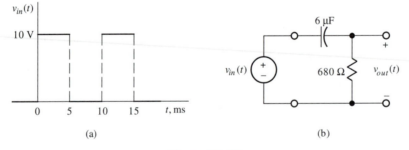

(a)　　　　　　　　　　　　　　　(b)

Figure P4.57

(b) After a long period of time with the pulses continuing, what are the amplitudes of the dc and first-harmonic components in the output spectrum?

4.58. Figure P4.58 shows a standard noninverting op-amp amplifier circuit.

(a) Represent the entire amplifier in a system diagram with main amplifier, summer, β network, and sampler. What are A, β, and the sign of the summer in this amplifier?

(b) Find the loop gain for this amplifier.

(c) Find the exact gain of the amplifier (no approximation) and compare with the gain derived in the text, Eq. (4.109).

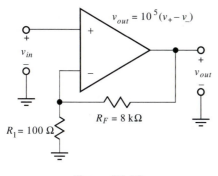

Figure P4.58

4.59. A voltage is given by the expression

$$v(t) = 12 + 6\sin(200\pi t) - 3\cos(300\pi t)\,\text{V}$$

(a) What is the fundamental frequency in hertz?
(b) Find the total power in volts².
(c) Find the rms value of this voltage.
(d) If passed through a low-pass filter that reduces the 100-Hz component by 3 dB, how much in dB would the 150-Hz component be reduced?

4.60. Using the circuit in Fig. P4.60, design an amplifier that produces an output of $20\log(v_{in}/1\text{ V})$ that is, the output is the input voltage in dB compared with 1 V. Assume that the diode characteristic is described by Eq. (2.17) (ignore the +1 term) for $\eta = 1.4$, $I_0 = 2 \times 10^{-10}$ A, and $V_T = 0.0259$ V. Specify R_F for the required gain and V_B to remove the constant term in Eq. (4.120). (*Note:* V_B could be obtained with a voltage divider connected to the negative power supply.)

4.61. For the circuit shown in Fig. P4.61 find the following:

(a) Find the sign of A for the feedback to be negative.

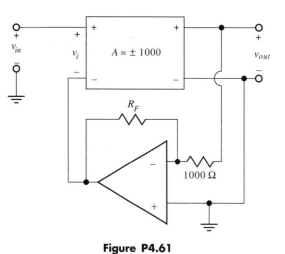

Figure P4.61

(b) We want a gain magnitude of 10 for the system. Find R_F.

4.62. For the circuit shown in Fig. P4.62, find the following:
(a) The loop gain.
(b) The gain with feedback.

4.63. The analysis of the passive low-pass filter in Sec. 4.2 ignores the loading at input and output. Consider now the circuit in Fig. P4.63, which contains a source resistance R_s and a load resistor R_L. The filter function $\mathbf{F}(\omega) = \mathbf{V}_{out}/\mathbf{V}_{in}$ will be identical in form to that in Eq. (4.36), except (1) the gain will no longer be unity in the region below the critical frequency, ω_c, and (2) the equation for ω_c is changed because it depends on the source and load resistances. Determine the revised expressions for ω_c and $\mathbf{F}(\omega)$.

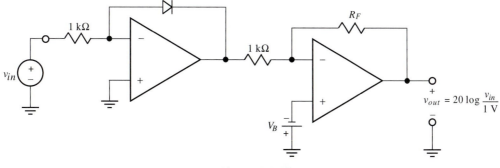

Figure P4.60

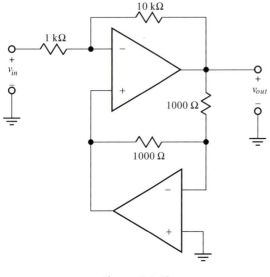

Figure P4.62

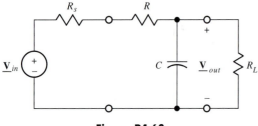

Figure P4.63

4.64. The small-signal amplifier in Fig. 2.50 has an input

impedance $Z_{in} = R_B \parallel r_\pi$ where $R_B = R_1 \parallel R_2$, which often is unacceptably low for a voltage amplifier because r_π is relatively small. A circuit in which feedback improves the input impedance is shown in Fig. P4.64(a). The small-signal equivalent circuit for the amplifier is shown in Fig. P4.64 (b). This circuit uses feedback to control impedance. The feedback is proportional to the emitter current and is subtracted from the input voltage at the base of the base–emitter *pn* junction.

(a) Show that the input impedance to the transistor base is raised from r_π to $r_\pi + (1+\beta)R_E$, where β is the current gain of the transistor. *Hint:* This circuit is similar to the emitter-follower circuit, especially in feedback effects.

(b) For the transistor ($r_\pi = 450\ \Omega$) and component values in Fig. 2.50, but with an $R_E = 10\ \Omega$ added to the base, calculate the input impedance and gain of the amplifier.

4.65. The power supply shown in Fig. P4.65 consists of a full-wave rectifier, an *LC* filter, and a resistive load. If the filter is designed correctly, the current in the inductor will be continuous, two diodes will always be conducting, and the rectifier output, $v_{FW}(t)$, will be a full-wave rectified waveform having a spectrum

$$v_{FW}(t) = V_P\left[\frac{2}{\pi} + \frac{4}{3\pi}\cos(2\omega_1 t)\right.$$
$$\left. - \frac{4}{15\pi}\cos(4\omega_1 t) + \cdots\right]$$

where ω_1 is the input frequency, 60 Hz in this case.

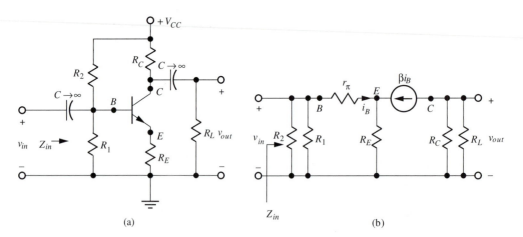

(a)

(b)

Figure P4.64

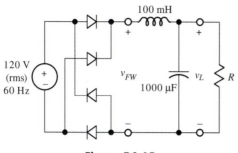

Figure P4.65

Because the current in the inductor is mostly dc and second harmonic, the condition for continuous current is that the peak value of the second harmonic current be less than the dc current.
(a) Determine the maximum value of R for continuous current in the inductor. Consider only dc and second harmonic. *Hint*: Treat the capacitor as a short circuit for ac.
(b) With the value determined in part (a), approximate the percent ripple in the load voltage, $V_{p-p}/V_{dc} \times 100\%$, where V_{p-p} is the peak-to-peak in the load voltage. Make reasonable approximations.

4.66. Figure P4.66 shows a standard op-amp circuit, except for the capacitor in the feedback circuit. The switch is closed at $t = 0$. Determine the output voltage as a function of time after the switch closure at the input. *Hint*: You may use the methods of Chap. 1, but you may have to use the methods of this chapter to write the DE to determine the time constant. Assume linear operation of the op amp.

4.67. The op-amp circuit shown in Fig. P4.67 is assembled with the switch closed and then the

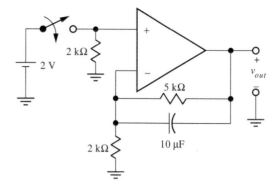

Figure P4.66

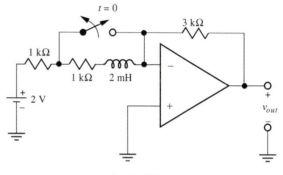

Figure P4.67

switch is opened at $t = 0$. Determine the output voltage to show the voltage before the switch is opened and the transient as a result of opening the switch. Make a sketch of v_{out}. *Hint*: You may use the methods of Chap. 1. You may have to use the methods of this chapter to write the DE to determine the time constant.

Answers to Odd-Numbered Problems

4.1. (a) $T_0 = 66.8°$, $T_1 = 17.7°$, $\theta_1 = -195°$, and $\omega_1 = 2\pi/12$ rad/ms; (b) $50.3°$.

4.3. $v_{HW}(t) = V_p [1/\pi + (1/2) \sin(\omega_1 t) - (2/3\pi) \cos(2\omega_1 t)]$, $n = 4, 8, 12, ...$, the same as before, but if $n = 2, 6, 10, ...$, change the signs to $-$.

4.5. (a) Discrete; (b) 117 V^2; (c) the dc would change to -5 V; (d) 450 Hz.

4.7. (a) 150 Hz; (b) 100 V^2; (c) 4.24 V; (d) 100 V^2.

4.9. 76.6 V^2.

4.11. Proof depends on $\sin(n\pi) = 0$ for n even, -1 for $n = 1, 5, 9, ...$, and $+1$ for $n = 3, 7, 11, ...$.

4.13. 40.5%.

4.15. 0.2 ms.

4.17. (a) ± 2.236 V; (b) 100 µs; (c) 2.10 W.

4.19. 3.18 V dc, 0.133 V peak.

4.21. -0.969 dB.

4.23. (a) High pass; (b) 1270 Hz; (c) 112.5 V^2; (d) 12.5 V^2.

4.25. Standard low-pass filter with resistance of 187 Ω.

4.27. $v_{out} = 10$ V, $v_i = 0.083$ V, $v_f = 1.000$ V, $v_{in} = 1.083$ V.

4.29. (a) -6.58; (b) 0.075 approximate, 0.074 exact.

4.31. (a) -100; (b) 0.0990.

4.33. (a) $Gain = -\dfrac{A_i R_L}{1 + (R_L/R_F)A_i}$; (b) 556 Ω;

(c)

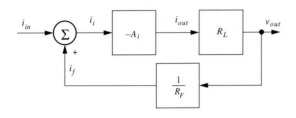

(d) $L = -\dfrac{R_L}{R_F}A_i = -8.99$; (e) $A_f \Rightarrow R_F$.

4.35. (a) 0.660 V; (b) 9.16 ms.

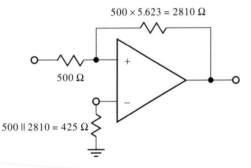

4.37. 15 dB $= 10^{15/20} = 5.623$, $R_1 = 500\Omega$, $R_F = 2.81\,\text{k}\Omega$, inverting amplifier.

4.39. (a) 6.02 dB; (b) 3R.

4.41. $v_{out} = -v_{in}(R_F/R_1) + V.$

4.43. (a) $v_{out} = v_{in}\dfrac{-A}{1 + R_1/R_F + AR_1/R_F}$;

(b) the same answer; (c) $A \gg 1$ and $AR_1/R_F \gg 1$ such that $A_f \Rightarrow -R_F/R_1$.

4.45. The gain is between -12.3 and -8.18.

4.47.

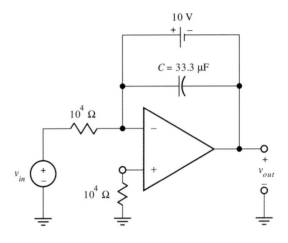

4.49. $v_a = 0$, $v_b = -3$ V.

4.51. (a) 8.6 kΩ; (b) 2.5; (c) -7.5; (d) 0.133 s.

4.53. (a) positive; (b) 1.5 V; (c) 0.6 V.

4.55. $v_{out} = 10(1 - e^{-t/15.9 \text{ ms}})$ for $0 < t < 10$ ms and $v_{out} = 4.67$ V for $t > 10$ ms.

4.57. (a) 7.06 V; (b) dc $= 0$, fundamental $= 5.93$ V peak.

4.59. (a) 100 Hz; (b) 166.5 V^2; (c) 12.9 V rms;
(d) -5.12 dB.

4.61. (a) -1000; (b) 99 Ω.

4.63. $\omega_c' = 1/[C \times R_L \,\|\, (R + R_s)]$ and

$$\mathbf{F}(j\omega) = \left(\frac{R_L}{R_L + R + R_s}\right) \times \frac{1}{1 + j(\omega/\omega_c')}.$$

4.65. (a) 113 Ω; (b) 7.4%.

4.67. $v_{out} = -3 + 3e^{-t/1\mu s}$ V.

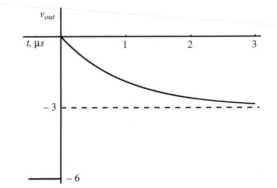

Instrumentation Systems

1. To understand how to perform worst-case and statistical error analysis
2. To understand how to recognize floating signals and be able to amplify difference-mode signals
3. To understand analog signal and noise spectra and how active filters can improve system performance
4. To understand the principles of cabling, grounding, and shielding to reduce interference
5. To understand digital signal sampling, conversion, and processing

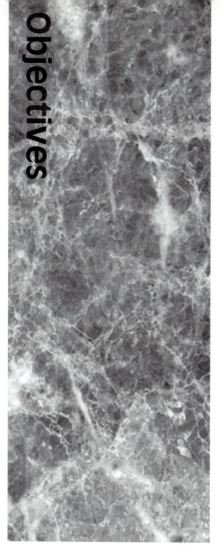

objectives

Instrumentation systems combine analog and digital techniques to acquire, condition, store, and process data. Issues such as error analysis and noise reduction are considered, and new applications for op amps are presented.

General Considerations

Data gathering and reduction were once simple but tedious tasks. Meters were read by eye and results were recorded by hand. At best, a continuously recording strip chart left a wiggly line of data to be examined for trends, interpolated by eye, and analyzed by hand calculations.

Computers have made data gathering and reduction quick, automatic, and sophisticated. The computer can monitor many inputs, adjust the gain of analog channels to keep data within a prescribed range, filter incoming signals to improve data quality, process and record data, furnish displays, and produce outputs.

Analog Information

An instrumentation system.   Figure 5.1 shows an instrumentation system that employs analog and digital processing. We now discuss the various subsystems as an introduction to the structure of this chapter.

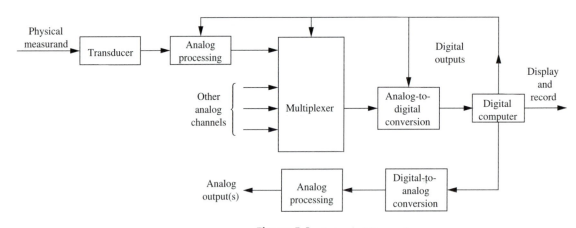

Figure 5.1   Instrumentation system.

The Frequency Domain

Transducer.   The transducer produces an electrical output indicative of some physical *measurand* such as pressure, temperature, or angular position. Transducers are discussed in this introductory section.

measurand

Analog processing.   Amplification and filtering are normally required to prepare the signal for conversion to digital form. Figure 5.1 shows a control signal coming from the computer into this subsystem because often the gain of an analog channel is controlled to keep the signal within a prescribed range. Analog processing is discussed in Sec. 5.2.

Multiplexer.   Typically, several analog channels are processed sequentially through a multiplexer, which is a digitally controlled switch. The multiplexer accepts parallel inputs from several analog channels and provides one analog output at a time for conversion to digital form. Multiplexers are discussed in Sec. 5.3.

Digital Information

Analog-to-digital (A/D) conversion.   The A/D converts the information from analog to digital form. Often, the time variations of the analog signal must be arrested with a sample-and-hold circuit while A/D conversion is taking place. We discuss two types of A/D converters in Sec. 5.3.

Digital computer.   The brains of the entire operation, and the immediate recipient of the information, is a digital computer. This might be a microprocessor dedicated to the instrumentation system or it might be a general-purpose computer that is structured to perform the required data-acquisition function simultaneously with other activities. For example, a desktop personal computer can be adapted to accept analog and digital data inputs, and standard programs are available to supervise the data-gathering activity. We do not investigate the processing and storage functions of the computer in this chapter except for a brief discussion of digital filtering.

Digital-to-analog (D/A) conversion.   Often, the computer must provide outputs in analog form. If, for example, the data monitor were part of a control system, the computer might furnish analog output signals as feedback to the controller of the process affecting the physical measurand. We discuss D/A converters in Sec. 5.3.

Processing of analog outputs.   Analog outputs often require filtering and amplification for controlling process functions. However, no new topics are introduced, so we have no further discussion of analog outputs.

Transducers

transducer, sensor

A *transducer*, also called a *sensor*, converts a physical quantity to an electrical signal. Most transducers produce analog signals, but some have digital outputs. Many types of transducers exist for most types of physical measurands; they differ in physical processes, noise, accuracy, linearity, ruggedness, output impedance, frequency response, and need for frequent calibration. After discussing bridge circuits and strain gages, we describe some common transducers used to measure position and angle, pressure, flow rate, and temperature.

Bridge circuits.   Many transducers employ a resistor whose resistance changes as a function of the measurand. The most common circuit used to convert a resistance change to a voltage change is the Wheatstone-bridge circuit, shown in Fig. 5.2. The bridge consists of two voltage dividers, and the output voltage is the difference in the voltages created by the voltage dividers. Application of the voltage-divider relationship gives the output voltage:

$$v_{out} = V \times \left(\frac{R_0}{R_1 + R_0} - \frac{R_2}{R_3 + R_2} \right) \tag{5.1}$$

balanced bridge   The bridge is *balanced* when the output voltage is zero, which requires

$$\frac{R_0}{R_1} = \frac{R_2}{R_3} \tag{5.2}$$

If the bridge is approximately balanced, small changes in R_0 from balance, ΔR_0, produce corresponding small changes in output voltage, given by

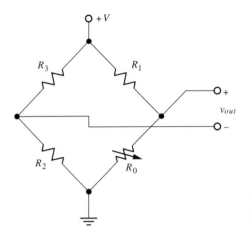

Figure 5.2 The Wheatstone-bridge circuit converts a resistance change in R_0 to a change in output voltage.

$$\Delta v_{out} = V \times \frac{R_1 \, \Delta R_0}{(R_0 + R_1)^2} \tag{5.3}$$

look-up table

but large changes in R_0 are related to the changes in output voltage by the nonlinear relationship given by Eq. (5.1). Furthermore, the change in the resistance of the transducer is often nonlinear. Thus, the measurement system may have to process this nonlinear information to determine the measurand. With a computer-based system, this conversion is often performed by means of a *look-up table* that stores the nonlinear function in the computer.

EXAMPLE 5.1 | **Balanced bridge**

Temperature is measured by a temperature-sensitive resistor called a thermistor. The nominal 10-kΩ thermistor is placed in a balanced bridge that has all resistors of 10 kΩ. How much does R_0 have to increase to give a $v_{out} = 1/50 \times V$?

SOLUTION:
Using Eq. (5.3) with $R_1 = R_0$, we find

$$\frac{1}{50} \times V = V \times \frac{\Delta R_0}{4R_0} \quad \Rightarrow \quad \Delta R_0 = \frac{R_0}{12.5} = 800 \ \Omega \tag{5.4}$$

WHAT IF? What if you use the exact relationship in Eq. (5.1)?[1]

---

[1] $\Delta R_0 = +833$ or $-769 \ \Omega$.

Strain gages. Another common measurement component is the strain gage, which can be used to indicate strain, force, pressure, or acceleration. A *strain gage* is a resistor that is attached to a mechanical member to share the elongation of the member. A bonded strain gage, shown in Fig. 5.3 (a), is a resistance that is bonded to the member with an adhesive. Elongation changes the resistance of the gage by stretching the wire. The bonded strain gage is placed in a bridge circuit to produce an output voltage. Bonded strain gages are temperature-sensitive, and normally a second strain gage that shares the thermal environment, but not the strain, is used in the same bridge to minimize thermal effects.

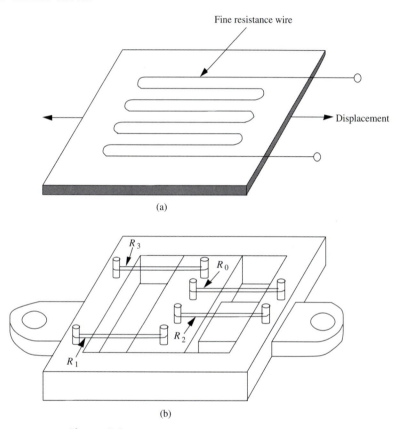

Figure 5.3 (a) Bonded strain gage; (b) unbonded strain gage.

An unbonded strain gage is more sensitive than the bonded variety. As shown in Fig. 5.3(b), the entire bridge circuit is placed in a fixture that strains all four resistors, and temperature effects tend to cancel. More sensitive yet is the semiconductor strain gage, in which a special semiconductor material is bonded to the mechanical member. However, the semiconductor strain gage must be used in a temperature-controlled environment because its resistance is influenced strongly by temperature.

Position transducers: potentiometers. Figure 5.4 shows (a) translational and (b) rotational potentiometers. The potentiometer consists of a fixed resistor with a movable contact that responds to physical movement, thus changing the resistance ratio and

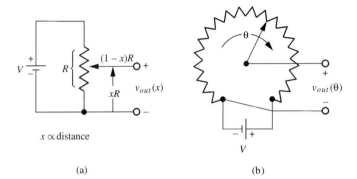

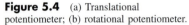

Figure 5.4 (a) Translational potentiometer; (b) rotational potentiometer.

(a)

(b)

hence the output voltage of the voltage divider. The resulting change in resistance could be used in a bridge circuit or could produce a voltage directly, as shown in Fig. 5.4. The unloaded output voltage in Fig. 5.4(a) is

$$V_{out}(x) = V \times \frac{xR}{xR + (1-x)R} = xV \qquad (0 < x < 1) \qquad (5.5)$$

Potentiometers are available in which the output voltage relates to the physical movement through linear or logarithmic functions. Rotational potentiometers are available in single- or multiturn versions.

Position transducers: linear-variable differential transformers. The linear-variable differential transformer uses variable coupling between the primary and two secondaries of a transformer to create ac voltages that depend on the position of a magnetic slug. The diodes and resistors shown in Fig. 5.5 rectify the ac voltages and produce a dc output voltage that, after suitable filtering, gives an indication of the position of the magnetic slug. With the slug in the center, equal voltages are created across the two resistors and the output voltage is zero. With the slug off center, the voltages become unequal and produce an output voltage whose polarity indicates the direction of movement.

Pressure transducers: diaphragm type. Pressure can be converted to force or displacement through the use of a diaphragm, bellows, or spiral tube. Therefore, pressure can be measured with a strain gage or some other means for monitoring force or displacement.

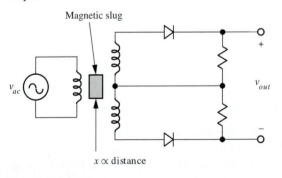

Figure 5.5 The linear-variable differential transformer.

Pressure transducers: integrated-circuit pressure cells. Pressure cells based on semiconductor sensors are manufactured complete with bridge circuit and amplifiers. In effect, the input to the cell is a pressure, and the output is an electrical signal.

Fluid-flow transducers. Fluid flow in a pipe can be monitored through the differential pressure across an orifice or screen mesh. Fluid flow can also be measured through a turbine placed in the stream. The rotation rate of the turbine can be measured digitally to determine flow rate.

Temperature transducers. Many types of temperature sensors are available because almost all physical processes are affected by temperature. Sensors differ according to range and precision; a different transducer would be used to measure temperatures in the range of 5000°F than to monitor the temperature for an air-conditioning system. In the following, we describe two common types of temperature transducers that produce electrical outputs.

thermistor

Temperature transducers: thermistors. A *thermistor* is a resistor, made of a semiconductor, whose resistance depends on temperature. The change of intrinsic-carrier concentration with temperature produces a resistance that decreases strongly with increasing temperature. A typical temperature range for a thermistor is −100° to +300°C, and the resistance change is nonlinear. For monitoring of small temperature changes, the thermistor may be placed in a bridge circuit, but for large temperature changes, other circuit arrangements may be required.

EXAMPLE 5.2 | **Thermistor**

A thermistor has a resistance given by

$$\ln\left(\frac{R(T)}{R_0}\right) = \alpha\left(\frac{1}{T} - \frac{1}{T_0}\right) \qquad (5.6)$$

where $R(T)$ is the resistance at T in Kelvin, $R_0 = 500\ \Omega$ at $T_0 = 20°C$ (293 Kelvin), and $\alpha = 4000$ is a parameter of the semiconductor. Find the temperature for $R(T) = 1000\ \Omega$.

SOLUTION:
From Eq. (5.6),

$$\ln\left(\frac{1000}{500}\right) = 4000\left(\frac{1}{T} - \frac{1}{293}\right) \;\Rightarrow\; T = 278.8\ \text{K}\ (5.8°\ \text{C}) \qquad (5.7)$$

WHAT IF? What if you solve for $R(T)$ for these parameters?[2]

---

[2] $R(T) = 5.89 \times 10^{-4}\, e^{\,4000/T}\ \Omega$.

Temperature transducers: thermocouples.

A *thermocouple* is a junction between two dissimilar metals that produces a voltage that depends on the junction temperature. Thermocouples utilize this effect to monitor temperatures in a wide range, up to 2500°C. As shown in Fig. 5.6, two junctions are required, with a reference junction kept at constant temperature, often 0°C in an ice slush. Many types of metals are used, depending on the temperature range required, and typical temperature coefficients are +50 μV/°C.

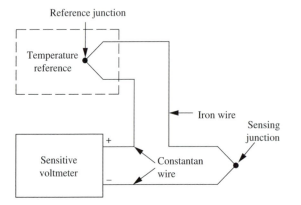

Figure 5.6 A thermocouple-based temperature measurement requires two junctions, one at a known temperature.

Summary.

In this section, we discussed the more common transducers used to measure displacement, force, pressure, fluid flow, and temperature. Sensors are available for measurement of virtually any physical variable. The technology of measurement is advanced and diverse; the best source of information is application literature from manufacturers.

Error Analysis

LEARNING OBJECTIVE 1.

To understand how to perform worst-case and statistical error analysis

Need for error analysis.

Errors are introduced into a measurement system by the transducer, by the noise and nonlinearities of the analog electronics, by A/D conversion, and by digital signal processing. Error analysis combines the errors from the various components and processes to estimate the total error in the measurement. Some of the contributing errors are random errors describable by statistical methods, and some of the errors are estimates of uncertainty in calibration and system properties. Such errors may be combined only by making assumptions; and error analysis, even when undertaken with total objectivity, can be controversial. This section presents basic methods of error analysis.

Exact and linearized error analysis.

The output of a measurement system depends on many factors, and in principle all can introduce errors. For a simple example, let us say we require a current source of magnitude, I, which we establish from a voltage source V, a resistance R, and the op-amp circuit shown in Fig. 5.7.

Sources of error.

We assume the magnitude of the error in V to be δV, the magnitude of the error in R to be δR, and we ignore other sources of error such as finite loop gain, op-amp power-supply fluctuations, and thermal noise. The interpretation of the errors in V and R depends on the method of analysis employed, as is discussed in what

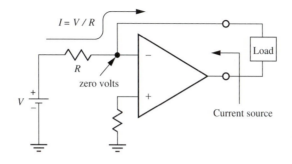

Figure 5.7 A current source simulated with an op amp.

follows. These errors are not known quantities but rather uncertainties that we are willing to estimate. For example, δV would depend on power supply ripple, noise introduced into the wiring, the influence of variations in line voltage and temperature on the voltage source, and whatever else might conceivably influence the input voltage. Likewise, δR might be estimated from temperature change and calibration uncertainty. Usually, the magnitude of each error is estimated, the sign assumed randomly plus or minus. Our purpose in this section is to show how such "known" errors may be combined to estimate an overall error.

In this context, Ohm's law takes the form

$$I \pm \delta I = \frac{V \pm \delta V}{R \pm \delta R} \tag{5.8}$$

where δI is the magnitude of the uncertainty in the output current. For purposes of illustration, we assume $V = 10\ \text{V}$, $\delta V = 200\ \text{mV}$, $R = 500\ \Omega$, and $\delta R = 1\%$. Equation (5.8) may be used to determine δI by either the *worst-case* or the *statistical* error analysis shown in what follows.

worst-case error analysis

Exact worst-case analysis. *Worst-case error analysis*, as the name implies, assumes that the errors are at their maximum and add in the worst possible way. The errors in this case, δV and δR, are interpreted to be the maximum errors in the various factors, and the signs are chosen such that the effects of all errors are cumulative.

EXAMPLE 5.3 | **Worst-case errors**

Calculate the worst-case error in the current-source output using the uncertainties given before.

SOLUTION:
Equation (5.8) gives a maximum current with maximum voltage and minimum resistance:

$$I + \delta I = \frac{V + \delta V}{R - \delta R} = \frac{10 + 0.2}{0.99 \times 500} = 20.606\ \text{mA} \tag{5.9}$$

and similarly a minimum output current of

$$I + \delta I = \frac{V - \delta V}{R + \delta R} = \frac{10 - 0.2}{1.01 \times 500} = 19.4059\ \text{mA} \tag{5.10}$$

Equations (5.9) and (5.10) may be solved for I and δI to yield

$$I = 20.006 \text{ mA} \qquad \text{and} \qquad \delta I = 0.600 \text{ mA} \tag{5.11}$$

which may be expressed as

$$I = 20.006 \pm 0.600 \text{ mA (worst-case)} \tag{5.12}$$

In the example, the average value given by Eq. (5.12) is very near to the nominal value of 20 mA calculated before, because the relative errors are small. We put "worst-case" by our stated error to make explicit the method of error calculation and interpretation.

Linearization. Equation (5.8) is a simple expression, but usually many factors are involved in an error analysis, and the large number of variables and the nonlinear character of the expressions can lead to mathematical difficulties. For this reason, a linearized analysis is often performed. A linearized approximation to Eq. (5.8) is legitimate if the errors are small on a percent basis. We may linearize Eq. (5.8) by power-series expansion to the form

$$I(1 \pm \varepsilon_I) = \frac{V(1 \pm \varepsilon_V)}{R(1 \pm \varepsilon_R)} \approx \frac{V}{R} \times (1 \pm \varepsilon_V \pm \varepsilon_R \pm \text{higher-order terms in } \varepsilon\text{'s}) \tag{5.13}$$

where $\varepsilon_I = \delta I/I$, $\varepsilon_V = \delta V/V$, and $\varepsilon_R = \delta R/R$ are the normalized errors. The first term on the right side does not involve the errors and is the nominal output current. We may drop the higher-order terms and cancel the zeroth-order term on both sides of Eq. (5.13) to obtain a form involving only the errors:

$$\pm \varepsilon_I = \pm \varepsilon_V \pm \varepsilon_R \tag{5.14}$$

Equation (5.14) is valid with the ε's as normalized errors or as percent errors. In this simple case, both sources of error receive equal weight in establishing the resultant error, but in more complicated systems, the weights can differ.

EXAMPLE 5.4 | **Percent errors**

If the power supply voltage has a value of 10 V with an uncertainty of ± 200 mV and a 1% tolerance 500-Ω resistor is used, what is the worst-case error in the current using a linearized analysis?

SOLUTION:
The nominal current is 20 mA. We interpret the ε's in Eq. (5.14) as describing percent errors; hence we have

$$\varepsilon_V = \pm 2\% \pm 1\% = \pm 3\% \tag{5.15}$$

where we have assumed that errors add. Thus, the worst-case error is 3% of 20 mA, or 0.6 mA worst case.

Summary. In our examples, the difference between exact and linearized analysis is small because the errors are small. Worst-case analysis is pessimistic because it assumes that all errors are at their maximum values and that all the effects of errors accumulate. For this reason, worst-case error analysis is used primarily when life, property, or product acceptability is dependent on the results.

statistical error analysis

Statistical error analysis. A *statistical* analysis of system errors considers that errors are independent random variables described by probability density functions. The underlying mathematical models are beyond the scope of our treatment. Use of statistical techniques with the full nonlinear expressions is difficult at best, and usually a linearized analysis is performed.

With this model, the quoted errors of the ε's are standard deviations and they combine in a Pythagorean addition. If we assume a standard deviation of 2% in V and 1% in R.

$$\varepsilon_I = \sqrt{(2)^2 + (1)^2} = 2.24\% \text{ (1 s.d.)} \tag{5.16}$$

Note the following:

- With Pythagorean addition, the total errors are strongly dominated by the larger contributors. Small errors do not matter unless there are many of them.

- The "(1 s.d.)" means one standard deviation and implies the statistical method for combining errors. For our example, the statistical method yields a current of 20.00 ± 0.45 mA (1 s.d.). Normally, Gaussian statistics are assumed, and the results imply that the true current lies within the stated error bounds with a probability of 68% and within twice the error bounds with a probability of 95%.

Statistical error analysis is optimistic because it assumes that the errors can be small and that error cancelations occur to some extent. Experience shows that this method is usually valid, and its simplicity makes it popular.

Check Your Understanding

1. The voltage applied to a Wheatstone bridge influences the output voltage unless the bridge is balanced. True or false?

2. Reversing both diodes in the linear-variable differential transformer does not change the output voltage. True or false?

3. A measurement system has three sources of error, which produce maximum errors of 1%, 1.8%, and 2.7%. What is the worst-case error possible in the output?

4. Repeat the previous problem if the errors are standard deviations. What is the standard deviation due to the combined errors?

Answers. **(1)** True; **(2)** false: the output changes sign; **(3)** 5.5% (worst-case); **(4)** 3.40% (1 standard deviation).

The purpose of the analog section of the instrumentation system is to provide the A/D converter with analog information as free from noise as possible and in the acceptable voltage range. This involves amplification, gain control, removal of unwanted dc voltage, filtering, and cabling the system to minimize interference. In this section, we first treat matters relating to amplification and gain control, then deal with analog filtering techniques, and end with a brief section on grounding and shielding of instrumentation cables.

**Power ground,
signal ground,
single-ended signal,
floating signal**

Instrumentation Amplifiers

Single-ended and floating signals.

In power systems we use grounds for safety; the *power ground* is the point of a circuit connected to earth by a large wire. By contrast, the *signal ground* of an electronic circuit is a connected set of conductors distributed to many points of the circuit to give a common reference voltage. Signal ground may or may not correspond to power ground. When the signal voltage appears between a wire carrying the signal and the signal ground, as shown in Fig. 5.8(a), the signal is *single-ended*. When the signal voltage appears between two wires independent of signal ground, as shown in Fig. 5.8(b), the signal is *floating*. The voltage source v_{cm} in Fig. 5.8(b) might be voltage introduced by the circuit of the transducer or might represent noise induced in both wires carrying the signal. The floating signal can be considered two single-ended signals, with the real signal being their difference.

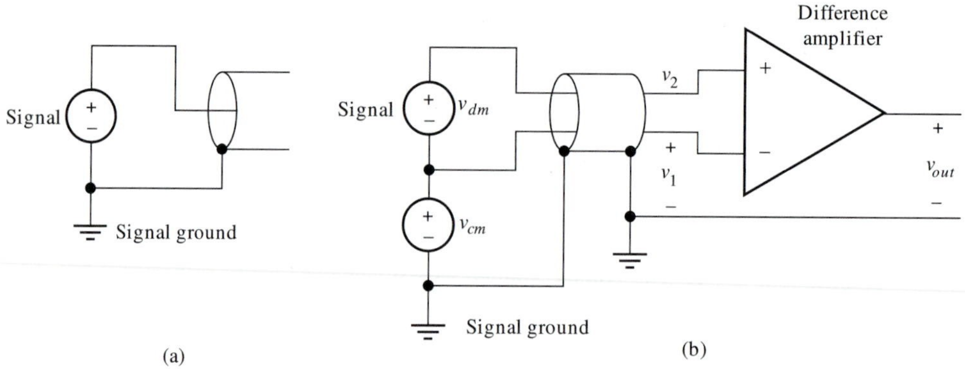

> **LEARNING OBJECTIVE 2.**
>
> **To understand how to recognize floating signals and be able to amplify difference-mode signals**

Figure 5.8 (a) Single-ended signal; (b) floating signal.

**difference
amplifier**

**Impedance
Level**

Amplifying floating signals.

With a single-ended signal, the standard op-amp amplifiers presented in Chap. 4 could be used. The noninverting amplifier would be favored because its high input impedance minimizes loading effects. A floating signal requires a *difference amplifier*, that is, an amplifier that subtracts the signals between the two input conductors and amplifies their difference. We presented a circuit that accomplishes this in Fig. 4.56, but that circuit suffers from a low input impedance. After discussing difference- and common-mode signals, we present an amplifier with high-input impedance for floating inputs.

Difference- and common-mode signals. Signals of the type shown in Fig. 5.8(b) are described by a difference-mode signal and a common-mode signal. As shown in Fig. 5.9, the *difference-mode* signal, v_{dm}, is the difference between the two input voltages, as shown in Fig. 5.8(b). The *common-mode* signal is the average between v_1 and v_2 and is the voltage that the difference amplifier is required to reject, or at least minimize. The common- and difference-mode voltages relate to the single-ended voltages as

$$v_{dm} = v_2 - v_1 \qquad v_2 = v_{cm} + \frac{v_{dm}}{2}$$
$$\Leftrightarrow \qquad\qquad\qquad (5.17)$$
$$v_{cm} = \frac{v_2 + v_1}{2} \qquad v_1 = v_{cm} - \frac{v_{dm}}{2}$$

Figure 5.9 The difference-mode signal is the difference between two voltages, and the common-mode signal is their average.

difference mode, common mode

common-mode rejection ratio, CMRR

Common-mode rejection. An amplifier with a floating input, as shown in Fig. 5.8(b), ideally should have equal but opposite gains for both inputs, but in practice has slightly unequal gains. Let the gain for the noninverting input be $+A_2$ and the gain for the inverting input be $-A_1$, where both As are positive. The output, therefore, is

$$v_{out} = +A_2 v_2 - A_1 v_1 = \underbrace{\left(\frac{A_1 + A_2}{2}\right)}_{\substack{\text{difference-mode} \\ \text{gain, } A_{dm}}} \times v_{dm} + \underbrace{(A_2 - A_1)}_{\substack{\text{common-mode} \\ \text{gain, } A_{cm}}} \times v_{cm} \qquad (5.18)$$

where the difference-mode gain, A_{dm}, is the average of A_2 and A_1 and the common-mode gain, A_{cm}, is the difference between A_2 and A_1. The *common-mode rejection ratio*, CMRR, is the ratio

$$\text{CMRR} = \left| \frac{A_{dm}}{A_{cm}} \right| \qquad (5.19)$$

A difference amplifier should have a large common-mode rejection ratio. A typical op amp would have a CMRR of 70 dB, which means that the gain for the difference-mode signal is $10^{+70/20} = 3162$ times the gain for the common-mode signal. In a given circuit application, the amplifier utilizing the op amp would have effects from the external circuit that would degrade the CMRR from that of the op amp alone. The magnitude of the common-mode voltage also has to lie within acceptable bounds for the op amp to operate satisfactorily.

EXAMPLE 5.5 | **Bridge outputs**

The Wheatstone bridge in Fig. 5.2 operates with $V = 12$ V, $R_1 = R_2 = R_3 = 10$ kΩ, and $R_0 = 12$ kΩ. With these values, the voltage across R_0 is $v_2 = 6.545$ V and the voltage across R_2

is $v_1 = 6.000$ V. We assume that this signal is amplified by a difference amplifier with a CMRR of 70 dB. What is the normalized error in the output due to the common-mode component?

SOLUTION:
The common-mode and difference-mode voltages can be determined from Eqs. (5.17), with the result $v_{dm} = 0.545$ V and $v_{cm} = 6.273$ V. Assuming a difference-mode gain of A_{dm}, the common-mode gain would be smaller by 70 dB: $A_{cm} = A_{dm}/3162$. The normalized error in the output due to the common-mode signal is, therefore,

$$\frac{6.273 A_{dm}/10^{70/20}}{0.545 A_{dm}} = 3.64 \times 10^{-3} = 0.364\% \qquad (5.20)$$

WHAT IF?   What if $R_0 = 11$ kΩ and the CMRR $= 72$ dB?[3]

Impedance Level **5**

Instrumentation amplifier.   The instrumentation amplifier shown in Fig. 5.10 has high and equal input impedances at both inputs and an output voltage that amplifies the difference between the input voltages. We recognize that the output op amp provides subtraction and amplification with a gain of R_F/R_1, as shown in Eq. (4.115). We may analyze the input stage by the methods of Sec. 4.4, p. 248. Because $v_+ \approx v_-$ for both amplifiers, the voltages at the top and bottom of R_2 are v_2 and v_1, respectively. The current in R_2 is thus

$$i = \frac{v_2 - v_1}{R_2} \qquad (5.21)$$

Because the op-amp inputs have negligible input current, the current through R_2 must also go through both R_3's, so the difference in output voltages is

$$v_2' - v_1' = iR_2 + 2iR_3 = (v_2 - v_1)\left(1 + \frac{2R_3}{R_2}\right) \qquad (5.22)$$

Hence, the input stage gives a gain of $(1 + 2R_3/R_2)$, in addition to providing high and equal input impedances to the sources of v_2 and v_1. The difference-mode gain of the instrumentation amplifier shown in Fig. 5.10 is thus

$$A_{dm} = \frac{R_F}{R_1}\left(1 + \frac{2R_3}{R_2}\right) \qquad (5.23)$$

The common-mode gain of the circuit depends on the CMRR for the output amplifier and the degree to which the R_F/R_1 ratios are matched for the two inputs to the subtractor circuit.

---

[3] The error would be 0.540%.

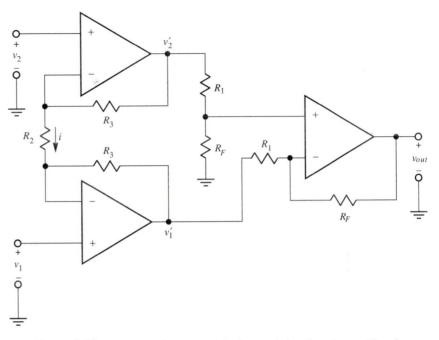

Figure 5.10 The instrumentation amplifier has coupled noninverting amplifiers for high-input impedance, followed by a difference amplifier.

<table>
<tr><td>**EXAMPLE 5.6**</td><td>**Instrumentation-amplifier design**</td></tr>
</table>

A strain gage has a maximum output signal of 10 mV and a maximum of 1 V is allowed by the A/D converter. Design an instrumentation amplifier with the required gain using $R_2 = 10$ kΩ and $R_1 = 10$ kΩ.

SOLUTION:
The required gain is $20\log(1 \text{ V}/10 \text{ mV}) = 40$ dB. We choose 20 dB/stage of amplification. This requires

$$10^{20/20} = 10.0 = 1 + \frac{2R_3}{R_2} \Rightarrow R_3 = 4.5R_2 = 45 \text{ k}\Omega$$

$$10^{20/20} = 10.0 = \frac{R_F}{R_1} \Rightarrow R_F = 10R_1 = 100 \text{ k}\Omega$$

(5.24)

WHAT IF?   What if the top $R_3 = 46$ kΩ and the bottom $R_3 = 44$ kΩ?[4]

---

[4] No problem: "$2R_3$" $= R_{3(\text{top})} + R_{3(\text{bottom})} = $ same.

Amplifier with bias removal. If the information appears as small changes in a dc voltage, we can use the amplifier shown in Fig. 5.11 to remove all or part of the dc voltage from the signal.[5] This circuit consists of a voltage-follower input stage, for high input impedance, followed by a subtractor. The dc input to the subtractor is derived from a potentiometer. The resistance of the potentiometer, R_p, should be somewhat less than R to avoid loading effects and maintain constant gain for the signal.

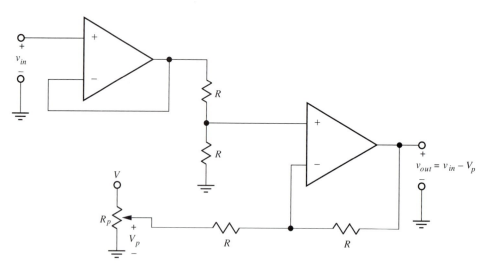

Figure 5.11 This amplifier can be used to remove a dc bias from a signal.

Programmable attenuator. An *attenuator* reduces the signal level by a prescribed amount. Attenuation is required when the signal level is too large for the A/D converter or some other component in the system. The inverting amplifier shown in Fig. 5.12 (a) has voltage gains of -1, -0.5, -0.1, -0.05, or -0.01, depending on the switch settings. The table in Fig. 5.12(b) shows the standard settings, with a "1" indicating that the switch is closed. The attenuation in dB is the negative of the gain in dB. The switches might be relay switches or FET switches, controlled by an overrange indication from the A/D converter.

Summary. In this section we discussed a number of op-amp circuits that are useful in analog conditioning of signals. We defined difference-mode and common-mode signals and introduced the common-mode rejection ratio for an amplifier. We now discuss active filtering of analog signals.

Analog Active Filters

active filter

The Frequency Domain

The analog section of an instrumentation system normally involves filtering. The filters typically are *active filters* employing op amps with frequency-dependent feedback networks. In this section, we first discuss sources of analog noise and then present representative active filter circuits for low-pass and high-pass filters.

---

[5] If the signal spectrum is not too low, a blocking capacitor can remove the dc and pass the signal. The circuit of Fig. 5.11 works for all signal frequencies.

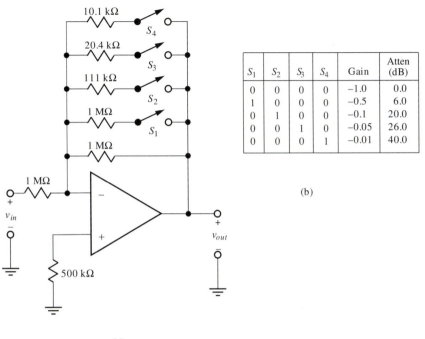

S_1	S_2	S_3	S_4	Gain	Atten (dB)
0	0	0	0	−1.0	0.0
1	0	0	0	−0.5	6.0
0	1	0	0	−0.1	20.0
0	0	1	0	−0.05	26.0
0	0	0	1	−0.01	40.0

(b)

(a)

Figure 5.12 This programmable attenuator reduces the signal level by factors up to 100, depending on the switch settings.

Signal and noise spectra. The transducer output signal has a time structure that may be described in the frequency domain by a spectrum. The transducer spectrum must be estimated as a basis for designing the filters of the analog section of the system. Noise from several sources is also present in the system. The purpose of the analog filter is to pass the transducer signal and to eliminate as much of the noise as possible. Because the signal and noise spectra typically overlap, the filter cannot fully eliminate the noise.

The spectrum of the signal is determined by the time variations in the measurand and the construction of the transducer. The specifications of the manufacturer of the transducer should give some indication of the time response.

EXAMPLE 5.7 | **Transducer bandwidth**

A temperature transducer is specified as having a 0.5-s time constant. What signal bandwidth might be expected from this transducer?

SOLUTION:
The specification of a time constant implies a first-order system like an RC low-pass filter, presumably due to the thermal inertia of the material. Because the 3-dB frequency for such a filter is $f_c = 1/(2\pi RC)$, with $RC = \tau$, the time constant, a good estimate of the transducer signal bandwidth, B, is

$$B \approx \frac{1}{2\pi\tau} = \frac{1}{2\pi \times 0.5} = 0.318 \text{ Hz} \qquad (5.25)$$

Variations of the temperature at higher frequencies would have little effect on transducer output.

Types of noise. The principal components of the noise in the system are wide-band noise, $1/f$ noise, and interference. Figure 5.13 represents the spectra of wideband and $1/f$ noise. The wide-band noise consists of thermal noise, shot noise, and partition noise. Thermal noise arises in all resistors and is thermodynamic in origin; Chap. 6 gives more details. Shot and partition noise components arise in transistors due to the discrete character of electrical carriers and become a problem at small current levels. Interference is man-made noise that can be reduced by proper circuit layout, shielding, and grounding techniques, as discussed in the next section.

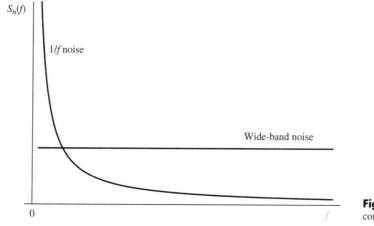

Figure 5.13 Two unavoidable noise components are $1/f$ and wide-band noise.

1/f noise. The $1/f$ noise component, which exists in all natural processes, shows the accumulated effects of small changes. For example, resistors and transistors are constantly changing in their microstructure due to heat, cosmic-ray damage, vibration, and a myriad of other effects. These combine to produce *drift*, that is, a slow meandering of signal values. Because the signal normally has a $1/f$ component of its own, which is part of its information content, the $1/f$ noise component cannot be eliminated by straightforward filtering.

Reducing 1/f noise. An important technique to minimize this noise component is the synchronous modulation/detection system shown in Fig. 5.14(a). The signal from the transducer is *chopped* or modulated by switching between the transducer output and a reference signal, as shown in Fig. 5.14(b). After amplification and analog filtering, the output component at the chopping frequency is detected by a phase-selective detector. The benefit of this scheme is that $1/f$ noise contributed by the electronic system is reduced because only these signal components near the modulation frequency are de-

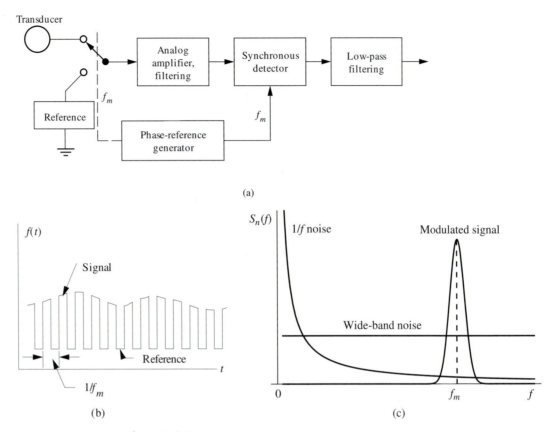

Transducer

Analog amplifier, filtering

Synchronous detector

Low-pass filtering

f_m

Reference

f_m

Phase-reference generator

(a)

$f(t)$

Signal

Reference

t

$1/f_m$

(b)

$S_n(f)$

$1/f$ noise

Modulated signal

Wide-band noise

0

f_m

f

(c)

Figure 5.14 (a) The lock-in detection system chops the signal at a frequency f_m and detects the information at that frequency; (b) the slow variations in the signal are modulated by the chopping; (c) the effect of the modulation is to move the information to a higher frequency so that $1/f$ noise can be eliminated by the filtering and detection process.

tected. In effect, the modulation moves the information to a high frequency so that the $1/f$ noise can be filtered, as shown in Fig. 5.14(c). In Chap. 6, we again encounter modulation and detection in communication systems.

active filter, passive filter

Active filters. An *active filter* combines amplification with filtering. The *RC* filters investigated in Chap. 4 are *passive filters* because they provide only filtering. An active filter uses an op amp to furnish gain but has capacitors added to the input and feedback circuits to shape the filter characteristics.

In Chap. 4 we derived the gain of an inverting amplifier in the time domain. In Fig. 5.15, we show the frequency-domain version. We may easily translate the earlier derivation into the frequency domain:

$$v_{in} \Rightarrow \underline{\mathbf{V}}_{in}(\omega), \qquad v_{out} \Rightarrow \underline{\mathbf{V}}_{out}(\omega)$$

$$A_V = -\frac{R_F}{R_1} \Rightarrow \underline{\mathbf{F}}_V(\omega) = \frac{-\underline{\mathbf{Z}}_F(\omega)}{\underline{\mathbf{Z}}_1(\omega)} \tag{5.26}$$

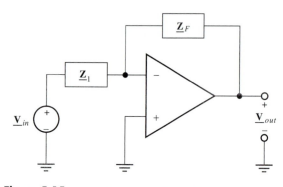

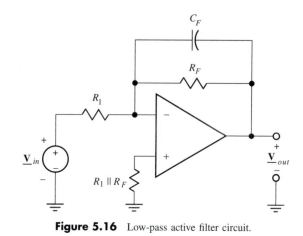

Figure 5.15 Active filter circuit. $\underline{\mathbf{Z}}_F$ and $\underline{\mathbf{Z}}_1$ shape the filter characteristic.

Figure 5.16 Low-pass active filter circuit.

The filter function, $\underline{\mathbf{F}}_V(\omega)$, is thus the ratio of the two impedances with an inversion, and in general gives gain as well as filtering.

Low-pass active filter. Placing a capacitor in parallel with R_F, Fig. 5.16, will at high frequencies tend to lower the magnitude of $\underline{\mathbf{Z}}_F$ and hence reduce the gain of the amplifier; consequently, this capacitor converts an inverting amplifier into a low-pass filter with gain. We may write

$$\underline{\mathbf{Z}}_F(\omega) = R_F \| \frac{1}{j\omega C_F} = \frac{1}{1/R_F + j\omega C_F} = \frac{R_F}{1 + j\omega R_F C_F} \qquad (5.27)$$

Because $\underline{\mathbf{Z}}_1 = R_1$, the filter function from Eq. (5.26) is

$$\underline{\mathbf{F}}_V(\omega) = -\frac{R_F}{R_1} \frac{1}{1 + j\omega R_F C_F} = A_V \frac{1}{1 + j(\omega/\omega_c)} \qquad (5.28)$$

where $A_V = -R_F/R_1$ is the gain without the capacitor, and $\omega_c = 1/R_F C_F$ is the cutoff frequency. The gain of the amplifier is approximately constant until the frequency exceeds ω_c, after which the gain decreases with increasing ω.

EXAMPLE 5.8 **Low-pass active filter**

A pressure transducer requires a low-pass filter with a gain of 20 dB and an input impedance of 1 kΩ. The time constant of the transducer is 100 ms. Design a suitable filter for this application.

SOLUTION:

To give the required input impedance, $R_1 = 1\text{k}\Omega$. The low-frequency gain is $A_v = -10^{20/20} = -10$, so $R_F = 10R_1 = 10$ kΩ. From Eq. (5.25), we estimate the required cutoff frequency to be $1/(2\pi \times 0.1) = 1.59$ Hz. This is the critical frequency; thus, from the discussion of Eq. (5.28), we calculate

$$C_F = \frac{1}{\omega_c R_F} = \frac{1}{2\pi \times 1.59 \times 10^4} = 10 \ \mu\text{F} \qquad (5.29)$$

The Bode plot of this filter function is shown in Fig. 5.17. The shape is identical to that of the low-pass filter in Fig. 4.25, but there is an increase in gain due to the op amp.

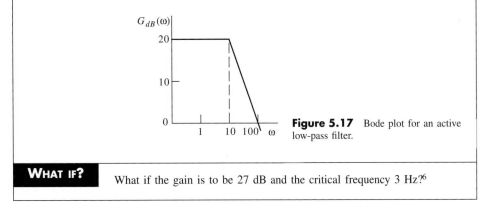

Figure 5.17 Bode plot for an active low-pass filter.

WHAT IF?   What if the gain is to be 27 dB and the critical frequency 3 Hz?[6]

High-pass active filter.   The high-pass filter shown in Fig. 5.18 uses a capacitor in series with R_1 to reduce the gain at low frequencies. For this filter, $\mathbf{Z}_F = R_F$ and $\mathbf{Z}_1 = R_1 + 1/j\omega C_1$; hence, the filter function is

$$\mathbf{F}_V(\omega) = -\frac{R_F}{R_1 + 1/j\omega C_1} = A_V \times \frac{j\omega/\omega_c}{1 \times j(\omega/\omega_c)} \qquad (5.30)$$

where $A_V = -R_F/R_1$ is the gain without the capacitor, and $\omega_c = 1/R_1 C_1$ is the cutoff frequency, below which the amplifier gain is reduced. The Bode plot of this filter characteristic for $R_F/R_1 = 10$ is shown in Fig. 5.19. This Bode plot is identical to that of the high-pass filter given in Fig. 4.27 except for the increase in gain due to the op amp.

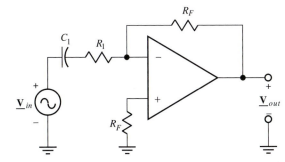

Figure 5.18 High-pass filter circuit.

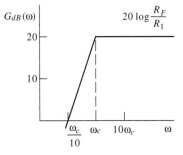

Figure 5.19   Bode plot for active high-pass filter for $R_F = 10R_1$.

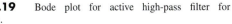

[6]$R_F = 22.4 \ \text{k}\Omega$ and $C_F = 2.37 \ \mu\text{F}$.

Butterworth-filter characteristics. The general low-pass Butterworth-filter characteristic is

$$|\mathbf{F}(f)| = \frac{1}{\sqrt{1 + (f/f_c)^{2n}}} \tag{5.31}$$

where n is the order of the filter and f_c is the critical frequency. For $n = 1$, we have the characteristic of the low-pass filter discussed in Chap. 4. The filter characteristics for $n = 1$ and $n = 2$ are shown in the asymptotic Bode plots of Fig. 5.20. The second-order filter characteristic drops at -40 dB/decade after the critical frequency is passed.

Low-pass, two-pole Butterworth filter. The circuit shown in Fig. 5.21 gives a second-order Butterworth-filter characteristic. This circuit combines positive and negative feedback, and its principles of operation are complex. We will confirm the characteristic by using the op-amp analysis techniques developed in Chap. 4.

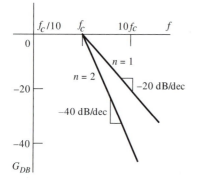

Figure 5.20 Bode plots for first- and second-order Butterworth low-pass circuits.

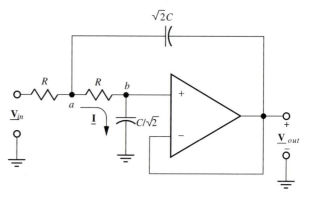

Figure 5.21 Second-order Butterworth low-pass-filter circuit.

Analysis of Butterworth-filter circuit. Applying the op-amp principles of Sec. 4.4, p. 248, we require that the input voltage to both op-amp inputs be essentially $\mathbf{V}_{out}$. Thus, the current $\mathbf{I}$ shown in Fig. 5.21 is

$$\mathbf{I} = j\frac{\omega C}{\sqrt{2}}\mathbf{V}_{out} \tag{5.32}$$

Hence, the voltage at point a is

$$\mathbf{V}_a = \mathbf{V}_{out} + R\mathbf{I} = \mathbf{V}_{out}\left(1 + \frac{j\omega RC}{\sqrt{2}}\right) \tag{5.33}$$

We can now write KCL for node a, which is a standard nodal equation:

$$\frac{\left(1 + j\frac{\omega RC}{\sqrt{2}}\right)\mathbf{V}_{out} - \mathbf{V}_{in}}{R} + \frac{\left(1 + j\frac{\omega RC}{\sqrt{2}}\right)\mathbf{V}_{out} - \mathbf{V}_{out}}{1/j\sqrt{2}\omega C} + j\frac{\omega C}{\sqrt{2}}\mathbf{V}_{out} = 0 \tag{5.34}$$

Equation (5.34) has the solution

$$\underline{\mathbf{V}}_{out} = \frac{\underline{\mathbf{V}}_{in}}{1 - (\omega RC)^2 + j\sqrt{2}\,\omega RC} \tag{5.35}$$

For the Bode plot, we require the square of the magnitude of the filter gain:

$$|\underline{\mathbf{F}}(\omega)|^2 = \left|\frac{\underline{\mathbf{V}}_{out}}{\underline{\mathbf{V}}_{in}}\right|^2 = \frac{1}{[1 - (\omega RC)^2]^2 + (\sqrt{2}\,\omega RC)^2} = \frac{1}{1 + (\omega RC)^4} \tag{5.36}$$

which is the required Butterworth characteristic, Eq. (5.31), for $n = 2$ and

$$f_c = \frac{1}{2\pi RC} \tag{5.37}$$

EXAMPLE 5.9 | **Butterworth filters**

Design a second-order low-pass Butterworth filter with a 3-dB frequency of 8 Hz. The output impedance of the input transducer is less than $100\,\Omega$.

SOLUTION:

The critical frequency given by Eq. (5.37) is the 3-dB frequency. The design requires one choice. We may ensure that the filter has insignificant loading of the signal source by making R much greater than $100\,\Omega$. Accordingly, we choose $R = 10\,\mathrm{k}\Omega$ and Eq. (5.37) determines the value of C to be

$$C = \frac{1}{2\pi R f_c} = \frac{1}{2\pi (10^4)8} = 1.99\ \mu\mathrm{F} \tag{5.38}$$

This is not the value of either capacitor in the circuit in Fig. 5.21 but is rather the geometric mean between the two capacitors. The capacitors are thus approximately $2.8\,\mu\mathrm{F}$ and $1.4\,\mu\mathrm{F}$, and the required circuit is shown in Fig. 5.22.

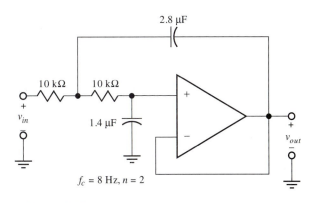

Figure 5.22 Second-order low-pass Butterworth filter circuit.

| WHAT IF? | What if you want the gain of your filter in dB at 10 Hz?[7] |

Other filters. Standard designs exist for high-pass Butterworth filters and wide- and narrow-band band-pass Butterworth filters, as well as several other filters having superior characteristics to the Butterworth in some aspects. Handbooks and application literature give practical circuits for many active filter circuits. Filter manufacturers produce filters of prescribed characteristics for high-frequency, extremely narrow-band band-pass, and narrow-band band-reject filters.

Cabling, Grounding, and Shielding Techniques

interference, EMI

Impedance Level

Sources of interference. Unwanted, man-made electrical signals constitute a class of noise called *interference*.[8] Interference may couple into a transducer circuit via an electric field, a magnetic field, or a common ground wire. Electric-field interference arises from high-voltage or high-frequency signals and couples into high-impedance portions of electronic circuits. Magnetic fields induce signals by coupling to loops of wire in a low-impedance electronic circuit. Both types of interference may be reduced by proper circuit layout, shielding, and grounding techniques.

LEARNING OBJECTIVE 4.

To understand the principles of cabling, grounding, and shielding to reduce interference

Reducing interference. Sensitive electronic equipment picks up interference in the vicinity of high-power electrical equipment such as motors, welders, and transformers, near radio and TV stations, or near power-electronic equipment such as motor controllers and large power supplies. Interference may be reduced by avoiding such locations if possible and shielding electronic circuits with grounded metallic cabinets and power lines with metal conduit. Circuits susceptible to magnetic interference may be shielded by magnetic foil. Low-level signals should be protected by twisted-pair conductors within a braided outer conductor.

signal ground

Grounding. The signal ground is a network of wires, all tied together, to provide a reference potential of zero volts to all parts of the circuits. Grounding problems arise because currents flow in ground wires, and because the ground wires have resistance and inductance, voltage differences are created between different portions of the grounding system. Such differences can create false signals at low-level inputs. One precaution is to connect all grounds within a system to a common point through separate wires so that ground currents do not share wires, but this is usually unnecessary. A more practical technique is to provide a separate grounding system for high-level signals and signals that have rapid transitions that is isolated from the grounding system for low-level instrumentation signals.

Long cable runs between instruments or between a transducer and its analog electronics can cause special problems in grounding. When connected through the cable in an attempt to provide a common ground between two separate systems, as shown in Fig. 5.23, as much as 100 mV of voltage can exist between "ground" *a* and "ground" *b*

---

[7] The gain is −5.37 dB at 10 Hz.

[8] Or *EMI* for electromagnetic interference.

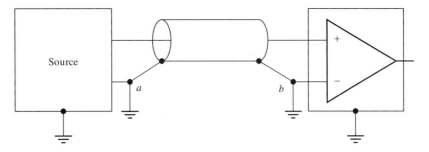

Figure 5.23 Electronic circuits with grounds connected through the connecting cable. This approach can cause interference problems if the two "ground"s are not at the same potential.

due to interference signals. This connection is unacceptable unless signals are much larger than that level of interference.

The cabling connection shown in Fig. 5.24(a) allows separate grounds. The input to the receiving chassis is floating, with the cable shield as one input. The connection in Fig. 5.24(b) uses a common ground, but the signal is carried by a twisted-pair shielded cable and thus floats. The signal-carrying wires are twisted to reduce magnetic coupling.

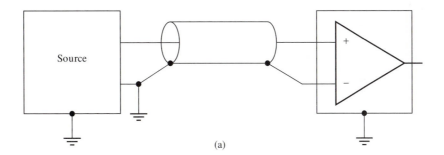

(a)

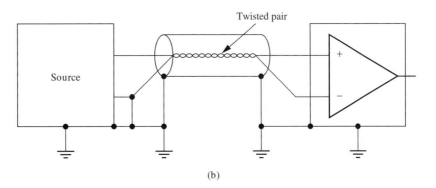

(b)

Figure 5.24 (a) No common ground is provided in this connection; (b) here a common ground is provided, but the signal is separate from the ground connection.

Check Your Understanding

1. Grounding one part of a floating input produces a single-ended input. True or false?

2. The voltages of a floating input are $+1.5$ V and -0.8 V. Find the common-mode and difference-mode components.

3. A difference amplifier has a voltage gain of 15,000 and a common-mode rejection ratio of 65 dB. What is the gain for the common-mode component of the input?

4. What would be the gain of the programmable attenuator in Fig. 5.12 if all switches were closed?

5. Show that the product of the voltage gain and the bandwidth of the low-pass filter in Fig. 5.16 is independent of the resistor in the feedback path.

6. The grounding system used in Fig. 5.23 would be acceptable for logic signals with a 5-V amplitude. True or false?

7. Are twisted-pair cables used to reduce electric-field or magnetic-field interference?

Answers. **(1)** True; **(2)** $V_{cm} = 0.35$ V, $V_{dm} = 2.3$ V; **(3)** 8.44; **(4)** $0.00629 = -44.0$ dB; **(5)** gain $\times B = 2\pi/R_1 C_F$; **(6)** true; **(7)** magnetic-field interference.

5.3 DIGITAL SIGNAL PROCESSING

Analog-to-Digital Conversion

LEARNING OBJECTIVE 5.

To understand digital signal sampling, conversion, and processing

A/D, ADC, ADC precision

Introduction. An *analog-to-digital converter* (ADC or A/D) changes an analog signal to a digital signal for processing in digital form. Figure 5.25(a) shows an ADC with analog input, a *GO* input to initiate conversion, outputs to indicate the status of the conversion, and three output bits. If the ADC is linear, the digital output will be related to the analog input as indicated in Fig. 5.25(b). The *precision* of the ADC is the number of output bits, in this case 3. The *range* of the ADC is given by the maximum and minimum input voltages that can be converted with at most a one-half least-significant-bit (LSB) error.

EXAMPLE 5.10 | **Errors in an ADC**

The full-scale (FS) input voltage of a 3-bit ADC is 10 V. As shown in Fig. 5.25(b), the range of the ADC is from 625 mV (1/16 FS) to 8.125 V (13/16 FS).[9] Find the worst-case error and rms error from digitizing the analog signal.

SOLUTION:
The worst-case error is 1/16 FS, or 0.625 V, as shown in Fig. 5.25(b). The rms error requires computing the rms value of a sawtooth waveform with a peak value of 0.625 V. We integrate over one-half the error cycle:

---
[9] If we use 111 for an overrange indication.

$$V_e^2 = \frac{1}{FS/16} \int\limits_{-FS/32}^{+FS/32} (0.625)^2 \left(\frac{v_e}{FS/32}\right)^2 dv_e = \left(\frac{0.625}{3}\right)^2 \qquad (5.39)$$

where V_e is the rms error and v_e is the error voltage, the difference between the sloping and stair-step functions in Fig. 5.25(b). Thus, the rms error is $0.625/\sqrt{3} = 0.361$ V.

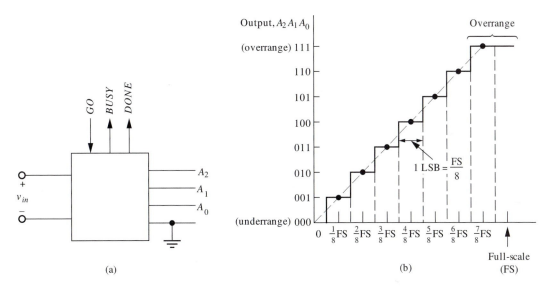

Figure 5.25   a) The A/D converter takes an analog input and converts to a parallel digital output; (b) the output binary word depends on the analog input.

**underflow,
overflow,
ADC range**

Underflow and overflow.   Voltages below 625 mV would give an output of 000, indicating underrange, or *underflow*, and voltages above 9.375 V would give an output of 111 plus an overrange, or *overflow*, indication. The underrange and overrange indications can be used to control the gain of the analog channel to keep the analog voltage within the range of the ADC. The *resolution* of the ADC is the FS voltage divided by the number of output states, in this case, $10\,V/2^3 = 1.25$ V, and corresponds to 1 bit. Typical commercial ADCs have 8- and 12-bit precision, indicating 256 and 4096 states, respectively.

**flash converter,
underflow,
overflow**

Two-bit flash A/D converter.   Figure 5.26(a) shows a 2-bit flash ADC; "flash" because it yields instantaneous output. The resistors in series set up three reference voltages, and the analog input voltage is compared simultaneously with all references, with the results shown in the truth table of Fig. 5.26(c). The lowest comparator indicates underflow and the highest indicates overflow. The middle three comparators indicate which of the four possible ranges contains the input, and the logic circuit in the dash-line box converts this information to a binary output code. The A/D conversion takes place almost instantaneously. This type of ADC, also called a *parallel encoder*, is available commercially in 4- and 8-bit versions. Figure 5.26(b) shows that the ranges for this converter are set up with an offset from a linear function passing through the origin, unlike the ideal in Fig. 5.25(b).

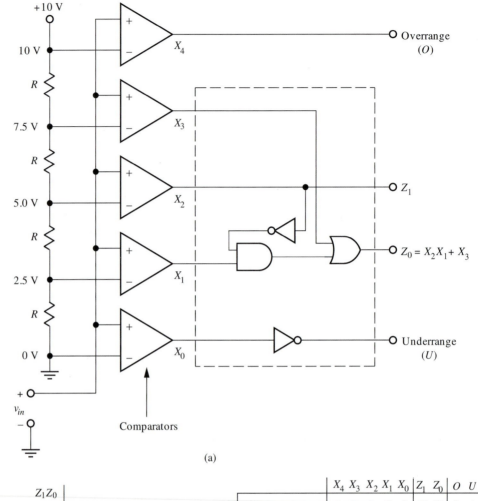

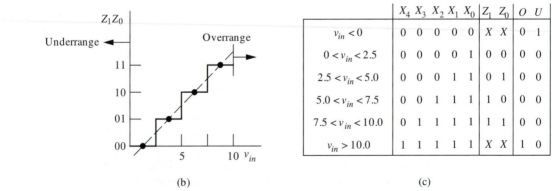

	X_4	X_3	X_2	X_1	X_0	Z_1	Z_0	O	U
$v_{in} < 0$	0	0	0	0	0	X	X	0	1
$0 < v_{in} < 2.5$	0	0	0	0	1	0	0	0	0
$2.5 < v_{in} < 5.0$	0	0	0	1	1	0	1	0	0
$5.0 < v_{in} < 7.5$	0	0	1	1	1	1	0	0	0
$7.5 < v_{in} < 10.0$	0	1	1	1	1	1	1	0	0
$v_{in} > 10.0$	1	1	1	1	1	X	X	1	0

(b)   (c)

Figure 5.26   (a) The 2-bit flash ADC converter compares the input with ranges of voltage; (b) the output–input characteristic of the ADC; (c) the truth table of the ADC, including overrange and underrange indications.

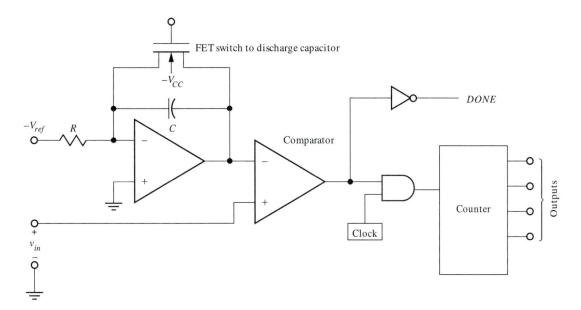

Figure 5.27 The single-slope converter counts clock pulses until an increasing ramp voltage exceeds the input voltage.

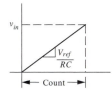

Figure 5.28 The count is proportional to the input voltage.

Single-slope ramp analog-to-digital conversion.

A single-slope ramp converter uses an integrator to generate a linearly increasing voltage to compare with the unknown voltage. The circuit shown in Fig. 5.27 uses an op amp to integrate an input reference voltage. This gives output ramp function shown in Fig. 5.28. While the ramp is increasing, the AND gate allows the counter to count clock pulses, but when the comparator indicates that the ramp voltage has exceeded the analog input, the count stops, $DONE = 1$, and the counter output is displayed as indicated in Fig. 5.27. The capacitor is then discharged electronically, and the ADC is ready to begin its conversion cycle again.

EXAMPLE 5.11 **ADC**

An 8-bit ADC has a range from 0 to 10 V and a 100-kHz clock frequency. Find the time it takes to make the conversion and the required ramp slope.

SOLUTION:
Eight bits corresponds to a count of $2^8 = 256$ from the clock, which requires $256 \times T = 256 \times 10^{-5} = 2.56$ ms. The slope in Fig. 5.28 is $(10\text{ V})/(2.56\text{ ms}) = 3906$ V/s. For $V_{ref} = 8$ V, $RC = 2.05$ ms.

WHAT IF? What if a 1-MHz clock is used on a 12-bit converter?[10]

---

[10]Converts in 4.10 ms; $RC = 3.28$ ms.

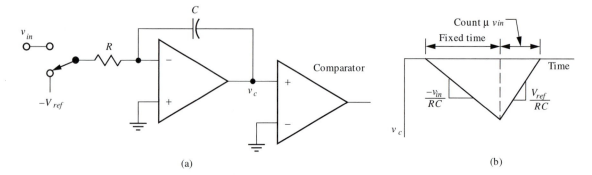

(a)

(b)

Figure 5.29 (a) Dual-slope ramp conversion input stage; (b) the number of counts while the voltage is coming to zero indicates the analog input voltage.

Dual-slope A/D ramp conversion. Dual-slope conversion has several advantages over single-slope conversion. Figure 5.29(a) shows the input to the integrator connected first to the unknown analog input voltage, and the output voltage increases in the negative direction a fixed amount of time, normally a multiple of the power-line period to reject this form of interference. The input then is switched to a known reference voltage of opposite polarity, and clock pulses are counted until the integrator voltage passes through zero, Fig. 5.29(b). The advantage of this scheme is that the unknown is compared to the voltage reference, independent of variations in the clock frequency. This is the method employed in most digital voltmeters.

Sampling

multiplexer

Typically, transducers monitoring physical processes produce data rates well below the capability of a computer. Thus, many inputs can be monitored simultaneously. Numerous analog inputs may be sampled at discrete times sequentially through a *multiplexer*, which is essentially an electronic multipole switch. A sample-and-hold circuit may be required to arrest change in an input while A/D conversion is being performed. Finally, the discrete samples of the time-varying inputs may be processed as a time sequence by the computer by techniques known as digital filtering. Processes associated with the sampling of the analog inputs are discussed in this section.

Analog multiplexers. The analog multiplexer, AMUX, shown in Fig. 5.30(a) accepts eight analog inputs. A 3-bit address $B_2B_1B_0$ selects the input to connect to the output. By cycling through all inputs, the ADC can sequentially convert all inputs to digital form for processing by a computer.

FET switch. Figure 5.30(b) shows how the AMUX uses an FET switch to connect and disconnect an input from the output. The FET is used as a voltage-controlled switch. With zero voltage on the gate, the drain–source resistance is many megohms and the input voltage is blocked. With sufficient voltage applied to the gate, the FET is placed in the ohmic region, and the drain–source resistance is 25 to 100 Ω, allowing the analog input to appear across the 47-kΩ resistor and the output. The internal logic in the AMUX decodes the channel address and connects one channel at a time with a "break-before-make" connection to ensure that no two input channels are shorted together through the switches.

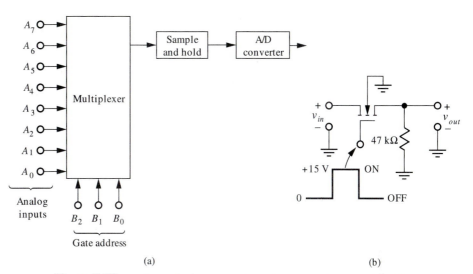

(a) (b)

Figure 5.30 (a) This multiplexer has eight analog inputs. The input selected by the incoming address is connected to the sample-and-hold circuit for analog-to-digital conversion. (b) The FET switch is an open circuit except when its gate is at +15 volts, connecting input to output.

EXAMPLE 5.12 AMUX loss and cross coupling

The output impedances of the analog inputs to a 8-input AMUX is $100\,\Omega$. The FET OFF resistance is $10\,\text{M}\Omega$ and its ON resistance is $60\,\Omega$. Find the loss of the AMUX when the FET is ON.

SOLUTION:

One channel is ON and seven channels are OFF. The equivalent circuit is shown in Fig. 5.31. The output voltage compared with the open-circuit voltage at the analog input is given by a voltage divider:

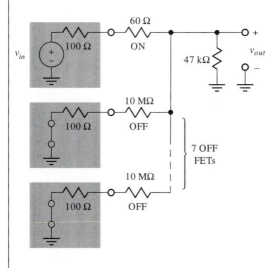

Figure 5.31 Circuit with one FET ON and the other seven OFF.

$$\frac{v_{out}}{v_{in}} = \frac{47 \text{ k}\Omega \| (10 \text{ M}\Omega/7)}{47 \text{ k}\Omega \| (10\text{M } \Omega/7) + 100 + 60} = 0.996(-0.03 \text{ dB}) \tag{5.40}$$

WHAT IF?   What if the FET is OFF? Find the isolation between inputs assuming another switch ON.[11]

sample and hold

Sample-and-hold circuit.
Figure 5.32 shows a *sample-and-hold* circuit, which is used to arrest time variations in the input signal during ADC processing. The circuit uses two voltage-follower amplifiers to buffer input and output. An FET switch is activated long enough for the capacitor to charge to the input voltage, and then the capacitor holds the voltage while A/D conversion is taking place. The magnitude of the capacitance is a compromise between the requirements for rapid charging and long voltage-retention time: The small capacitor required for rapid charging places a limit on how long the capacitor can hold the voltage to a prescribed tolerance, considering the leakage current through the FET and into the op amp.

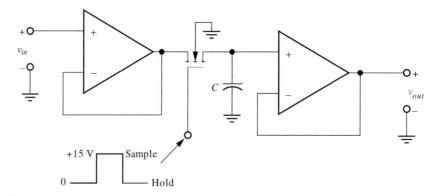

Figure 5.32   The sample-and-hold circuit uses a capacitor to remember the voltage after the FET switch is gated ON briefly.

EXAMPLE 5.13   **Capacitance design**

In the sample-and-hold circuit of Fig. 5.32, the FET ON resistance is $25\,\Omega$ and the requirement is for charging to 99% of the input value in $1\,\mu$s. Find the required capacitance.

SOLUTION:
Assuming $v_C(0) = 0$ V, the transient is of the form

---

[11] The isolation is 100 dB.

$$v_C(t) = V_\infty(1 - e^{-t/\tau}) \tag{5.41}$$

where V_∞ is the output of the buffer and $\tau = RC = 25C$, assuming negligible output impedance from the buffer. Thus,

$$e^{-1\,\mu s/\tau} = 0.01 \quad \Rightarrow \quad \tau = 0.217\,\mu s \quad \Rightarrow \quad C = 8.7\,nF \tag{5.42}$$

WHAT IF? What if the output impedance of the buffer is $5\,\Omega$ and a 10 nF capacitor is used? What is the error in capacitor voltage for $1\,\mu s$ ON time?[12]

Discrete samples. Let us consider that we have a data-acquisition system with several inputs. We use an AMUX to sample the inputs with the sample-and-hold circuit holding the sampled voltages long enough for A/D conversion. The end result of this operation is a sequence of samples representing the time structure of each input, as shown by Fig. 5.33(a). The discrete sampling of the time structure of the input raises two questions: "What is an appropriate sampling rate?" and "How should we process the samples to improve the information content of the data?"

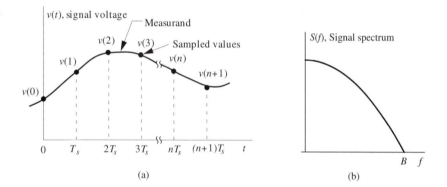

Figure 5.33 (a) The continuous waveform is sampled at a frequency $f_s = 1/T_s$, producing a sequence of discrete values, $v(0)$, $v(1)$, ...; (b) the spectrum of the signal waveform before sampling. The Nyquist criterion states that the sampling frequency must be at least twice the signal bandwidth to avoid loss of information and prevent confusion.

Nyquist sampling criterion. The question of how often to sample is answered by the *Nyquist criterion*, which states that no information is lost through the sampling process if the sampling frequency f_s satisfies the following criterion:

$$f_s > 2B \tag{5.43}$$

aliasing

where B is the bandwidth of the waveform being sampled, as shown in Fig. 5.33(b). Thus, you must sample two or more times during each period of the highest frequency

---

[12] 3.6%.

in the signal. If the Nyquist criterion is violated by sampling too slowly, information is lost and the information content of the sampled values becomes confused through *aliasing*, the mixing of high frequencies into the signal band. For a wideband signal, the bandwidth prior to sampling should be limited to prevent aliasing.

EXAMPLE 5.14 **Nyquist sampling**

A temperature transducer has a 0.5-s time constant. The signal bandwidth of this transducer was determined on p. 295 to be 0.318 Hz. What is an appropriate time between samples to monitor the temperature information out of the transducer?

SOLUTION:
The calculated bandwidth is not an exact guide because some signal spectrum lies above this critical frequency. To prepare this signal for sampling, we might use a second-order Butterworth low-pass filter with a critical frequency of 0.4 Hz, and then sample at 0.9 Hz to be safe. Thus, a sample every 1.11 seconds would get all the information that passes the filter.

Digital filtering. Digital filtering concerns the processing of the samples. Of course, the samples may be used directly, but often signal characteristics are improved by processing the samples. To illustrate this technique, we consider two sample-processing schemes for the filter indicated by Fig. 5.34.

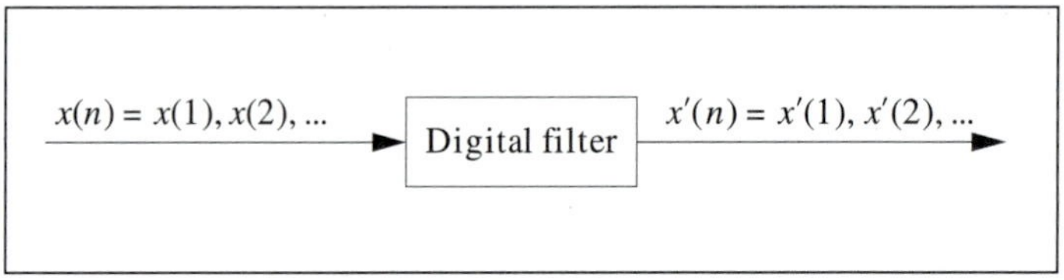

$x(n) = x(1), x(2), ...$ → Digital filter → $x'(n) = x'(1), x'(2), ...$

Figure 5.34 The digital filter accepts discrete input samples and uses a numerical procedure to produce output discrete data. The "filter" is a software algorithm in the computer.

FIR, nonrecursive

FIR and IIR filtering. The first is called a *finite impulse response* (*FIR*, or *nonrecursive*) filter, because it calculates the output, $x'(n)$, from previous values of the input, $x(n), x(n-1), \ldots$. Consider, for example, the FIR filter represented by the calculation

$$x'(n) = \frac{x(n) + x(n-1)}{2} \qquad (5.44)$$

IIF, recursive

The output is the running unweighted average of the current and previous sample of the input. By contrast, an infinite impulse (*IIF*, or *recursive*) filter calculates $x'(n)$ from previous inputs, $x(n), x(n-1), \ldots$, and previous outputs, $x'(n-1), x'(n-2), \ldots$. An example is

$$x'(n) = \frac{x'(n-1) + x(n)}{2} \qquad (5.45)$$

which averages the input with the previous output to give the new output.

We may compare the response of the FIR and the IIR filters described by Eqs. (5.44) and (5.45) by examining their responses to a sudden increase of the input. Figure 5.35(b) shows the responses calculated in the table in Fig. 5.35(a) for the filter algorithms given in Eqs. (5.44) and (5.45). We filled in between the samples with lines to aid in identification of the responses. The IIF algorithm has a greater smoothing effect because it has a longer memory of the past history of the input.

		0	1	2	3	4
Input	$x(n)$	0	1	1	1	1
FIR	$x_1'(n)$	0	0.5	1	1	1
IIR	$x_2'(n)$	0	0.5	0.75	0.88	0.94

(a)

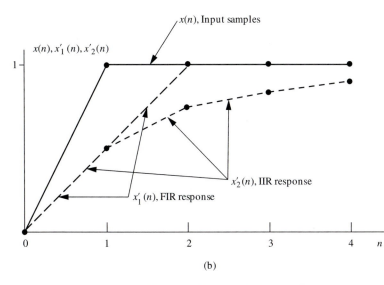

(b)

Figure 5.35 (a) Table showing input and output of a digital filter with the filter algorithms given in Eqs. (5.44) and (5.45). The input is a sudden increase in the input value. (b) Plots of input and output samples.

Digital filters and their effects are an established technique of digital-signal processing. Digital filters exist to simulate analog filters such as the Butterworth filters and to eliminate discrete frequencies, 60-Hz hum, for example.

Digital-to-Analog Conversion

The instrumentation system shown in Fig. 5.1 includes the possibility of an analog output from the computer because often the purpose of the data-gathering function is to control the processes being monitored. For example, the temperature of chemical reagents might be monitored to maintain a prescribed temperature through heaters. In this section we discuss digital-to-analog converters (D/A converters or DACs).

DAC, D/A

Function. The *digital-to-analog converter* (DAC or D/A) accepts a digital input and produces an analog output. The input–output characteristics are the same as those shown in Fig. 5.25 (b) for the ADC, except that the vertical axis represents the input and the horizontal axis represents the output. Depending on details of the construction and specifications, the output might be offset from zero in the manner of Fig. 5.26(b). We will show circuits for two representative 3-bit DACs.

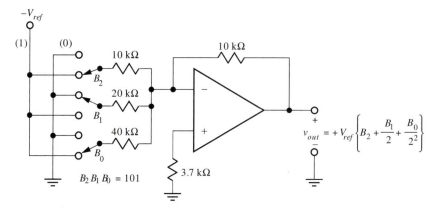

$$v_{out} = +V_{ref}\left\{B_2 + \frac{B_1}{2} + \frac{B_0}{2^2}\right\}$$

Figure 5.36 This 3-bit digital-to-analog converter circuit sums voltages.

Voltage-summing DAC.

The circuit shown in Fig. 5.36 uses a binary-weighted sum of inputs controlled by three single-pole, double-throw switches representing the digital word. The principle of operation of a summing amplifier circuit is discussed on page 253. Here each input is either $-V_{ref}$ or 0 V, depending on the switch setting. The switch positions shown correspond to the binary word $B_2B_1B_0 = 101$, and the output is $1.25V_{ref}$. Clearly, 000 produces zero output and 111 produces $1.75V_{ref}$. The least significant bit is thus $0.25V_{ref}$ for this DAC.

Ladder DAC.

The circuit shown in Fig. 5.37(a) is based on the properties of the R-$2R$ ladder network, which are indicated in Fig. 5.37(b). We use the principle of superposition to examine the effect of the LSB input (B_0). Note that the equivalent resistance of the ladder is $2R$ into the series resistors and R into the parallel resistors going to the inputs. Thus, the equivalent resistance values are the same, regardless of the length of the ladder. At each input, the input resistance is $3R$, so the input current from the source at B_0 is $I_0 = -V_{ref}/3R$. The $R-2R$ ladder network halves the current at each node. By the time the input current at B_0 reaches the op-amp input, it is halved three times, a factor of 8, such that the current into the op amp is $-V_{ref}/24R$. As explained on page 249, this current goes through the feedback resistor of $6R$ and produces an output voltage of $+V_{ref}/4$. The input of each higher-order bit produces a larger voltage by a factor of 2 because the current is halved one less time. Thus, the input at B_2 produces an output voltage of $+V_{ref}$, and generally the ladder can be extended to a larger number of bits for an 8- or 12-bit DAC.

EXAMPLE 5.15 **R-2R ladder DAC**

Find the input current to the inverting amplifier in Fig. 5.37(a) for $V_{ref} = 10$ V and $R = 2$ kΩ.

SOLUTION:
The properties of the R–$2R$ ladder are such that the impedance is $2R$ to ground each direction from the node above the $2R$ resistor connected to $-V_{ref}$. Thus, the source current at B_0 is

$$I_0 = \frac{-V_{ref}}{2R + 2R\|2R} = \frac{-V_{ref}}{3R} = \frac{-10}{6\ \text{k}\Omega} = -1.67\ \text{mA} \tag{5.46}$$

The input current at B_2 is the same. The input to the inverting amplifier is $I_0/2 + I_0/8 = -1.04$ mA, and $v_{out} = +1.04$ mA $\times 6 \times 2$ k$\Omega = 12.5$ V.

WHAT IF?

What if you calculate the most current that the R-$2R$ ladder can draw from the V_{ref} power supply in Fig. 5.37(a)?[13]

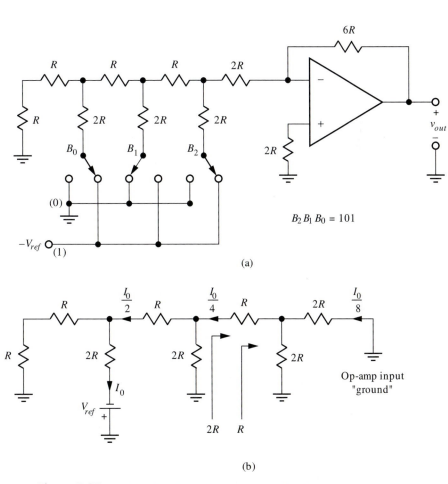

(a)

$B_2 B_1 B_0 = 101$

(b)

Figure 5.37 (a) This digital-to-analog converter is based on the current-division properties of the R-$2R$ ladder circuit; (b) the R-$2R$ ladder network divides the current by a factor of 2 at each node.

---

[13] Each input sees $3R$, so $I_{ref} = -5$ mA for $B_2B_1B_0 = 111$.

Check Your Understanding

1. The increment of voltage corresponding to the least significant bit for an 8-bit ADC would be one-half the increment of voltage for a 4-bit ADC. True or false?

2. The underrange signal on the ADC in Fig. 5.26(a) would be a digital one for a negative input signal. True or false?

3. The dual-slope ramp conversion ADC is immune to minor variations in the clock frequency. True or false?

4. A sample-and-hold circuit may not be required if the ADC is fast enough in its conversion speed. True or false?

5. If the output signal from a temperature-indicating device contains no significant information beyond 1 Hz in its output spectrum, what is the maximum sampling period that can be used for this device without losing information?

6. A digital filter filters digital signals. True or false?

7. What is the increment in output voltage corresponding to a 1-bit change in the least significant bit for a 8-V full-scale ADC with a precision of 8 bits?

8. Show that the op-amp bias current sees the same impedance at the $+$ and $-$ input terminals in Fig. 5.37(a).

Answers. **(1)** False; **(2)** true; **(3)** true; **(4)** true; **(5)** 0.5 s; **(6)** false; **(7)** 31.25 mV; **(8)** $6R \parallel (2R + R) = 2R$.

CHAPTER SUMMARY

This chapter builds on the material in Chap. 3, "Digital Electronics", and Chap. 4, "Analog Electronics." We introduce transducer technology and error analysis, but focus on the processing of analog and digital signals. Active filters improve the signal-to-noise ratio by eliminating noise out of the signal bandwidth. Filters must limit signal bandwidth to eliminate confusion and loss of information due to sampling. The sampling process involves analog multiplexing and a sample-and-hold circuit to enable analog-to-digital conversion. Finally, we show how digital signals are converted to analog form for control and display purposes.

Objective 1: To understand how to perform worst-case and statistical error analysis. Measurements should have stated errors that combine the various error contributions of components of the instrumentation system. Errors may be combined on a worst-case or statistical basis. The meaning of stated errors should be explicit.

Objective 2: To understand how to recognize floating signals and be able to amplify difference-mode signals. A signal is floating when the information voltage is present between two wires, neither of which is grounded. Practical systems can within limits reject the average voltage and amplify the difference voltage.

Objective 3: To understand analog signals and noise spectra and how active filters can improve system performance. Several types of noise spectra are present in a system. Filtering can be effective when the signal and noise have different spectra. Active filters use negative feedback in an op-amp circuit to give gain and filtering.

Objective 4: To understand the principles of cabling, grounding, and shielding to reduce interference.

Construction of instrumentation systems requires attention to the nature of the noise environment and the nonideal nature of connections between system components. Sensitive electronic equipment must be shielded from electric- and magnetic-field interference, and may require separate grounding systems for low-level signals.

Objective 5: To understand digital signal sampling, conversion, and processing.

The Nyquist sampling theorem gives the minimum sampling rate for an analog signal based upon the signal bandwidth. Sampled analog signals can be stored in sample-and-hold circuits for conversion to digital form. Sampled signals may be filtered by numerical algorithms.

Chapter 6 deals with another application of electronics, communication systems. Nonlinear effects, which constitute noise in instrumentation systems, are the key to the frequency-shifting operations common to all communication systems.

GLOSSARY

1/f noise, p. 296, the frequency-domain representation of the accumulated effects of small changes. The corresponding time-domain concept is "drift."

Active filters, p. 294, filters based on op amps with frequency-dependent feedback networks, often combining amplification with filtering.

ADC precision, p. 304, the number of output bits of an ADC.

ADC range, p. 305, the maximum and minimum input voltages that can be converted with at most a one-half least significant bit (LSB) error.

ADC resolution, p. 305, the full-scale voltage divided by the number of output states.

Aliasing, p. 311, an undesirable result of sampling a signal too slow, through which information is lost.

Analog-to-digital converter (ADC or A/D), p. 304, a circuit that converts an analog signal to a digital signal for processing in digital form.

Attenuator, p. 294, a device that reduces the signal level by a prescribed amount, usually expressed in dB.

Chopped signal, p. 296, a signal that is modulated by switching between a signal source and a constant reference signal.

Common-mode rejection ratio, CMRR, p. 291, the ratio of the amplification of the difference-mode signal to the common-mode signal in an amplifier or system.

Common-mode signal, p. 291, the average of two single-ended signals.

Difference amplifier, p. 290, an amplifier that subtracts the signals between the two input conductors and amplifies their difference.

Difference-mode signal, p. 291, the difference between the two single-ended signals.

Digital-to-analog converter (DAC or D/A), p. 313, a circuit that accepts a digital input and produces a corresponding analog output.

FIR digital filter, p. 312, a finite impulse response (or nonrecursive) filter calculates output values as a weighted sum of past input values.

Floating signal, p. 290, a signal that appears between two wires, neither of which is signal ground.

IIF digital filter, p. 312, an infinite impulse (or recursive) filter calculates output values as a weighted sum of past input and output values.

Interference, p. 302, man-made electrical noise.

Look-up table, p. 282, a means for representing a nonlinear function, used in the interpretation of transducer signals.

Multiplexer, p. 308, an electronic multipole switch, used to cyclically connect the back end of an instrumentation system to several input signal sources.

Nyquist criterion, p. 311, the rule that gives the minimum sampling rate to capture all the information in a signal in terms of its bandwidth.

Overflow, p. 305, an indication that the input signal of an ADC is too large to be converted.

Parallel encoder, p. 305, An ADC that compares the input signal with internal references and gives output almost instantaneously.

Sample-and-hold circuit, p. 309, a circuit used to arrest time variations in a signal for ADC processing.

Signal ground, pp. 290, 302, in an electronic circuit, a connected set of conductors distributed to many points of the circuit to give a common reference voltage, which may or may not correspond to power ground.

Single-ended signal, p. 290, a signal that appears between one wire and signal ground.

Single-slope ramp ADC, p. 307, an ADC that compares the input voltage with an internally generated ramp function while a counter measures the time at which the two voltages are equal, yielding a digital output.

Statistical error analysis, p. 289, a an analysis of system errors based on the assumption that errors combine randomly.

Strain gage, p. 283, a resistor that is attached to a mechanical member to share the elongation of the member, used to monitor the mechanical strain.

Thermistor, p. 285, a resistor, made of a semiconductor, whose resistance depends on temperature, used to monitor temperature.

Thermocouple, p. 286, a junction between two dissimilar metals, which produces a voltage that depends on the junction temperature, used to monitor temperatures.

Transducer, p. 281, also called a sensor, a device that converts a physical quantity to an electrical signal.

Underflow, p. 305, an indication that the input signal is too small to be converted by an ADC.

Worst-case error analysis, p. 287, a method of error analysis that assumes that all errors are at their maximum and add in the worst possible way; thus the effects of all errors are assumed to be cumulative.

PROBLEMS

Section 5.1: Introduction to Instrumentation Systems

5.1 Find the resistor values for a Wheatstone bridge that uses a voltage of 8 V and a 10-Ω change from balance in one resistor produces a 10-μV change in the output voltage. Assume all resistors the same at balance.

5.2. A thermistor is used in the bridge circuit shown in Fig. P5.2. The characteristic of the thermistor is

$$\ln\left[\frac{R(T)}{R_0}\right] = 4000\left(\frac{1}{T} - \frac{1}{T_0}\right)$$

where $R(T)$ is the resistance at temperature T in kelvin, and R_0 is the resistance at temperature T_0, in this case 10 kΩ at 20°C. Find and plot the output voltage of the bridge as a function of temperature in the range $0 < T < 100$°C. *Note:* °C + 273 = kelvin.

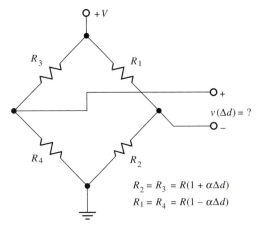

$$R_2 = R_3 = R(1 + \alpha\Delta d)$$
$$R_1 = R_4 = R(1 - \alpha\Delta d)$$

Figure P5.3

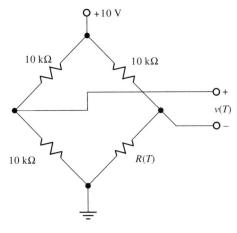

Figure P5.2

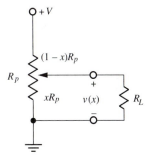

Figure P5.4

5.3. The circuit in Fig. P5.3 represents an unbonded strain gage. All four resistors are affected by changes in the length of the device, Δd, with opposite resistors charged in opposite directions. Determine the output voltage as a function of Δd.

5.4. The potentiometer transducer in Fig. P5.4 gives an output voltage proportional to the displacement x only if there is no load ($R_L = \infty$); otherwise, loading effects make the output nonlinear. Determine the ratio of the load resistance, R_L, to the potentiometer resistance, R_p, such that the maximum error is 1% of full scale. *Hint:* Find the maximum value of the output resistance and make this 1% of the load resistance.

5.5. A voltage in a first-order transient circuit decays from 10 to 0 V with an RC time constant. If the resistance can vary by 5% and the capacitor by 10%, find the following:

(a) The percent of variation in the time constant using an exact and a linearized analysis.

(b) The percent of variation in the voltage at $t = RC$, where RC is the nominal time constant. Use the exact analysis.

5.6. A measurement of y is derived from two measurements, $y = x_1 + x_2$, where $x_1 = 6(1 \pm 3\%)$ and $x_2 = 3(1 \pm 5\%)$.

(a) Find y, assuming errors in the components are maximum errors. Use worst-case conditions.

(b) Find y, assuming errors in the components are 1 standard-deviation errors. Use statistical error analysis.

5.7. The resistive voltage divider in Fig. P5.7 is built with 10% tolerance resistors.

(a) Determine the exact limits of the output voltage if $R_2 = 2R_1$ (nominal values).

(b) Explain why the limits do not give a 20% error due to each 10% error.

5.8. Assume that you have a number of measurements of the same physical quantity, all of the same accuracy. Show that the standard deviation of the average is reduced by $1/\sqrt{n}$ from the standard deviation of a single measurement, where n is the

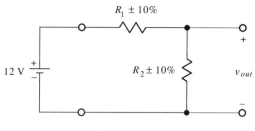

Figure P5.7

number of samples that are averaged. To improve the accuracy by a factor of 10, how many independent measurements must be made?

Section 5.2: Analog Signal Processing

5.9. For the instrumentation amplifier in Fig. 5.10, assume that the two R_3's are not identical but differ by a normalized error of ε_R. That is, assume $R_{3(\text{bottom})} = R_{3(\text{top})}(1 \pm \varepsilon_R)$.

(a) Using the op amp analysis technique from Chap. 4, determine the error in the output due to this asymmetry.

(b) Does this affect the common-mode rejection ratio or just the difference-mode gain of the amplifier?

5.10. Determine the output of the noninverting amplifier in Fig. 4.50 if the amplifier has finite gain (A_d) and a finite common-mode rejection ratio ($A_c = A_d/$ CMRR). Express your answer is terms of A_d, CMMR, R_1, and R_F.

5.11. An amplifier has a gain of 50 dB for the difference input and a common-mode rejection ratio of 72 dB. The input signals are 1.012 and 1.006 V. What is the percent of error in the output signals due to the common-mode component?

5.12. Find the gain in dB of the attenuator in Fig. 5.12 when the input digital signal to the switches is $S_1 S_2 S_3 S_4 = 1010$.

5.13. What is the most attenuation in dB that can be produced by the attenuator in Fig. 5.12, and what is the switch configuration to achieve this attenuation? *Hint*: More than one switch can be closed.

5.14. Design an attenuator similar to the one in Fig. 5.12 that gives 0, 2, 4, or 6 dB of attenuation. The input impedance should be 1 MΩ.

5.15. The circuit in Fig. 5.11 has a gain of unity for the signal only if the loading due to the potentiometer is

ignored. Find the ratio R_p/R such that the gain of the signal channel is always between 0.97 and 1.00.

5.16. Design an op-amp-based amplifier with the following characteristics:
- Amplifies the difference between two signals
- Filters the signal with the filter function shown in Fig. P5.16
- Has an input impedance of 1 kΩ at both inputs
- Uses one 0.1-µF capacitor
- Uses no more than two op amps

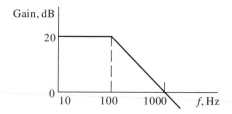

Figure P5.16

5.17. The sound energy in the voice has a power spectrum with most of the energy between 500 and 3000 Hz. Design an active filter to pass only this band of frequencies and also give a gain of 15 dB. Use no capacitor larger than 1 µF. Combine the first-order high-pass and low-pass characteristics.

5.18. For the active filter circuit shown in Fig. P5.18, find the following:

(a) Find the filter function $\mathbf{F}\,(j\omega)$.

(b) If the input voltage is 2 V dc, what is the output voltage?

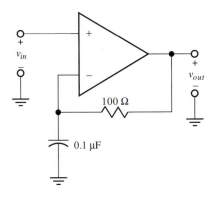

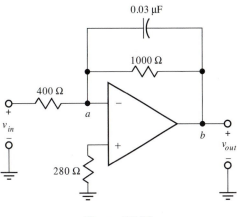

Figure P5.18

Figure P5.20

(c) Is there a critical frequency? If so, what is it in hertz, and does the gain begin to decrease or increase at the critical frequency?

5.19. The Bode plot of the gain of a single-ended active filter is shown in Fig. P5.19.
 (a) Draw an active filter circuit to give this characteristic.
 (b) If the circuit capacitor is $12\,\mu F$, find the resistors.
 (c) At what frequency is the amplitude of the output voltage equal to the input voltage if the signal is a sinusoid?
 (d) If $v_{in}(t) = 5\sqrt{2}\cos(2400\pi t)$ is the input signal, which is the output signal in the time domain?

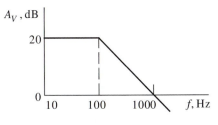

Figure P5.19

5.20. For the op-amp circuit shown in Fig. P5.20 find the following:
 (a) If a 1-V battery is connected from ground to the input, with the + of the battery connected to ground, what is the voltage of points a and b relative to the ground?
 (b) Now instead of a battery, the input is a 1-V pulse. What is the duration of the shortest pulse that the amplifier can pass without serious distortion?

5.21. Design an active filter having a gain of +22 dB, a phase shift of 180°, an input resistance of $860\,\Omega$ in the passband, and a high-pass characteristic with a cutoff frequency of 1250 Hz.

5.22. Derive the filter function, $\mathbf{F}(\omega) = \mathbf{V}_{out}/\mathbf{V}_{in}$, for the high-pass filter shown in Fig. 5.18. Confirm the formula for the cutoff frequency, ω_c. What is the input impedance in the band of frequencies above the cutoff frequency, ω_c?

5.23. Design an active high-pass filter with a cutoff frequency of 5000 Hz, an input impedance of $1000\,\Omega$ in the high-frequency region, and a gain of +20 in the passband. *Hint*: Use a passive high-pass filter at the input of a noninverting amplifier.

5.24. An op-amp amplifier is required to pass pulses of 1-ms width but to eliminate unwanted frequencies higher than those required for the pulse. The gain must be −3. Assume that you have a $0.0033\text{-}\mu F$ capacitor available. Give the circuit and component values.

5.25. In an audio system, an amplifier is required that will give a gain of 10 (+ or −, it does not matter) and have upper and lower cutoff frequencies to exclude signals outside the normal audio region, 30 to 20,000 Hz. Combine high-pass and low-pass active filters and design a circuit that will accomplish this. Assume that the largest capacitor you have available is $200\,\mu F$.

5.26. A second-order Butterworth low-pass filter with gain has a cutoff frequency of 100 Hz. The gain at 10 Hz is 20 dB. What would be the gain in dB at 10 kHz?

5.27. (a) Design a second-order low-pass Butterworth filter with a cutoff frequency of 1000 Hz. The largest capacitor may be no more than 0.1 μF.

(b) Determine the gain (amplitude and phase shift) at 500, 1000, and 5000 Hz.

Section 5.3: Digital Signal Processing

5.28. An 8-bit ADC has a range from 0 to a full-scale value of 8 V.
  (a) What nominal increment in input voltage causes a 1-bit change at the output?
  (b) What range of input voltages corresponds to the output of 00001101? Use no-offset characteristic similar to that shown in Fig. 5.25(b).

5.29. Verify the logic for Z_0 and Z_1 in the 2-bit flash ADC in Fig. 5.26(a) by making Karnaugh maps for Z_0 and Z_1 with X_1, X_2, and X_3 as inputs.

5.30. Redesign the flash ADC in Fig. 5.26(a) to have a characteristic passing through the origin. In other words, what would be the voltage-reference levels (and the resistors in the voltage-divider circuit) such that the transition to 01 is at $v_{in} = 0.125V_{ref}$, to 10 at $v_{in} = 0.375V_{ref}$, and so on.

5.31. An 8-bit ADC of the type shown in Fig. 5.27 takes 10 ms to make its conversion.
  (a) What is the clock frequency?
  (b) If the input range is 0 to 8 V and the reference voltage is 10 V, what is the RC product for the integrator?

5.32. The sample-and-hold circuit in Fig. 5.32 was designed in the text to have an 8.7-nF capacitor. If the combined leakage current due to the FET, op-amp input bias current, and capacitor leakage were 1 nA, how long does the capacitor hold the voltage to 99% of a 10-V capacitor voltage when the FET is turned OFF?

5.33. The FET switches used in the multiplexer in Fig. 5.30(a), one of which is shown in Fig. 5.30(b), have a resistance of 25 Ω when ON and a resistance of 10 MΩ when OFF. Assume that inputs A_0 through A_7 come from sources with a 100-Ω output impedance. Determine the coupling in decibels from the channel that is ON to the other seven channels that are OFF. *Note*: The eight channels are switched by eight FETs, all of which connect to the same 47-kΩ resistor for an output.

5.34. The sample-and-hold circuit in Fig. 5.32 uses an input voltage-follower amplifier with an output impedance of 2 Ω and an output voltage-follower amplifier with an input impedance of 20 MΩ. The FET resistance is 25 Ω when ON and 10 MΩ when OFF. The FET is ON for 1 μs and OFF for 1 μs for the ADC to make its conversion. Under these conditions, what is the optimum capacitor to maximize the voltage at the end of the 2-μs cycle, and what is the final voltage as a percent of the input voltage to the sample-and-hold circuit?

5.35. Determine the response of the two digital filters shown in Fig. 5.35(a) to the sudden increase at the input if the filters use three terms: for the FIR filter, $x'(n) = [x(n) + x(n-1) + x(n-2)]/3$, and for the IIR filter, $x'(n) = [x(n) + x'(n-1) + x'(n-2)]/3$.

5.36. Show that the FIR digital filter in Eq. (5.44) will remove a 60-Hz component in the data if the sampling period is 1/120 s.

5.37. A transducer signal is sampled every 5 s. The output spectrum is sufficiently broad to cause confusion if the transducer bandwidth is not limited prior to sampling. Determine the cutoff frequency of a simple low-pass filter to ensure that 99% of the power to the sampler lies within the Nyquist limit. Assume a worst-case transducer spectrum that is white noise.

General Problems

5.38. The highway department wishes to study the roughness of certain highways. For this purpose, a vehicle is instrumented with a downlooking sonar ranging unit and the associated electronics and data-recording equipment, as suggested by Fig. P5.38. The truck moves at 50 mph and road height variations are thereby converted to a continuous time signal. The sonar spot size is 3 inches in diameter, so roughness at smaller scales is lost. Of course, the vehicle shock-absorber/spring system acts as a high-pass filter, so the really low frequencies are lost. The output voltage of the sonar unit is analog with voltages in the range of 0 to 100 mV. The sonar unit is followed by a low-pass filter, an amplifier, a sample-and-hold circuit, and an 8-bit ADC that

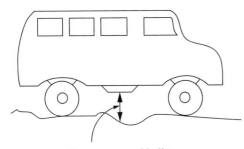

Sonar measures this distance

Figure P5.38

requires a positive input voltage in the range of 0 to 5 V. The output of the ADC goes into a computer, which actively controls the sample-and-hold circuit and the ADC.

(a) What would be an appropriate cutoff frequency for the filter such that noise outside the data spectrum is eliminated?

(b) Design the amplifier. Assume an input impedance of 1 kΩ and an ideal op amp.

(c) What is the minimum sampling rate for the system to record all the data?

(d) What voltage difference does 1 bit represent at the output of the ADC?

Answers to Odd-Numbered Problems

5.1. All 2 MΩ.

5.3. $v(\Delta d) = -\alpha\Delta d \times V$.

5.5. **(a)** Exact: $+15.5\%$, -14.5%; linearized: $\pm15\%$; **(b)** $V = 3.656 \pm 15.1\%$.

5.7. **(a)** $v_{max} = 8.516$ V and $v_{min} = 7.448$ V, $v = 7.982 \pm 6.69\%$; **(b)** $R_2 s$ in the numerator and denominator are correlated; also $R_1 = R_2$ should not preserve percent accuracy.

5.9. **(a)** $v_1' - v_1' = (v_1 - v_2) \times [1 + R_{3(top)}/R_2 \times (2 \pm \varepsilon_R)]$; **(b)** does not affect the common-mode gain, just the difference-mode gain.

5.11. 4.22%.

5.13. -44.0 dB.

5.15. $R_p/R = 0.255$.

5.17.

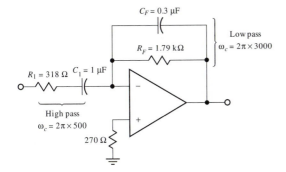

5.19. **(a)** Standard low-pass filter; **(b)** $R_F = 5.3$ kΩ, $R_1 = 530$ Ω; **(c)** 1000 by inspection; **(d)** $0.587 \cos(2400\pi t - 85.2°)$ V.

5.21.

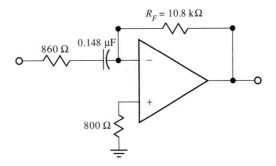

5.23.

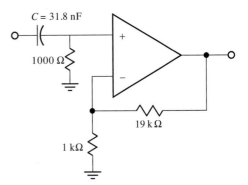

5.25.

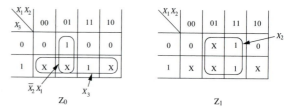

5.27. **(a)** "C" $= 70.7$ nF and $R = 2250\ \Omega$;
(b) $0.970 \angle -43.3°$, $0.707 \angle -90°$,
$0.03997 \angle -163.6°$.

5.29.

$X_1 X_2$ X_3	00	01	11	10
0	0	1	0	0
1	X	X	1	X

$\overline{X}_2 X_1$ X_3

Z_0

$X_1 X_2$ X_3	00	01	11	10
0	0	X	1	0
1	X	X	1	X

X_2

Z_1

5.31. **(a)** 25,600 Hz; **(b)** 0.0125 s.

5.33. -100 dB.

5.35. FIR is 0, 0.333, 0.667, 1, 1, 1, …; IIR is 0, 0.333, 0.444, 0.592, 0.679, 0.757, 0.812, 0.856, etc.

5.37. 1.57 mHz.

6

Communication Systems

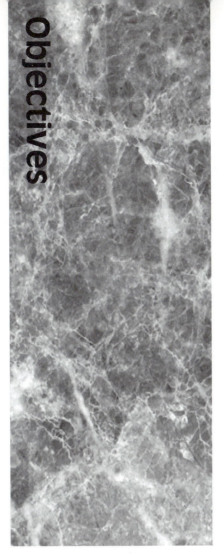

objectives

1. To understand the effects of nonlinear circuits on the spectra of signals
2. To understand the superheterodyne radio circuit and be able to relate RF, LO, and IF frequencies
3. To understand the properties of free and guided electromagnetic waves as they relate to communication systems
4. To understand the properties of antennas and their role in communication systems

Modern communication systems continue to proliferate: cellular phones, computer networks, TV satellites, and optical links for telephone service. Differing in details, these and other communication systems all use the same basic techniques of modulation, transmission, reception, and demodulation in a superheterodyne radio circuit. We also consider antennas and radio-wave propagation.

communication systems

Introduction to communication systems.
As a pipeline moves liquid from one location to another, so a *communication system* moves information from one location to another. A communication system consists of a source of information with some sort of encoder, a recipient of information with the appropriate decoder, and a medium connecting the two. Some examples are obvious: the telephone system, commercial TV broadcasting, and computer networks for stock-market transactions. Some are not so obvious: an internal bus in a computer, a police radar for measuring speed, and a system for collecting tolls from moving trucks on a toll road.

Electric signals have replaced smoke signals, Pony Express, and carrier pigeons in communication technology for several reasons. Electric signals travel with speeds approaching the speed of light. Electric signals in the form of radio waves go almost anywhere. At certain broadcast frequencies, the radio waves follow the Earth's curvature and also reach distant points by bouncing off the ionosphere. Electric signals lend themselves to a variety of ingenious coding schemes such as amplitude modulation, frequency modulation, and various digital codes.

Contents of this chapter.
In this section, we focus on the system used in commercial amplitude-modulation (AM) radio. First, we look at how information is encoded onto and decoded from AM radio signals. The explanation requires expansion of our understanding of the frequency domain to include nonlinear effects. Then we consider the various forms of electromagnetic waves used in radio systems. We examine several common types of antennas, both in transmitting and receiving. Finally, we consider the details of two communication systems.

General Principles of Nonlinear Devices in the Frequency Domain

Linear circuits.
Linear circuits create no new frequencies. If you excite a linear circuit at a certain frequency, it will respond only at that frequency. The analysis of ac circuits using phasors and impedance is founded entirely on this principle.

Nonlinear circuits with one input frequency.
Nonlinear circuits create new frequencies. We discovered in Chap. 2 that a diode, a nonlinear device, can rectify a signal, thus producing, in addition to the frequency of the input sinusoid, frequency components at dc and all the even harmonics of the input frequency. These new frequencies are created by the nonlinear action of the diode.

frequency multipliers

The spectrum of a rectified sinusoid is one instance of the general principle suggested in Fig. 6.1. For a nonlinear circuit with a sinusoidal input, the output of the circuit in general contains all the harmonics of the input. The amplitudes of the various output harmonics depend on the details of the circuit, but in principle, all harmonics are produced.

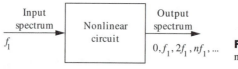

Figure 6.1 Nonlinear circuits create new frequencies.

Using these new frequencies. This property of nonlinear circuits finds direct application in many practical devices. One familiar example is a power supply, where the dc output is the zeroth harmonic of the input. Another application occurs in the generation of stable high-frequency sinusoids for precision communication systems or spectroscopic measurement systems. Often high-frequency signals of the required stability are generated by taking the output of a highly stable oscillator with a low frequency, say, 5 MHz, and putting this signal through a chain of *frequency multipliers* until it reaches the required high frequency, say, 3240 MHz. In such a scheme, the output of the multiplier chain is harmonically related to the low-frequency source and hence exhibits the same relative stability.

harmonic distortion, intermodulation distortion

On the other hand, the creation of new frequencies may be undesirable. A good audio amplifier should have low *harmonic distortion*, meaning that it should not generate harmonics of the audio signal. The amplifier should also have low *intermodulation distortion*. This term refers to the creation of new frequencies through the interaction of separate frequencies in the audio spectrum. Nonlinearities in the amplifier can cause both of these undesirable effects on the signal spectrum.

Nonlinear circuits with two input frequencies. The creation of harmonics in a nonlinear circuit, suggested in Fig. 6.1, is a special case of a more general property of nonlinear circuits suggested by Fig. 6.2. Here is a nonlinear circuit with two input frequencies, and the output in general contains all the harmonics of both input frequencies *plus* all the sum and difference frequencies of those harmonics. For example, if we had inputs of 30 and 100 Hz, the output would contain not only the harmonics of 30 (60, 90, 120, . . .) and the harmonics of 100 (200, 300, . . .), but the output would also contain the various sum and difference frequencies, such as $100 - 30$, $100 + 30$, $200 - 30$, $200 + 30$, $100 - 2 \times 30$, $100 + 2 \times 30$, $200 - 2 \times 30$, ... Radio systems use such new frequencies to encode communication signals.

Figure 6.2 Nonlinear circuits create all harmonics plus all sum and difference frequencies.

EXAMPLE 6.1 | **Output spectrum**

If two frequencies, $f_1 = 5$ MHz and $f_2 = 7$ MHz, are the inputs to a nonlinear circuit, what harmonics (n and m) give rise to an output frequency of 3 MHz?

SOLUTION:
The potential output frequencies are

$$f = nf_1 + mf_2, \text{ where } n, m = 0, \pm 1, \pm 2, \ldots, \tag{6.1}$$

where f represents all possible output frequencies and f_1 and f_2 are the input frequencies. By trial and error, we find that $n = 2$ and $m = -1$ works, and also $n = -12$ and $m = 9$.

WHAT IF? | What if it is 2 MHz you want?[1]

---

[1] $n = -1$ and $m = +1$ or $n = +6$ and $m = -4$ give 2 MHz.

Importance of filtering.   Although a multitude of frequencies can appear in the output of a nonlinear circuit, not all these frequencies are useful. The circuit designer must ensure that only the desired frequencies are strong in the output and that undesired components are minimized. Filters are used to remove unwanted frequency components in the output spectra of nonlinear devices. Only through careful control of the filtering properties of such circuits can the benefits of nonlinear action be achieved.

The design and operation of nonlinear circuits are complicated by the creation of new frequencies and the filtering of these frequencies to enhance desired and diminish unwanted components. In the following, we examine the applications of a few nonlinear circuits, emphasizing especially their role in communication systems. Our goal is to develop a sufficient basis for discussing the operation of AM and FM radio and the telephone system.

modulation, carrier

Analog Information

The Time Domain

modulation index

Modulation and Demodulation

Amplitude modulation.   In communication engineering, *modulation* is the process by which information at a low frequency is coded as part of a high-frequency signal. The higher frequency is called the *carrier* because it carries the information signal. In this section, we examine the amplitude-modulation (AM) scheme that is widely used in commercial broadcasting of radio and TV signals. Initially, we explain the nature of amplitude modulation, emphasizing the frequency-domain viewpoint throughout. Then we discuss why modulation is required in communication systems. Finally, we present a circuit that accomplishes amplitude modulation.

AM signals in the time domain.   Let us first consider the nature of amplitude modulation and how modulation affects the spectrum of the carrier. Equation (6.2) gives the equation of a modulated carrier in the time domain,

$$\underbrace{v_{AM}(t)}_{\text{AM signal}} = \underbrace{V_c[1 + m_a \cos(\omega_m t)]}_{v_e(t), \text{ the envelope}} \underbrace{\cos(\omega_c t)}_{\text{carrier}} \tag{6.2}$$

where ω_m is the angular frequency of the modulating sinusoid, ω_c is the frequency of the modulated carrier, and V_c is its amplitude. The parameter m_a, the *modulation index*, indicates the degree of modulation. The modulation index is proportional to the amplitude of the modulating signal, which is indicated by V_m in Fig. 6.3, subject to the restriction that m_a never exceeds unity; that is, $m_a \leqslant 1$.

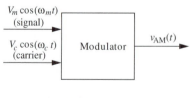

$$v_{AM}(t) = V_c[1 + m_a \cos(\omega_m t)] \cos(\omega_c t)$$

Figure 6.3   A modulator circuit has two inputs and one output.

The character of the modulated carrier and the role of the modulation index are shown in Fig. 6.4, which has modulated carriers for $m_a = 0.5$ and $m_a = 1.0$. Let us examine the role of the various factors characterizing the AM signal. The carrier is the sinusoid whose amplitude is being modulated. The carrier frequency controls the spacing of the zero crossings and individual peaks of the composite waveform. If we were to keep everything else the same yet double the carrier frequency, everything would look the same except there would be twice as many peaks and zero crossings. In other words, the hills and valleys of the amplitude would look the same, but the underlying sinusoid would have a higher frequency.

The amplitude of the carrier, V_c, characterizes the overall strength of the signal. With a broadcast signal, the amplitude of the received carrier received depends on the power of the station transmitter, the distance to the receiver, and the gains of the transmitting and receiving antennas.

The frequency of the modulating signals, ω_m, determines the time between the hills and valleys of the envelope, $v_e(t)$, as defined in Eq. (6.2). If, for example, the modulating frequency were doubled, the envelope of the carrier would vary up and down twice as fast.

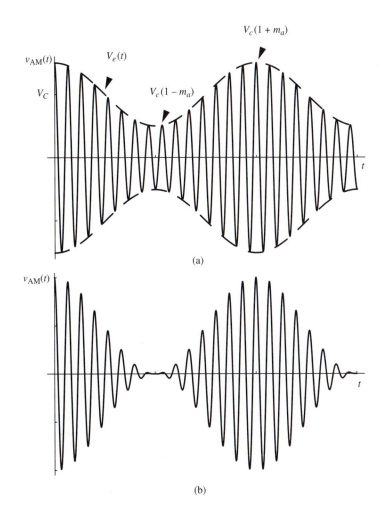

Figure 6.4 Two AM waveforms: (a) m_a = 0.5; (b) $m_a = 1.0$.

Finally, the modulation index controls the degree of modulation, the ratio of hills to valleys in the envelope. This parameter is limited to values less than unity, so the envelope factor in Eq. (6.2), $v_e(t)$, never goes negative. The modulation index is proportional to the amplitude of the modulating signal. The design of the modulator must ensure that the maximum amplitude of the modulating signal can be accommodated without exceeding the allowed value of unity.

$$m_a = KV_m, \quad \text{but} \quad m_a \leqslant 1 \tag{6.3}$$

where K is a constant.

EXAMPLE 6.2 | ## AM signal

An AM signal has zero crossings every 0.1 μs, a maximum amplitude of 12 V, and a minimum amplitude of 8 V, separated by 50 μs. Find the carrier frequency, the modulation index, and the modulation frequency.

SOLUTION:
Half the carrier period is 10^{-7} s, so the carrier frequency is 5×10^6 Hz, or 5 MHz. The carrier modulation index can be derived from the ratio of maximum and minimum voltages:

$$\frac{1 + m_a}{1 - m_a} = \frac{12}{8} \quad \Rightarrow \quad m_a = 0.2 \tag{6.4}$$

Finally, the modulation period must be twice the time between maximum and minimum amplitude; hence,

$$T_m = 2 \times 50 \ \mu s \quad \Rightarrow \quad f_m = \frac{1}{100 \ \mu s} = 10 \ \text{kHz} \tag{6.5}$$

WHAT IF?   What if you want the carrier voltage?[2]

AM modulator circuit. The circuit of Fig. 6.5 acts as a simple modulator circuit. Ignore for the moment the RC filter between a–b and c–d, and consider the part of the circuit that consists of two generators, a diode, and a resistor. The generator marked $v_c(t)$ is the carrier and has a high frequency. The generator marked $v_m(t)$ is the modulating signal, and it has a lower frequency and a smaller amplitude than the carrier.

Unfiltered output. Consider now the action of the diode on the signal at a–b. When the instantaneous carrier voltage, $v_c(t)$, exceeds the instantaneous modulating voltage, $v_m(t)$, the diode turns ON and current flows clockwise around the circuit. The output voltage in this case is equal to the instantaneous modulating voltage plus a small

---

[2] 10 V peak.

CHAPTER 6 COMMUNICATION SYSTEMS

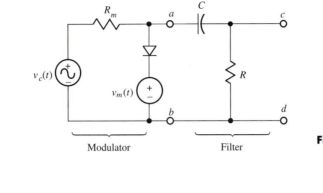

Figure 6.5 AM modulator circuit.

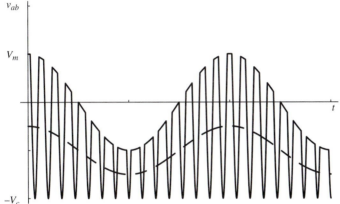

Figure 6.6 Unfiltered output of an AM modulator.

voltage required to turn ON the diode. On the other hand, when the instantaneous carrier voltage is smaller than the instantaneous modulating voltage, the diode is OFF and no current flows. During this time, there is no voltage drop across the resistor; hence, the voltage at *a–b* is equal to the carrier voltage. The result is shown in Fig. 6.6.

The Frequency Domain

Filtered output. The signal at *a–b*, shown in Fig. 6.6, differs from a pure AM signal in two respects:

1. Only the top is modulated, the bottom being flat.

2. The carrier waveform is not round on top and is thus not a pure sinusoid with varying amplitude.

We can solve the first problem by filtering out the dc and the component at ω_m, the dashed line in Fig. 6.6. These components we can eliminate with a high-pass filter that passes the carrier while rejecting the modulating frequency. This is the function of the high-pass filter between *a–b* and *c–d*.

The other problem, that the tops of the modulated carrier are distorted, implies that some higher harmonics of the carrier are also created in the modulator circuit. These also can be eliminated with a filter. With proper filtering, the output of the modulator in Fig. 6.6 looks like the ideal AM waveform in Fig. 6.4.

Spectrum of an AM signal. The spectrum of an AM signal can be determined through expansion of Eq. (6.2) with the trigonometric identity

$$\cos a \cos b = \tfrac{1}{2}[\cos(b+a) + \cos(b-a)] \tag{6.6}$$

When we expand Eq. (6.2) and apply Eq. (6.6), letting $a = \omega_m t$ and $b = \omega_c t$, we obtain

$$v_{AM}(t) = V_c \cos(\omega_c t) + m_a V_c \cos(\omega_m t) \cos(\omega_c t)$$

$$= V_c \cos(\omega_c t) + \frac{m_a V_c}{2}\left[\cos\underbrace{(\omega_c + \omega_m)}_{\substack{\text{sum}\\\text{frequency}}} t + \cos\underbrace{(\omega_c - \omega_m)}_{\substack{\text{difference}\\\text{frequency}}} t\right] \tag{6.7}$$

sidebands

This reveals a spectrum with three frequency components: the carrier at ω_c, an *upper sideband* at $\omega_c + \omega_m$, and a *lower sideband* at $\omega_c - \omega_m$. The spectrum before and after modulation are shown in Fig. 6.7. We broke the frequency scale in Fig. 6.7 because the carrier frequency is normally much higher than the modulating frequency.

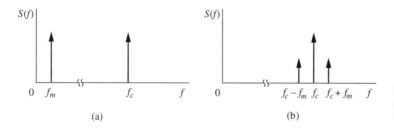

Figure 6.7 Modulation in the frequency domain: (a) before modulation; (b) after modulation.

EXAMPLE 6.3 | AM spectrum

The emergency warning broadcast tone is 853 Hz. Broadcast on a carrier of 1200 kHz, what would be the upper and lower sidebands?

SOLUTION:
Simply add and subtract the modulation frequency from the carrier frequency: 1,200,853 Hz and 1,199,147 Hz, respectively.

WHAT IF? What if the tone of A, 440 Hz, is broadcast on a carrier of 820 kHz? Where are the sidebands?[3]

Modulation with a broadband signal. If your AM radio received the AM signals in Fig. 6.4, and if the modulating frequency were in the audio band, you would hear a steady tone from the radio speaker, and you might suppose some sort of test was underway. In normal broadcasting, the program material would be music and voice signals, not steady tones. The modulated signal would thus be of the form

---

[3] The sidebands are 819,560 and 820,440 Hz.

$$v_{AM}(t) = V_c \left[1 + m_a s(t)\right] \cos(\omega_c t) \tag{6.8}$$

where $s(t)$ represents the program material and hence would generally have a broadband spectrum. In Eq. (6.8), $s(t)$ must remain smaller than unity so that the product $m_a s(t)$ never exceeds unity at any time.

Spectrum of broadband signals.   When the program information consists of a broadband signal such as music or voice, the individual frequencies comprising the spectrum of the audio signal are shifted to the sidebands. By law, the spectrum of the audio signal of standard AM broadcasting is limited to frequencies between 100 and 5000 Hz. The effect of modulating with such a signal is indicated by Fig. 6.8. The sidebands now become continuous spectra. The upper sideband is identical in shape with the audio signal, whereas the lower sideband is the mirror image. The bandwidth of the radio-frequency (RF) spectrum is twice the audio bandwidth; thus, 10 kHz of RF bandwidth is required for each AM station.

RF

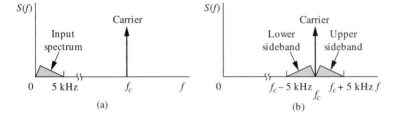

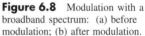

Figure 6.8 Modulation with a broadband spectrum: (a) before modulation; (b) after modulation.

If you could examine the spectrum of radio waves received by an AM radio antenna, you would see carriers for each station in your area, each bracketed by its upper and lower sidebands. Because the AM broadcast band covers frequencies from 535 to 1605 kHz, and because carriers can be spaced at 10-kHz intervals, there are channels for 106 AM stations, although obviously most channels are unoccupied in a given geographic area.

single sideband

Standard AM and single-sideband modulation. Because AM detectors are simple, standard AM modulation with both sidebands is used for commercial broadcasting. However, this AM system is inefficient in its use of the available bandwidth and the transmitter power. Bandwidth is wasted because although both sidebands and carrier are broadcast, all the information is contained in one sideband. Power is wasted because the carrier and one of the sidebands carry no information not already contained in the other sideband. For these reasons, sophisticated radio communication systems, such as a telephone company uses for long-distance messages, employ *single-sideband* modulation for economy in both power and bandwidth. Single–sideband modulation, as the name suggests, transmits only one of the sidebands over the communication path, and the receiver must generate the carrier and missing sideband to detect the information.

| EXAMPLE 6.4 | **Telephone supergroup** |

A telephone company uses frequencies between 312 kHZ and 552 kHz for a supergroup. The frequency band carries phone channels that have a 4-kHz bandwidth, single-sideband modulation. How many simultaneous phone conversations can be carried by a supergroup?

SOLUTION:

With single-sideband modulation, the RF bandwidth is equal to the information bandwidth, which is 4 kHz for a phone channel. The number of channels is, therefore, $(552 - 312)/4 = 60$ channels.

WHAT IF?   What if double-sideband modulation were used?[4]

Benefits of modulation.   Modulation shifts the information from the audio band to the RF radio band. We modulate to solve antenna problems, filtering problems, and confusion problems. The last problem is the easiest to understand: Clearly, we cannot allow all radio stations to broadcast their signals in the same frequency band because then our radios would receive all stations simultaneously. Hence, modulation allows different portions of the RF spectrum to be assigned to different radio stations, who place their signals into their allotted frequency bands. We choose stations by tuning our radio receivers to these bands.

The antenna problems solved by modulation are addressed in the section on antennas later in the chapter. We will consider how modulation solves filtering problems later in this section, after we have discussed AM detectors and mixers. Having shown how the information in the audio band is shifted to higher frequencies for broadcast, we now turn to the inverse operation, how your radio shifts the signal back to the audio band so that you can hear it.

Radio Receivers

Introduction.   The radio receiver is also based on the principle of spectrum shifting with nonlinear devices. We begin by discussing an AM detector or demodulator, but the heart of the radio is the mixer, another nonlinear device. The AM radio circuit exemplifies the importance of the frequency-domain viewpoint in communication engineering.

AM detector.   Figure 6.9 shows a simple circuit that will extract the audio information from the AM signal envelope. This is a half-wave rectifier circuit, with a filter capacitor. The diode allows the capacitor to charge to the peak voltage of the input, and the resistor represents the load.

To function as an AM detector, this circuit requires an RC time constant intermediate between the period of the carrier and that of the audio information, that is,

---

[4] Then the bandwidth of the supergroup could handle 30 channels.

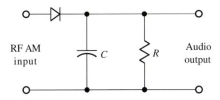

Figure 6.9 AM detector circuit.

$$\frac{2\pi}{\omega_c} < RC < \frac{2\pi}{\omega_m} \tag{6.9}$$

With this provision, the detection of the envelope of the carrier takes place as suggested by Fig. 6.10. The output of the detector follows the peaks of the input and hence approximates the envelope of the AM signal. The period of the carrier is drawn relatively large, resulting in a jagged appearance for the output; but in practice, the period of the carrier is so short that the output is smooth. The small amount of jaggedness, which is inevitable, constitutes a high-frequency noise component in the output, easily removed with a low-pass filter. Disregarding this noise component, we see that the effect of the detector in the frequency domain is to shift the information in the sidebands back to the audio band, as shown in Fig. 6.11.

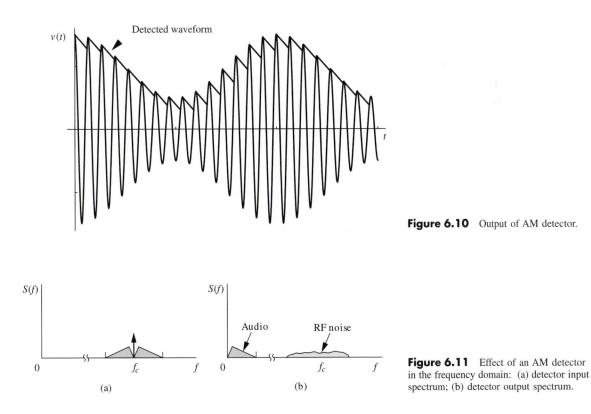

Figure 6.10 Output of AM detector.

Figure 6.11 Effect of an AM detector in the frequency domain: (a) detector input spectrum; (b) detector output spectrum.

EXAMPLE 6.5

AM detector

An AM radio receives a minimum carrier frequency of 540 kHz and a maximum modulation frequency of 5 kHz. Find a suitable RC time constant for the AM detector.

SOLUTION:
From Eq. (6.9),

$$\frac{1}{540 \times 10^3} < RC < \frac{1}{5 \times 10^3} \tag{6.10}$$

One approach is to use the geometric mean between the two limits:

$$RC = \sqrt{1.85 \ \mu s \times 200 \ \mu s} = 19.2 \ \mu s \tag{6.11}$$

WHAT IF?   What if the simple average is used?[5]

---

demodulation, mixing

Demodulation.   Figures 6.8 and 6.11 reveal that the detector inverts the modulation process. For this reason, the word *demodulation* is often used to describe this process. We now have all the makings of a radio: We can shift the audio information to the sidebands of a carrier for broadcast and we can shift the information spectrum back to the audio region for conversion to acoustic waves with loudspeakers. However, standard radios use *mixing*, another nonlinear process, to shift the spectrum of the received signal to an intermediate frequency between the RF and the audio. A *mixer*, like a modulator, is a nonlinear circuit with two inputs and one output, as suggested by Fig. 6.12(a). Indeed, modulators and mixers operate on similar principles yet differ in their function in communication systems.

Mixer.   Figure 6.12(b) shows a circuit that acts as a mixer. The input called $\underline{V}_2$ at ω_2 is the RF carrier and sidebands. The input $\underline{V}_1$ at ω_1 is the local-oscillator (LO) signal, to be explained in what follows. The frequency of the mixer output is placed intermediate between the RF and the audio, and hence is called the intermediate frequency (IF). We resonate the inductor and capacitor at the intermediate frequency; hence, the capacitor's impedance at RF will be so small that the RF current will be controlled by the diode characteristic [Eq. (2.16), assume $\eta = 1$].

$$i_D = I_0(e^{v_D/V_T} - 1) = I_0\left[1 + \frac{v_D}{V_T} + \frac{1}{2}\left(\frac{v_D}{V_T}\right)^2 + \cdots - 1\right]$$

$$= a_1 v_D + a_2 v_D^2 + \cdots \tag{6.12}$$

Impedance Level

where a_1 and a_2 are constants. Because the impedance level of the LC filter at RF is small, the voltage across the diode is approximately the sum of the two voltage sources;

---

[5] The answer is $RC = 101 \ \mu s$, but this is not a good method.

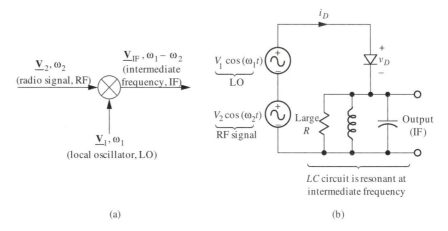

(a)   (b)

Figure 6.12 (a) Mixer input and output frequencies; (b) mixer circuit. The LC filter presents a small impedance at the radio frequency but a large impedance at the intermediate frequency.

$$v_D \approx V_1 \cos \omega_1 t + V_2 \cos \omega_2 t \qquad (6.13)$$

and therefore the power-series form of Eq. (6.12) has many terms involving the two input frequencies.

Mixer output frequencies. Table 6.1 shows some of the terms that will appear in the mixer output, and we have entered in the last column the frequency components that will be produced. The output provides an example of the principle suggested in Fig. 6.2, that a nonlinear circuit with two input frequencies will in general create all the harmonics of those input frequencies, plus the sums and differences of those harmonics. Had we extended our power series in Eq. (6.12) to higher-order terms and used trigonometric identities to expand these terms, we would have verified the presence of all higher harmonics, with their sums and differences.

TABLE 6.1 Some of the Frequencies Created in the Mixer

Time-Domain Expansion	*Frequency Component*
$i_D = a_1[(V_1 \cos (\omega_1 t) + V_2 \cos (\omega_2 t)] + a_2[V_1 \cos (\omega_1 t) + V_2 \cos (\omega_2 t)]^2 + \ldots$	
$= a_1 V_1 \cos (\omega_1 t)$	ω_1
$\quad + a_1 V_2 \cos (\omega_2 t)$	ω_2
$\quad + a_2 V_1^2 \cos^2 (\omega_1 t) \; \{= a_2(V_1^2/2)[1 + \cos (2\omega_1 t)]\}$	dc, $2\omega_1$
$\quad + 2a_2 V_1 V_2 \cos (\omega_1 t) \cos (\omega_2 t) \; [= a_2 V_1 V_2 \cos (\omega_1 t + \omega_2 t)$	$\omega_1 + \omega_2$
$\qquad\qquad\qquad\qquad\qquad\qquad\qquad + a_2 V_1 V_2 \cos (\omega_1 t - \omega_2 t)]$	$\omega_1 - \omega_2$
$\quad + a_2 V_2^2 \cos^2 (\omega_2 t) \; \{= a_2(V_2^2/2) [1 + \cos (2\omega_2 t)]\}$	dc, $2\omega_2$

The intermediate frequency. Only one frequency is desired; all others are noise to be eliminated by filters. We require the difference frequency, $\omega_1 - \omega_2$, listed in the second column in Table 6.1. This component represents a shift of the information in the carrier and sidebands to the intermediate frequency. Figure 6.13 shows the effects of mixing and demodulation. The higher frequency, f_1, is called the *local-oscillator* (LO) frequency, because this frequency is generated within the radio by an oscillator circuit. The mixer combines this frequency with the carrier and sidebands and shifts the entire RF spectrum to the *intermediate frequency* (IF). The IF signal is amplified and filtered and then detected by a normal AM detector; the audio signal is thus recovered. This audio signal is further amplified, low-pass filtered, and furnished to a loudspeaker.

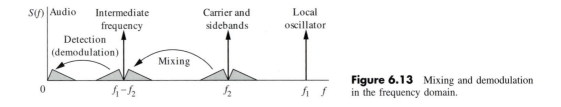

Figure 6.13 Mixing and demodulation in the frequency domain.

EXAMPLE 6.6 **AM system**

Consider a standard AM radio receiving a signal broadcast on a carrier of 1200 kHz. The standard IF for the AM radio is 455 kHz; find the LO frequency.

SOLUTION:

If the LO is placed above the RF, the LO frequency would be the sum of the RF and IF frequencies; hence, the local oscillator frequency must be $1200 + 455 = 1655$ kHz. The resulting AM radio system is shown in Fig. 6.14.

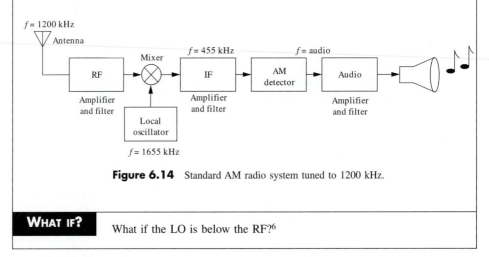

Figure 6.14 Standard AM radio system tuned to 1200 kHz.

WHAT IF?   What if the LO is below the RF?[6]

---

[6] This is a bad idea, but if you insist, 745 kHz.

Filtering. We now give the reasons for shifting the information to an IF rather than using direct demodulation of the RF signal. As a preliminary, we review the filtering that we have already encountered in the circuit. We require low-pass filtering at the output of the AM detector to separate the audio signal from the detector noise. This filtering is usually incorporated into the audio amplifier circuit and is relatively easy to accomplish because of the wide difference in frequency between the audio spectrum and the noise spectrum.

Similarly, we need filtering at the mixer output to eliminate all but one of the frequencies created by the nonlinear action of the mixer diode. This filtering is likewise easily accomplished because again there is a large difference between the desired and the undesired frequency components in the mixer output. As we will see, this filtering is accomplished by the IF amplifier.

Reasons for using an intermediate frequency.

The principal filtering task of the radio, and the most demanding, is the filtering of the RF spectrum to eliminate all the other radio stations that are broadcasting simultaneously with the station of interest. If Fig. 6.13 portrayed the true effect of mixing, it would show the signals from many AM radio stations side by side in the RF spectrum, and the radio has to select one and eliminate the rest. How is this filtering accomplished?

The obvious solution is to put a narrow-band filter in the RF amplifier to select the desired station and reject the rest. The two reasons why this is impractical are: (1) the requirements on such a filter would strain the limits for practical inductors and capacitors; and (2) the requirement for a *tunable* filter would further complicate the design of a narrow-band RF filter.

Let us consider the second difficulty first. To construct a radio, we need to filter out all the radio stations except one; but we also need to be able to tune the radio to any of the available stations. We have not discussed tunable filters hitherto, nor will we here except to assert that they are difficult to design. The other problem is that, even without the requirement that the filter be tunable, it would be challenging to design an RF filter selective enough to reject all but one of the available radio stations. The combination of these two difficulties has driven radio designers to use intermediate frequencies, because mixing to an IF lessens both difficulties.

IF amplifier.

In Fig. 6.14, we have followed the mixer with an IF amplifier/filter. In the AM radio, this amplifier would have high gain only for frequencies within ± 5 kHz of the IF frequency of 455 kHz and hence would act as a filter eliminating all frequencies out of this band of frequencies. The IF amplifier is "fixed-tuned," that is, its passband remains the same as you tune the radio; and because its operating frequency is relatively low, it can provide effective filtering and high gain. Consequently, not only does the IF amplifier/filter eliminate the undesired frequencies created by the mixer, it eliminates the radio stations adjacent to the desired station. Thus, it performs the same function as an RF filter, but it filters at a lower frequency and does not need to be tunable.

Selecting stations.

The radio is tuned by varying the local-oscillator (LO) frequency. When you turn the tuning knob on a radio, the principal effect inside the radio is to vary the frequency of the LO. We saw before that to receive a radio station broadcasting at 1200 kHz, we had to set the LO frequency to 1655 kHz. Clearly, if we increase the LO to 1665 kHz, the station broadcasting at 1210 kHz would now be mixed into the bandpass of the IF amplifier/filter, and the station at 1200 kHz would be eliminated. By this technique, one radio station is selected and all others are rejected.

Image band. To the previous filtering actions of the radio circuit, we must add yet one more. In Fig. 6.14, you will note that the box representing the RF amplifier is also called a filter. This does not contradict our earlier assertion that practical problems prevent us from designing a narrow-band RF filter. Some filtering of the RF spectrum is necessary to eliminate the image band of the mixer. To explain the image of the mixer and why we need one more filter to get rid of it, we must refer back to our explanation of mixer operation. You will note in Table 6.1 that the term relating to the mixer action was of the form $\cos[(\omega_1 - \omega_2)t]$. This is the logical way to write this term if ω_1, the LO frequency, is higher than the RF frequency and if only one RF frequency comes into the mixer. But if the RF frequency were higher than the LO frequency, we would write the mixer term as $\cos[(\omega_1 - \omega_2)t]$. So if, for example, our LO were tuned to 1655 kHz and there were an RF input at $1655 + 455 = 2110$ kHz, then this frequency would also be mixed into the IF passband.

image band

Image rejection. We conclude that the mixer would not distinguish between RF frequencies above and below the LO frequency and hence would mix both into the IF amplifier/filter passband. Without additional filtering, the radio would receive two stations simultaneously, one above and one below the LO frequency. This second, undesirable frequency is called the *image band* of the mixer, and hence a broad-band RF amplifier/filter is required to eliminate any station broadcasting in the image band before it reaches the mixer.

EXAMPLE 6.7 | **Image band**

A microwave receiver is tuned to a carrier of 4226 MHz, has an IF of 60 MHz with a 4-MHz IF bandwidth, and an LO frequency below the carrier. What is the image frequency band for this receiver?

SOLUTION:
The LO would be at $f_{RF} - f_{IF} = 4166$ MHz. The center of the image band would be below the LO by the IF, so $f_{image} = f_{LO} - f_{IF} = 4166 - 60 = 4106$ MHz. Hence, the image band would be 4106 ± 2 MHz $= 4104$ to 4108 MHz.

WHAT IF? What if the LO is above the RF?[7]

---

superheterodyne receiver

The superheterodyne receiver. The mixer and IF amplifier characterize the *superheterodyne receiver*, which combines amplification, filtering, and spectrum shifting. Obviously, much amplification is required because of the small signals picked up by the antenna, and we indicated amplification in Fig. 6.14 at RF, IF, and audio frequencies.

---

[7] 4344 to 4348 MHz.

Spectrum shifting is performed by the mixer, to lower the spectrum of the information from the RF band to the IF band; and spectrum shifting is also accomplished by the AM detector to move the information back into the audio frequency band. Filtering is done at RF to eliminate signals in the mixer image band; filtering is done by the IF amplifier to eliminate unwanted mixer products and adjacent radio signals; and filtering is done at the audio frequencies to eliminate noise created by the detector.

The Frequency Domain

The radio provides an excellent example of the importance of the frequency-domain viewpoint in electrical engineering. Of the three essential processes involved in radio circuits, two (filtering and spectrum shifting) are accomplished in the frequency domain. We now present two additional examples using superheterodyne techniques

Frequency-modulation (FM) radio.

The frequency of a radio wave is less vulnerable to noise than the amplitude, so frequency modulation (FM) is used when high-quality sound or other information is required. Commercial monaural FM broadcasting uses an RF bandwidth of 150 kHz, and stereo uses 200 kHz for sum (Left + Right) and difference (Left − Right) channels. The FM band is from 88 to 108 MHz, with stations given 200 kHz in which to broadcast 15 kHz of audio bandwidth. The noise-suppression properties of FM therefore come at a cost of greater bandwidth.

Feedback

FM receivers.

Figure 6.15 shows the basic FM receiver. This is a superheterodyne receiver with an IF of 10.7 MHz and an IF bandwidth of 200 kHz. The detection scheme puts the FM signal, shifted to the IF, through a frequency-sensitive circuit that converts the FM to AM for demodulation with rectifier diodes similar to AM radio. A limiter circuit prior to detection eliminates AM noise in the signal. A signal proportional to the IF frequency is fed back to the LO to lock the LO frequency to the station frequency, thus eliminating drift.

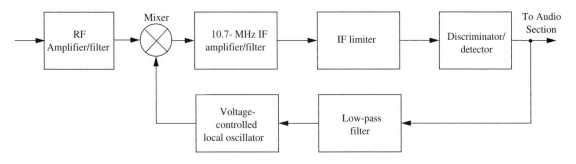

Figure 6.15 Basic FM system receiver. The LO is part of a feedback loop to lock onto the station frequency.

Base band

Analog Information

Telephone systems.

The traditional telephone circuit is a loop of wire from the central exchange to each subscriber. The voice spectrum of approximately 250–3300 Hz is sent over the wire in the base band, that is, without frequency shifting. Switching signals were obtained by opening and closing the loop at a rate of approximately 10–12 pulses/second and relays counted the pulses at the exchange to route the call. The ring signal was about 90 to 100 V ac at approximately 20 Hz.

The modern system uses a tone system to transmit switching information, and semiconductor switching computers have replaced the relays. The "ring" signal is commonly a tone that activates an audible device in the telephone receiver.

Digital Information

The local exchange switches up to 10^4 lines with the signal at the base band with a 4-kHz bandwidth. Between exchanges, the voice information is modulated with signal sideband plus carrier into groups of 12 adjacent channels between 60 and 108 kHz. These groups, 12 channels, are then placed into a supergroup of 60 channels between 312 and 552 kHz, and finally 10 supergroups, 600 channels, are placed into a master-group between 564 and 3084 kHz. For long distance, the voice channels are modulated to yet higher frequencies for transmission by coaxial cable, line-of-sight microwave, or satellite link. Currently digital systems using optical fibers are replacing traditional analog systems.

Noise in Receivers

noise, interference

Importance of noise. When we spoke of op amps in Chap. 4, we asserted that gain is cheap. With radio receivers, gain is not as cheap as with op amps, but we can still achieve as high a gain as we wish by adding more stages of amplification. We might suppose, therefore, that we can successfully detect any input RF signal, no matter how weak, simply by having adequate gain in the receiver. This is not true, however, because the sensitivity of a radio receiver is limited by random noise. *Noise*, by definition, is any undesired signal in the system, whether natural or man-made. Man-made noise is called *interference*; Chap. 5 showed techniques to reduce its effects.

blackbody radiation, shot noise, flicker noise

Sources of natural noise. Natural noise in communication systems comes from two sources. Some noise comes into the receiver from the antenna; for example, in an AM receiver, we can hear a sporadic crackling noise from lightning during storms. Other sources of noise that enter the system through the antenna are atmospheric gases (thermal noise) and extraterrestrial objects such as the sun and our galaxy (cosmic noise).

The second source of noise is the receiver itself. Just as any body radiates *blackbody radiation* due to the thermal motion of its atomic constituents, so also resistors radiate their own type of thermal noise into the circuits of which they are a part. We investigate resistor noise in what follows. Transistors add, in addition to thermal noise, *shot noise* due to the discrete charges that carry the current across the *pn* junctions, and *flicker* or $1/f$ noise, due to slow changes in device properties from aging, temperature, and other physical changes.

Equivalent Circuits

The Frequency Domain

Thermal noise. The thermal radiation of resistors plays an important role in the description of noise because resistors provide a simple calibration for noise signals. Figure 6.16(a) shows a resistor, R, and indicates the noise voltage generated by the thermal motion of the carriers in the resistor. The Thévenin equivalent circuit in Fig. 6.16(b) separates the noise voltage from the resistance, which here is noise-free. The power spectrum of the noise voltage in volts²/hertz is shown in Fig. 6.16(c),

$$S_n(f) = 4kTR \text{ V}^2/\text{Hz} \tag{6.14}$$

white noise

with k = Boltzmann's constant of 1.38×10^{-23} J/K. We note the following:

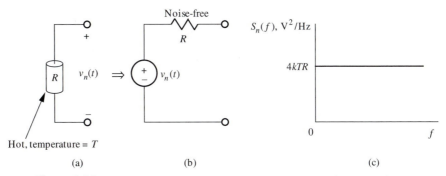

Figure 6.16 (a) A noisy resistor; (b) Thévenin equivalent circuit of a noisy resistor; (c) power spectrum of thermal noise.

■ The noise spectrum is *white noise*, meaning that all frequencies are present equally, at least throughout the radio spectrum. Thus, a resistor produces a noise power that is proportional to the bandwidth of the circuit.

■ The intensity of the power spectrum is proportional to the temperature of the resistor. For sensitive amplifiers, we may measure the noise of the system by comparing system noise to the noise from a resistor of known temperature.

■ The intensity of the power spectrum is proportional to the resistance. For this reason, large values of resistance are avoided in sensitive electronic amplifiers. The *available power* out of a circuit is the maximum power that the circuit is capable of delivering into a matched load. The available power spectrum from the circuit in Fig. 6.16(b) is

available power

$$S_{av}(f) = \frac{S_n(f)}{4R} = kT \text{ W/Hz} \tag{6.15}$$

where $S_{av}(f)$ is the available power spectrum in watts/hertz. Thus, the total power a noisy resistor will contribute to a matched circuit is

$$P = kTB \text{ W} \tag{6.16}$$

where B is the bandwidth in hertz. Equation (6.16) is useful in determining the signal-to-noise ratio in a communication circuit, once we have defined the noise figure and system temperature of an amplifier.

system temperature

System temperature. The noise originating in the resistors and semiconductors in a radio amplifier is generally broadband or white noise, at least over the bandwidth of the radio. Thus, we may represent the noise *as if* it originated in a resistor at the input of the radio. Figure 6.17(a) shows the true situation: An amplifier with a power gain of G and a bandwidth of B has internal noise sources that produce N_s watts of noise at its output. In Fig. 6.17(b) we show an equivalent amplifier, assumed to be noise-free, with the same noise at its output *attributed* to a noisy resistor at the input. The *system temperature* is defined as the temperature the resistor required in Fig. 6.17(b) to produce the same amount of output noise as in Fig. 6.17(a). Using Eq. (6.16), we can express the output noise as

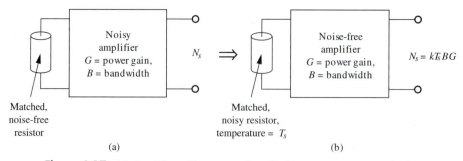

(a)   (b)

Figure 6.17 (a) Amplifier with output noise; (b) the system temperature is the temperature of the input resistor to give the same output noise.

$$N_s = kT_sGB \;\Rightarrow\; T_s = \frac{N_s}{kGB} \;\; \text{K} \tag{6.17}$$

where T_s is the system temperature and N_s is the output noise produced by the amplifier. All the internally generated noise is *referred to the input* of the receiver and then expressed as a temperature. The system temperature describes the sensitivity of the amplifier; the lower the system temperature, the more sensitive the amplifier.

operating noise figure

Noise figure.   The *operating noise figure* is defined as the input signal-to-noise ratio divided by the output signal-to-noise ratio. Figure 6.18 shows an amplifier with input signal and noise, S_{in} and N_{in}, and output signal and noise, S_{out} and N_{out}. The *operating noise figure* is defined to be

$$F = \frac{S_{in}/N_{in}}{S_{out}/N_{out}} = \frac{N_{out}}{N_{in}G} = \frac{N_{in}G + N_s}{N_{in}G} = 1 + \frac{N_s/G}{N_{in}} = 1 + \frac{T_s}{T_{in}} \tag{6.18}$$

where F is the operating noise figure of the amplifier and is often expressed in decibels, $F_{dB} = 10 \log F$, and T_{in} is the input noise expressed as a temperature. The first form on the right side of Eq. (6.18) is the definition of operating noise figure. The second form introduces the gain, $G = S_{out}/S_{in}$, and the third form breaks the output noise into two components, the amplified input noise and the internally generated noise. The last form shows that the noise figure depends on the internally generated noise, referred to the input of the amplifier, compared with the input noise. Thus, the operating noise figure depends in part on the input noise and not on the amplifier only. The operating noise figure is useful in performing signal-to-noise calculations in a communication system.

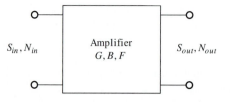

S_{in}, N_{in} Amplifier G, B, F S_{out}, N_{out}

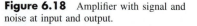

Figure 6.18 Amplifier with signal and noise at input and output.

EXAMPLE 6.8

Operating noise figure

The signal-to-noise ratio for a TV signal from a satellite antenna is $+50$ dB, and the operating noise figure of the amplifier in the radio receiver is 20 dB. Find the output signal-to-noise ratio of the receiver.

SOLUTION:

When we convert the definition of the noise figure in Eq. (6.18) to dB by taking 10 times the log of each side of the equation, we have

$$\left.\frac{S}{N}\right|_{out,\,dB} = \left.\frac{S}{N}\right|_{in,\,dB} - F_{dB} \;\Rightarrow\; \left.\frac{S}{N}\right|_{out,\,dB} = 50 \text{ dB} - 20 \text{ dB} = 30 \text{ dB} \qquad (6.19)$$

standard noise figure

Standard noise figure. For purposes of comparing amplifiers, we use the *standard noise figure*, in which the input noise is assumed to be the thermal noise of a matched resistor at a temperature of 290 K. For this case, $N_{in} = k(290)\,GB$; and from Eq. (6.18), we can express the standard noise figure as

$$F = 1 + \frac{T_s}{290} \qquad (6.20)$$

EXAMPLE 6.9

Amplifier comparison

One amplifier has a standard noise figure of 1.25 and another has a system temperature of 75 K. Which is the more sensitive amplifier?

SOLUTION:

To compare the amplifiers, we must express their noise contributions in the same terms. The first amplifier has a system temperature of

$$T_s = 290(F - 1) = 72.5 \text{ K} \qquad (6.21)$$

and thus produces slightly less noise than the second amplifier.

WHAT IF?

What if you use the 75-K amplifier in a satellite receiver that is looking at the cold sky, assuming an noise input of 20 K? What is the operating noise figure?[8]

---

[8] 6.77 dB.

EXAMPLE 6.10	**Output noise**

We assume the 72.5-K amplifier is used in a receiver with a gain of 120 dB, a bandwidth of 4 MHz, and an output impedance of $50\,\Omega$. What would be the output noise power due the amplifier, and what would be the peak-to-peak noise voltage into a $50\text{-}\Omega$ load?

SOLUTION:

The output available power can be determined from Eq. (6.17) as

$$N_s = kT_sGB = 1.38 \times 10^{-23} \times 72.5 \times 10^{120\,\text{dB}/10} \times 4 \times 10^6 \qquad (6.22)$$
$$= 4.00 \times 10^{-3}\ \text{W}$$

This would be the power into a matched load of $50\,\Omega$, and thus the rms voltage would be

$$\frac{V_{rms}^2}{50} = 4.00\ \text{mW} \Rightarrow V_{rms} = 0.447\ \text{V} \qquad (6.23)$$

The output voltage would be a random signal with an approximate correlation time of $\tau = 1/B = 0.25\ \mu\text{s}$. A good rule of thumb for such random voltages is that the peak-to-peak voltage is five to six times the rms value; thus, we would expect a peak-to-peak output voltage of about 3 V.

Check Your Understanding

1. If the input spectrum to a circuit contains a fundamental and third harmonic and the output contains in addition a second harmonic, is the circuit linear or nonlinear?

2. The input of a nonlinear system contains frequencies of 30 and 100 Hz. Which of the following can be in the output: 0, 70, 95, 100, 115, 210, 270 Hz?

3. Is the most critical filtering in a superheterodyne radio done by the RF amplifier, mixer, IF amplifier, or audio amplifier?

4. In a communication system, a carrier of 500 kHz is amplitude-modulated with a 5-kHz tone. What is the required radio-frequency bandwidth to pass the signal?

5. In a radio, is the RF amplifier required to remove the LO, IF, or image band of the mixer?

6. If in a radio, the RF frequency is 22 MHz, the IF frequency is 5 MHz, and the LO frequency is 27 MHz, what is the image frequency?

7. An amplifier has a standard noise figure of 6 dB. What is its system temperature?

8. Calculate the available noise power from a 300-K, 5-kΩ resistor in the bandwidth from 0 to 10 MHz.

Answers. **(1)** Nonlinear; **(2)** 0, 70, 100, 210, and 270 Hz; **(3)** IF amplifier; **(4)** 10 kHz; **(5)** image band of the mixer; **(6)** 32 MHz; **(7)** 865 K; **(8)** 4.14×10^{-14} W.

Introduction. Modern communication systems use electromagnetic (EM) waves as the medium for transmitting information. Guided EM waves are used in telephone and cable-TV systems, and radiated EM waves are broadcast by radio and TV stations and narrowcast by radar and space communication systems. In this section, we survey the physical properties of EM waves as they relate to communication systems, emphasizing the transmission of EM power from one antenna to another.

History. In early experiments in electricity, a connection between light, electric phenomena, and magnetic phenomena was unanticipated. It was known that moving charges produce a magnetic field (Ampère's circuital law) and that a changing magnetic flux produces an electric field (or a voltage, which is Faraday's law).[9] But no one related these phenomena to light. One of the major triumphs of mathematical physics occurred when James Clerk Maxwell (1831–1879) realized that mathematical consistency required that a changing electric field produce a magnetic field. When Maxwell added the needed term to the then-known equations, thus formulating Maxwell's equations, he showed that coupled electric and magnetic fields exist in the form of waves. Moreover, his predicted velocity of these waves corresponded to the known velocity of light. Thus, Maxwell unified electrical science and showed that light consists of electromagnetic waves. Maxwell's predictions were soon confirmed experimentally by Heinrich Hertz.

The electromagnetic spectrum. Maxwell discovered what we call the electromagnetic spectrum, summarized in Fig. 6.19. The figure incorporates the relationship between frequency and wavelength:

$$\lambda = \frac{c}{f} = \frac{3 \times 10^8}{f} = \frac{300}{f_{\text{MHz}}} \tag{6.24}$$

where λ is the wavelength in meters, c is the velocity of light in meters/second, f is the frequency in hertz, and f_{MHz} is the frequency in megahertz. Figure 6.19 lists the ways that electromagnetic waves are guided from point to point. We discuss waveguiding structures in what follows. Also shown are the ways in which free waves are affected by earth and ionosphere for long-distance communication, which we also discuss. The last column shows some of the applications of electromagnetic waves in the various frequency-wavelength ranges.

Free and guided electromagnetic waves. Figure 6.19 distinguishes between free waves and guided waves. Guided waves are one-dimensional waves guided by wires, coaxial cables, and the like. Free waves are launched by antennas and spread out in two- or three-dimensional space. We must investigate both types of waves to understand communication systems. We begin with guided waves.

---

[9] These fields and laws are discussed in courses in electrical physics.

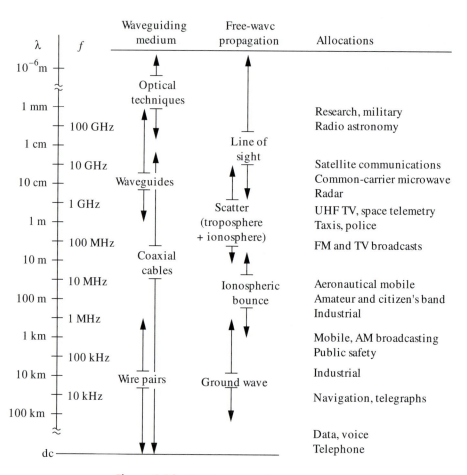

Figure 6.19 The electromagnetic spectrum.

Guided Electromagnetic Waves

transmission lines

Introduction. Along many rural roads, you see wires strung on high poles. The power line is usually on top, with the telephone cable below. These two examples of transmission lines exemplify two major applications, communication and distribution of electric power. The basic principles are the same for both types of lines, but in this section, we survey the major concepts of transmission lines used in communication systems.

A *transmission line* guides waves of EM energy from source to load. Such lines may be parallel wires, as described before, coaxial cables such as distribute TV and other communication services in urban areas, or an internal bus in a computer. Any "circuit" in which the distributed capacitance and inductance of the conductors becomes a factor in circuit performance is a transmission line. Put another way, any circuit whose physical dimensions are not greatly smaller than a wavelength at the highest significant frequency must be considered a transmission line. The analysis of waves on transmission lines follows from basic circuit theory, but space limitations prohibit development here.

The following are properties of transmission lines:

- Wave *velocity* on an overhead transmission line is the velocity of light, 3×10^8 m/s in air. In coaxial cables, the waves are slowed down by the plastic that separates inner and outer conductors. The formula for the wave velocity is

$$v = \frac{1}{\sqrt{LC}} \text{ m/s} \tag{6.25}$$

where L is the distributed inductance of the line in henrys/meter and C is the distributed capacitance of the line in farads/meter. For example, a common co-axial cable, RG-58 in Fig. 6.20, has $L = 0.253$ µH/m and $C = 101$ pF/m, so the wave velocity is 1.98×10^8 m/s. The wavelength at any frequency can be calculated from Eq. (6.24) if instead of c we substitute the velocity on the transmission line.

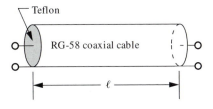

Inductance = 0.253 µH/meter $\times$ ℓ
Capacitance = 101 pF/meter $\times$ ℓ

Figure 6.20 RG-58 coaxial cable. The cable also has a small resistance, which causes loss.

- Power loss on the line is caused by distributed resistance and dielectric losses. For example, the loss of RG-58 is 5 dB/100 ft at 100 MHz.

- *Dispersion* describes the tendency of waves at different frequencies to travel at different velocities on the line. Dispersion causes distortion in the spectra of communication signals sent over transmission lines, but filters at the receiving end can compensate partially.

- The *characteristic impedance* gives the ratio of the voltage to the current waves that move along the line. The characteristic impedance, Z_0, of the transmission line is given by

$$Z_0 = \sqrt{\frac{L}{C}} \text{ } \Omega \tag{6.26}$$

For example, the characteristic impedance corresponding to the values of L and C given above is $50.0 \ \Omega$.

- *Reflection* occurs when the impedance of the load at the receiving end of the line differs from the characteristic impedance of the line. Reflections are avoided if possible because power is lost to the load, and often the reflected waves interfere with the source of the signal. When the load has the same impedance as the characteristic impedance of the line, the load is matched to the line, and hence no reflection occurs, as indicated in Fig. 6.21.

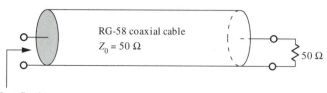

RG-58 coaxial cable
$Z_0 = 50\ \Omega$

$50\ \Omega$

No reflections

Figure 6.21 A matched transmission line with load. No reflection occurs at the load.

EXAMPLE 6.11 | **Transmission-line parameters**

A 10-meter piece of transmission line has a measured capacitance of 705 pF. A pulse traverses the length of the line in 50 ps. Find the inductance/meter for the line.

SOLUTION:
The capacitance must be 70.5 pF/m, and the velocity of the pulse on the line must be

$$v = \frac{10\ \text{m}}{50 \times 10^{-9}\ \text{s}} = 2.00 \times 10^8\ \text{m/s} \qquad (6.27)$$

From Eq. (6.25), we find

$$L = \frac{1}{v^2 C} = \frac{1}{(2.00 \times 10^8)^2 \times 70.5 \times 10^{-12}} = 0.355\ \mu\text{H/m} \qquad (6.28)$$

WHAT IF? What if you need to match this transmission line with a resistor? What value should you use?[10]

Free Electromagnetic Waves

What we mean by "free." In this section, we deal with electromagnetic waves that are free of man-made guiding structures. Such unconfined waves spread out in space. They may spread out in three dimensions, like waves radiated from a satellite antenna, or they may spread out in two dimensions, like waves radiated by a broadcast antenna and guided by the surface of the Earth. The distribution of energy flow in such waves is affected by the source of the energy, usually an antenna, and by the matter that the waves encounter, for example, by reflection from the ionosphere.

Free electromagnetic waves are used in almost all communication systems except local telephone and cable TV systems. For this reason, we need to understand the character of such waves; how they are launched; how they interact with Earth, obstacles, and the ionosphere; and how they are received. In this section, we focus on the character of the waves; in the next section, we look at how the waves are launched and received by antennas.

---

[10]71 Ω.

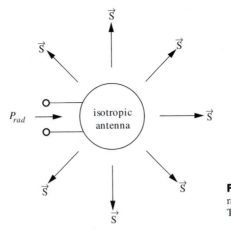

Figure 6.22 An isotropic antenna radiates energy in all directions equally. The antenna does not focus the energy.

Spherical waves.

The simplest case is that of waves that travel away in all directions from a source. Such waves consist of coupled electric and magnetic fields traveling near the speed of light. We may describe the waves by their polarization and power density, the power per unit area in the wave in watts per square meter. We begin with the isotropic (all-directional) antenna. Such an antenna is defined[11] to radiate equally in all directions, as suggested in Fig. 6.22. Because the energy travels outward in straight lines, and because energy is conserved, the power passing a spherical surface of radius R must account for all the power radiated by the antenna.[12] Thus, the power density is

Conservation Energy

$$S_{iso}(R) \times 4\pi R^2 = P_{rad} \implies S_{iso}(R) = \frac{P_{rad}}{4\pi R^2} \text{ W/m}^2 \tag{6.29}$$

polarization

where P_{rad} is the total power radiated, $4\pi R^2$ is the area of a sphere of radius R, and $S_{iso}(R)$ is the power density at a distance R from the source of the radiation. The power density is a vector quantity, having a direction in space away from the source. Equation (6.29) indicates that spherical waves weaken due to spreading as they travel away from their source.

Polarization.

Electromagnetic waves consist of electric and magnetic fields that, at any point, also have directions in space. Figure 6.23 shows the situation: The wave direction and power density are indicated by $\vec{S}$. The electric field, $\vec{E}$, and the magnetic field, $\vec{H}$, are directed at right angles to the direction of $\vec{S}$ and at right angles to each other. Thus, at a location we can describe a radio wave in terms of its direction of travel and its polarization, for example, E-field vertical relative to the Earth's surface.

Polarization is important because most antennas produce and receive only one polarization. Thus, if a satellite transmits vertical polarization and your local antenna re-

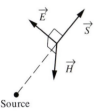

Source

Figure 6.23
Geometry of a wave traveling away from a source. The wave direction and power density are indicated by $\vec{S}$. The electric and magnetic fields are indicated by $\vec{E}$ and $\vec{H}$, respectively.

---

[11] It has been proved that an isotropic antenna is incompatible with Maxwell's equations and thus can never be constructed. It is a useful concept, however.

[12] There is a delay, of course, between the time of radiation and the time of passing the sphere. Equation (6.29) assumes steady power levels.

ceives only horizontal polarization, the radio waves arriving at your location will not be collected by the antenna. Commercial AM broadcast signals are vertically polarized, but FM and TV signals are required to use horizontal polarization. Circular polarization uses both vertical and horizontal polarizations, with a 90° phase shift between the two. Due to multiple reflection, however, transmitted signals can become in part depolarized, so the orientation of the receiving antenna is often not critical. At microwave[13] frequencies, aligning antenna and wave polarization is very important.

Surface waves.   At lower frequencies such as are used for AM broadcasting, radio waves can be guided by the surface of the Earth. These surface waves follow the curvature of the Earth and do not depend upon line-of-sight between transmitting and receiving antennas. In addition to the loss due to the spreading of the wave, a surface wave also diminishes due to resistive losses in soil.

Plane waves.   Radio waves originate from finite sources and hence have curvature as they spread out in space. However, for practical purposes, the radius of curvature is usually large enough to treat them as plane waves. In determining the intensity of such waves, we must consider the distance to the source, but in considering the local nature of the waves, we usually consider them as plane waves.

Radio-wave phenomena.   Earlier, we mentioned two effects that can happen to radio waves: they can be reflected and they can become depolarized. Among the more important effects on radio waves due to their interaction with matter are the following:

reflection, depolarization, refraction, diffraction, scattering, Doppler shift, attenuation

- *Reflection* can occur from the Earth's surface, bodies of water, buildings, and, at low frequencies, the ionosphere. Reflection is used in radar to locate the range and direction of a target. Sometimes reflectors are elevated on towers in microwave relay stations, as shown in Fig. 6.24. At low frequencies, the reflection of radio waves from the ionosphere allows long-distance communication.

- *Depolarization* occurs when a wave loses its pure polarization in the horizontal or vertical direction and becomes a partial mix of the two.

- *Refraction* refers to the bending of a wave toward the direction of denser matter. For example, radio waves in the lower atmosphere are refracted toward the Earth's surface because the atmosphere is more dense at the Earth's surface. This has the effect of extending the "radio horizon" slightly and is beneficial in line-of-sight microwave transmission systems.

- *Diffraction* occurs when waves spread into a shadow region behind an obstacle. Diffraction allows radio waves to be received beyond the radio horizon and behind buildings.

- *Scattering* occurs when radio waves bounce off a multitude of small objects and "scatter" in all directions. For example, a wave that was reflected from the surface of the ocean would be in part scattered by the irregular surface.

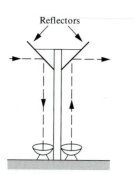

Reflectors

Figure 6.24 A microwave relay station using reflectors.

---

[13] Generally, above about 1000 MHz, but "microwave" refers more to the type of structures used to control the energy than to the absolute frequency.

- *Doppler shift* occurs when a wave reflecting from a moving target is shifted in frequency. This effect allows a radar to measure the speed of a target.

- *Attenuation* refers to the absorption of electromagnetic energy by matter. Rain and atmospheric gases cause attenuation at high microwave frequencies.

Summary. We investigated the behavior of free electromagnetic waves. We showed how waves diminish through spreading as they travel away from sources, and we described the various effects that can influence their travel through the atmosphere. We now describe the means for launching and receiving the waves.

Antennas

What is an antenna? An *antenna* is a structure that couples between a guided and a free electromagnetic wave. Most antennas can be used for both transmitting and receiving waves. The transmitting and receiving properties of an antenna are closely related, but are described in different terms. Figure 6.25 represents an antenna as a transmitting device. It receives power, P_{in}, from a circuit and radiates power in the form of radio waves, which we describe by their polarization and their power density as a function of distance from the antenna, $S(R)$. As a circuit element, the antenna is described by its input impedance, $\mathbf{Z}_{in}$. The radiation efficiency, η, is the ratio of the radiated power to the input power:

$$\eta = \frac{P_{rad}}{P_{in}} \tag{6.30}$$

Antenna gain. Parabolic reflectors, such as shown in Fig. 6.26(a), are used in radar and satellite communication systems because they focus the radiated power into a narrow region of space. Simple wire antennas, such as shown in Fig. 6.26(b), focus the power only slightly and are used for broadcast applications. The focusing ability of an antenna is described by the antenna gain, which is defined as

$$G = \eta \, \frac{S(R)}{S_{iso}(R)} \tag{6.31}$$

where G is the antenna gain. The antenna gain involves an efficiency factor and a focusing factor. The efficiency factor, η, describes how much of the power delivered to the antenna is radiated; $1 - \eta$ is the fraction lost in resistance and other losses. The focusing factor is the ratio of the power density radiated by the antenna divided by the power

antenna

LEARNING OBJECTIVE 4.
To understand the properties of antennas and their role in communication systems

antenna gain

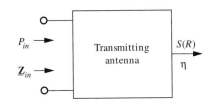

Figure 6.25 A transmitting antenna receives power from a circuit and radiates an electromagnetic wave.

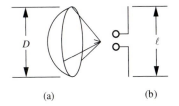

Figure 6.26 (a) A parabolic reflector-type antenna; (b) a wire-type antenna.

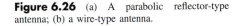

density of an isotropic antenna. The *antenna gain* thus is the power density of the antenna compared to that of a lossless antenna that does not focus at all. A simple wire antenna has a maximum gain in the range 1.5 to 2.0. A large parabolic reflector may have a gain exceeding 1000. The antenna is normally positioned to focus the energy in a preferred direction, and thus we deal usually with the maximum gain of the antenna.

We may combine Eqs. (6.29), (6.30), and (6.31) to obtain an important equation relating power density to distance from the antenna:

$$S(R) = \frac{P_{in}G}{4\pi R^2} \qquad (6.32)$$

Equation (6.32) assumes spherical-wave spreading of the waves and also the absence of reflection, attenuation, and the like.

EXAMPLE 6.12 | **Transmitting antenna**

A parabolic antenna has a maximum gain of 22 dB. How much power do we have to deliver to the antenna to produce a power density of $2\,\mu W/m^2$ at a distance of 10 miles?

SOLUTION:
Because 10 miles $= 1.61 \times 10^4$ meters, and $G = 10^{22/10} = 158$, we find from Eq. (6.32)

$$P_{in} = \frac{4\pi(1.61 \times 10^4)^2 \times 2 \times 10^{-6}}{158} = 41.1 \text{ W} \qquad (6.33)$$

WHAT IF? | What if the distance were doubled?[14]

Equivalent Circuit

Receiving antennas. Figure 6.27 represents a receiving antenna. The input to the antenna is a plane wave of a given polarization and power density. As a circuit element, the antenna is characterized by its Thévenin equivalent circuit. Although in practice the equivalent circuit is not easy to determine, we may work directly with the available power from the antenna.

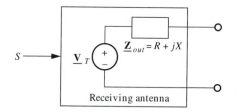

Receiving antenna

Figure 6.27 A receiving antenna. The input is a plane wave and the output is power furnished to a circuit.

---

[14] Then we need four times the power, 164.3 W.

As mentioned before, the available power of a circuit is the maximum power that the circuit is capable of delivering into a matched load. The load on the antenna is the input impedance of the radio receiver attached to the antenna, which must be $\mathbf{Z}_{out}^* = R - jX$ to receive the available power from the antenna. The ability of the antenna to collect power from a radio wave and deliver it to a circuit is characterized by the *effective area* of the antenna:

$$A_{eff} = \frac{P_{av}}{S} \tag{6.34}$$

where P_{av} is the available power in watts, S is the incident power density in W/m^2, A_{eff} is the effective area of the antenna in square meters, and we have assumed that the antenna is correctly oriented to receive the polarization of the incident radio wave. It can be shown[15] that the effective area of any antenna is related to its gain by

$$A_{eff} = \frac{\lambda^2}{4\pi} G \tag{6.35}$$

where λ is the wavelength in meters. Equations (6.32), (6.34), and (6.35) permit us to calculate the coupling of transmitter and receiver through a radio link.

EXAMPLE 6.13 | **Receiving antenna**

The power radiated in the previous example is received by an identical antenna at a frequency of 5020 MHz. Find the available power from the receiving antenna.

SOLUTION:
The wavelength from Eq. (6.24) is $300/5020 = 5.97 \times 10^{-2}$ meters, so the effective area by Eq. (6.35) is

$$A_{eff} = \frac{(5.97 \times 10^{-2})^2}{4\pi} \times 10^{22/10} = 4.50 \times 10^{-2} \text{ m}^2 \tag{6.36}$$

From Eq. (6.34), we can calculate the available power as

$$P_{av} = S \times A_{eff} = 2 \times 10^{-6} \times 4.50 \times 10^{-2} = 9.01 \times 10^{-8} \text{ W}$$

Communication equation. In the communication link pictured in Fig. 6.28, a transmitter delivers P_{in} of power to a transmitting antenna. A receiving antenna at a distance R accepts a portion of the radio wave and delivers an available power P_{rec} to a receiver with a matched input impedance. We consider only the line-of-sight wave and

---

[15] The proof of Eq. (6.35) from Maxwell's equations is difficult. A relatively simple proof can be obtained from thermodynamic equilibrium by considering resistors attached to each antenna. For the resistors to reach thermodynamic equilibrium through radiative heat transfer, Eq. (6.35) must be true. In other words, if Eq. (6.35) were not true, a perpetual motion machine theoretically could be constructed out of resistors and antennas.

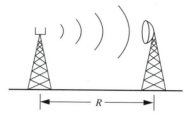

Figure 6.28 A radio transmission link with transmitting and receiving antennas.

thus neglect reflection, scattering, and the like. Combining Eqs. (6.32) and (6.34), we determine the received power, P_{rec}, to be

$$P_{rec} = \frac{P_{in}G_t}{4\pi R^2} \times A_{eff} \tag{6.37}$$

where G_t is the gain of the transmitting antenna and A_{eff} is the effective area of the receiving antenna. We may change the form of Eq. (6.37) by use of Eq. (6.35), with the result

$$P_{rec} = \frac{P_{in}G_tG_r}{(4\pi R/\lambda)^2} \tag{6.38}$$

space loss

where G_r is the gain of the receiving antenna. The coupling between transmitter and receiver depends, therefore, on the gains of the transmitting and receiving antennas and on the distance between antennas measured in wavelengths. The denominator in Eq. (6.38) is often called the *space loss* in the link and is often given in decibels. We illustrate its use in what follows; then we discuss two common types of antennas.

EXAMPLE 6.14 **Space loss**

What is the space loss in dB for the communication link described in the previous two examples?

SOLUTION:
The distance normalized to wavelengths is

$$\frac{R}{\lambda} = \frac{16,100}{5.97 \times 10^{-2}} = 2.69 \times 10^5 \tag{6.39}$$

The space loss in dB is

$$dB = 10 \log\left(4\pi \frac{R}{\lambda}\right)^2 = 10 \log\left(4\pi \times 2.69 \times 10^5\right)^2 = 130.6 \text{ dB} \tag{6.40}$$

WHAT IF? What if the distance were doubled?[16]

---

[16] 136.6 dB.

Reflector-type antennas. Figure 6.26(a) shows a reflector-type antenna commonly used in microwave communication systems. The metallic reflector is parabolic in shape, and a "feed" at the focus of the parabola radiates or receives the radio waves. Power is coupled between the transmitter (or receiver) and the feed through a transmission line. In the transmit mode, a spherical wave is radiated by the feed in the direction of the reflector, and the reflector redirects the spherical wave into one direction, as shown in Fig. 6.29. In the receive mode, the incoming plane wave is reflected from the parabolic reflector and converted into a spherical wave that converges on the feed. The feed collects power from the incoming spherical wave.

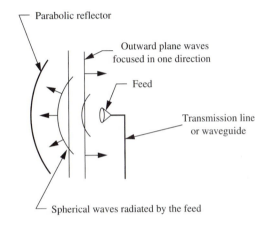

Figure 6.29 The parabolic reflector transforms the spherical waves radiated by the feed into plane waves focused in one direction. The plane waves eventually spread through diffraction into a beam.

A reflector antenna is most naturally described in terms of its effective area. A good rule of thumb is that about 50% of the incident energy that strikes the reflector is collected by the feed; the effective area is about

$$A_{eff} \approx \frac{1}{2} A_{geo} = \frac{1}{2}\left(\frac{\pi}{4}\right)D^2 = \frac{\pi D^2}{8} \tag{6.41}$$

where A_{geo} is the geometric area of the reflector and D is its diameter. Equation (6.35) gives the approximate gain of a reflector-type antenna as

$$G \approx \frac{4\pi}{\lambda^2} A_{eff} = \frac{\pi^2}{2}\left(\frac{D}{\lambda}\right)^2 \tag{6.42}$$

Thus, the gain of the antenna depends on the diameter measured in wavelengths. Equation (6.42) allows estimation of the gain of a microwave antenna based on its size and operating frequency.

EXAMPLE 6.15 **Antenna size**

Find the diameter of the antennas used in the examples on transmitting and receiving antennas. The antennas are parabolic reflectors with a 22-dB gain at a frequency of 5020 MHz.

SOLUTION:

The wavelength was calculated as 5.97 cm. Using Eq. (6.42), we find the diameter to be

$$10^{22/10} = \frac{\pi^2}{2}\left(\frac{D}{5.97 \text{ cm}}\right)^2 \Rightarrow D = 33.8 \text{ cm} \tag{6.43}$$

WHAT IF? What if the frequency were doubled? What would be the gain at the higher frequency?[17]

Wire-type antennas. Figure 6.26(b) shows a common wire-type antenna called a *dipole*. This antenna is used on TV receivers and is closely related to the telescoping antenna used on automobiles. For good antenna properties, the length of the antenna, ℓ, should be either one-half or a full wavelength of the electromagnetic wave.[18] Because the half-wave dipole is the smallest efficient antenna, this explains why audio signals must be modulated to higher frequencies for effective radiation. The half-wave dipole has a gain of about 1.5, and an input resistance of about 50 Ω. Thus, the effective area of this antenna is

$$A_{eff} = 1.5 \frac{\lambda^2}{4\pi} \tag{6.44}$$

array

Other antenna types. Multiple dipole antennas are often used in tandem arrangements called *arrays*. A common housetop TV antenna is an array antenna of several dipole antennas with slightly different lengths to increase the gain and bandwidth of the antenna. Often an AM station uses an array of two or more dipoles on separate towers to direct their signals toward populated areas.

An external antenna is avoided in an AM radio by using a small, multiturn coil wound on a ferrite rod. A horn antenna may be used for a microwave antenna, much as a megaphone is used by a cheerleader to focus his or her voice. The cornucopia antennas used in microwave relay systems combine a horn with a parabolic reflector, as indicated in Fig. 6.30(b).

Summary. We discussed the components of a communication system. Transmitting antennas radiate waves that are received by an antenna, which furnishes power to a radio receiver. The receiver amplifies and filters the signal, plus the inevitable noise, and

---

[17] 28 dB.

[18] For practical reasons, the length of an auto antenna is much less than the wavelength, which is about 300 m for AM signals. For this reason, auto antennas are inefficient.

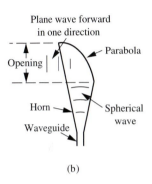

Plane wave forward
in one direction

Parabola

Opening

Horn

Spherical
wave

Waveguide

(b)

Figure 6.30 (a) A cornucopia horn; (b) antenna cross-section.

delivers the signal to the ultimate user of the information. We next consider two representative communication systems.

Check Your Understanding

1. A 0.5-μs pulse is transmitted down a RG-58 coaxial cable. How much space on the line does the pulse occupy at any given time?

2. One way that radio waves are bent is through refraction. True or false?

3. What is the wavelength of a radio wave with a frequency of 15 MHz?

4. An isotropic source radiates 10 W. What is the power density at a distance of 1 mile?

5. A wave with a power density of 2×10^{-9} W/m² falls on an antenna with an effective area of 10 m². How much power is available from the antenna to its receiver?

6. What is the space loss in decibels between two antennas that are 10^4 wavelengths apart?

7. What is the gain at 3000 MHz of a reflector-type antenna that has a diameter of 3 ft?

Answers. **(1)** 99.0 m; **(2)** true; **(3)** 20.0 m; **(4)** 0.307 μW/m²; **(5)** 2×10^{-8} W; **(6)** 63.3×10^{-12} or -102.0 dB; **(7)** 412.6, or 26.2 dB.

FM Broadcast Station

General description. Frequency modulation uses a wide bandwidth to achieve its remarkable noise-repression characteristics; specifically, a stereo FM station uses 200 kHz of RF bandwidth for an audio bandwidth of 50 to 15,000 Hz. The University of Texas at Austin operates an FM radio station, KUT, on a carrier frequency of 90.5 MHz. The station radiates approximately 15 kW of RF power.

Antenna system. The station employs an array of 12 helical antennas, radiating circular polarization. The Federal Communication Commission requires horizontal polarization for FM, as stated before, but permits circular polarization. Circular polarization is desirable because orientation of the receiver antenna becomes unimportant.

Frequencies exceeding about 20 MHz do not normally bounce off the ionosphere; hence, reception is limited to line-of-sight, extended somewhat by refraction and by diffraction over the radio horizon. The array of transmitting antennas directs power out toward the horizon. The gain of the array, considered as a single antenna, is about 6.4 (8.1 dB). The transmitter output is about 17 kW, of which 80% is radiated by the antenna; thus, the radiated power is 13.6 kW. The station announcers claim "100,000 watts," which presumably takes account of the antenna gain and normal media exaggeration.

Coverage. The antenna is located west of Austin on a tower that is 528 ft high. Assuming a perfect sphere, we calculate through simple geometry the distance, D, to the line-of-sight horizon to be

$$D = \sqrt{2R_{eq}h} \tag{6.45}$$

where h is the antenna height and R_{eq} is the equivalent radius[19] of the Earth, 1.3×3956 miles. Substitution into Eq. (6.45) yields a distance of about 32 miles to the radio horizon. The antenna tower is about 8 miles west of Austin on a high hill, which extends the radio horizon somewhat.

With Eq. (6.38), we may calculate the received power at the radio horizon. The wavelength is 300/90.5 m, and we assume a gain of 1.5 for the receiving antenna. Under these assumptions, the received power is 3.4×10^{-6} watts. Some noise would be broadcast with the signal; the FCC requires a 60-dB or better signal-to-noise ratio. Some noise would also be added to the signal due to thermal radiation of the Earth and atmosphere, but the main source of noise would be the receiver. We assume a standard noise figure of $F = 5$; hence, the noise power in the 200-kHz bandwidth of the receiver would be

$$N = k(F - 1)T_0B = 1.38 \times 10^{-23}(5 - 1) \times 290 \times 200 \times 10^3 \tag{6.46}$$
$$= 3.2 \times 10^{-15} \text{ W}$$

---

[19] The Earth's radius is increased by 30% to account for refraction.

where T_0 is the standard temperature (290 K) and the noise is referred to the input of the receiver. Thus, the signal-to-noise ratio is about 90.3 dB for a receiver at the radio horizon. This signal is much stronger than required for line-of-sight operation. The station is therefore extending coverage beyond the horizon.

Speed-Measuring Radar

Radar concepts. Radar is an acronym for *RA*dio *D*irection *A*nd *R*ange. A conventional radar sends out a pulse of power that reflects from a target such as an airplane and returns to the radar site. The direction to the target is determined from the directionality of the antenna, and the range to the target is determined from the time delay between transmission and reception of the pulse. Such radars are used for airplane traffic control, weather detection, and a variety of military and space applications.

A police radar measures neither direction nor range but speed. A continuous microwave signal is radiated from a directional antenna, bounces off a vehicle, and returns to the point of origin. The return signal is received with the same antenna and diverted to a receiver. The motion of the vehicle Doppler shifts the frequency of the return signal, allowing accurate determination of speed. In this section, we perform some representative calculations for such a radar.

Radar equation. Figure 6.31 pictures the situation. The power density at the vehicle would be given by Eq. (6.32), repeated here:

$$S = \frac{PG}{4\pi R^2} \text{ W/m}^2 \tag{6.47}$$

radar cross section   where G is the gain of the antenna, P is the power to the antenna, and R is the distance. The reflection from the target back toward the transmitter is described by the *radar cross section*, σ, which has the units of square meters. The radar cross-section includes two factors: the total power reflected and the directionality of that reflection toward the transmitter. Hence, the equivalent power reflected back toward the transmitter, P_{ref}, is

$$P_{ref} = \frac{PG\sigma}{4\pi R^2} \text{ W} \tag{6.48}$$

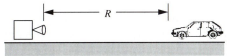

Figure 6.31   A speed-measuring radar.

This power weakens due to spreading and produces a power density at the transmitter/receiver of

$$S_{rec} = \frac{PG\sigma}{(4\pi R^2)^2} \text{ W/m}^2 \tag{6.49}$$

where S_{rec} is the power density at the transmitter/receiver. The power available from the antenna is

$$P_{av} = \frac{PG\sigma A_{eff}}{(4\pi R^2)^2} = \frac{\lambda^2 PG^2\sigma}{(4\pi)^3 R^4} \quad \text{W} \tag{6.50}$$

radar equation

where A_{eff} is the effective area of the transmitter/receiver antenna. The second form of Eq. (6.50) uses the relationship between effective area and gain from Eq. (6.35). Equation (6.50) is called the *radar equation* and relates the available power to the transmitted power, the radar cross section of the target, and the gain of the radar antenna.

Received power calculation. Modern police radars use a frequency of 24,150 MHz, which corresponds to a wavelength of 1.242 cm. The antenna is a horn with a 4-in. diameter. The gain of the antenna from Eq. (6.42) is about 330. The power source in the radar is a solid-state device, and the radiated power is a minimum of 10 mW. The radar cross section of an automobile, being an irregularly shaped object, is a statistical quantity because the incident energy reflects from many points on the surface and combines in a random fashion in various directions. We use a nominal value of 0.01 m². We assume a range of 100 yards (91.4 m). Substitution of these numbers into the radar equation, Eq. (6.50), yields an available power of

$$P_{av} = \frac{\lambda^2 PG^2\sigma}{(4\pi)^3 R^4} = \frac{(0.0124)^2(10 \times 10^{-3})(330.1)^2(0.01)}{(4\pi)^3(91.4)^4} \tag{6.51}$$

$$= 1.212 \times 10^{-14} \quad \text{W}$$

Assuming the receiver is matched to the antenna system, this is the received power, which we must compare with the noise in the system to determine the signal-to-noise ratio.

Signal-to-noise estimation. To determine the noise level in the receiver, we must know the noise figure and the bandwidth. Because the noise power out of the receiver is proportional to bandwidth, the minimum bandwidth is used. The minimum bandwidth depends in turn on the maximum speed that the system would expect to encounter in practice. We assume 150 mph (67.0 m/s) and use the Doppler equation:

$$\Delta f = f \times \frac{2v}{c} = 24.15 \times 10^9 \times \frac{2(67.0)}{3 \times 10^8} = 10.8 \text{ kHz} \tag{6.52}$$

This would be the bandwidth of our receiver. The noise in the receiver depends on the input noise from the antenna and the receiver noise figure, with the major contribution coming from the receiver. A worst-case standard noise figure for a receiver of this type is $F = 10$. Using Eqs. (6.20) and (6.17), we estimate the noise, referred to the receiver input, to be

$$N = k(F - 1)T_0 B = 1.38 \times 10^{-23}(10 - 1) \times 290 \times 10.8 \times 10^3 \tag{6.53}$$

$$= 3.89 \times 10^{-16} \text{ W}$$

The signal-to-noise ratio is thus about 31.2. This is not very impressive, and speeding tickets would be easy to discredit if this were the end of the story.

Equation (6.53) gives the signal-to-noise ratio if only one sample of the speed were obtained, but the radar would be able to collect many samples in a short time. Let us assume that the radar collects samples over 0.5 s before displaying the speed. This means that a counter determines the frequency of the return for 0.5 s, and then the results are displayed, scaled to indicate miles per hour. The correlation time of the receiver noise is approximately $1/B = 0.1$ ms, and hence the number of independent samples is approximately 5000. A well-known rule of statistics is that accuracy improves as the square root of the number of independent samples, which would be about 70 in this case. Thus, the signal-to-noise ratio of the speed measurement is about 2200, or 33.4 dB. This gives an accurate measurement of target speed.

CHAPTER SUMMARY

Telecommunications is at present a dynamic area of technical and economic growth. A world wide communication network for computers has emerged, and cellular pager and telephone coverage is spreading rapidly. Optical systems based on digital codes are replacing microwave systems based on analog techniques in all long-distance communication.

In this chapter, we considered the components of communication systems. We showed how nonlinear devices called modulators are used to shift communication signals to high frequencies for transmission, and we showed how similar principles are used in radio receivers to recover the information. These techniques have been adapted to optical systems and continue to be the basis of communication systems.

We described the properties of the free and guided EM waves used in communication systems. We defined the concepts used to describe the properties of transmitting and receiving antennas, and we used these to derive a communication equation for the power delivered to a radio receiver. Finally, we examined the details of two communication systems, emphasizing signal-to-noise calculations.

Objective 1: To understand the effects of nonlinear circuits on the spectra of signals. When two sinusoids are combined in a nonlinear circuit, all the harmonics of both and all sums and differences of the harmonics of both are created. This effect can be used to shift information in one band of frequencies to another. Modulation shifts information at low frequencies to high frequency for broadcasting or some other form of transmission. Mixing and demodulation shift the information back to lower frequency for application.

Objective 2: To understand the superheterodyne radio circuit and be able to relate RF, LO, and IF frequencies. The superheterodyne radio circuit combines several types of filtering with frequency shifting and amplification. The genius of the technique is the local oscillator–mixer–intermediate-frequency circuit.

Objective 3: To understand the properties of free and guided electromagnetic waves as they relate to communication systems. Free and guided electromagnetic waves are used to transport communication signals. Wave effects such as dispersion, diffraction, and scattering limit communication systems in various ways.

Objective 4: To understand the properties of antennas and their role in communication systems. Antennas couple between free and guided electromagnetic waves. In transmission, antennas are described by their gain and receiving antennas are described by their effective area. We derive a communication equation relating received power to transmitted power, antenna properties, and distance between transmitting and receiving antennas. Two practical systems are analyzed.

This chapter concludes the section dealing with electronic techniques. The next chapter introduces mathematical methods for system analysis that are used in all types of systems, especially control systems.

GLOSSARY

Antenna, p. 355, a structure that couples between a guided and a free electromagnetic wave.

Antenna gain, p. 355, the power density of an antenna compared to that of a lossless antenna that does not focus at all.

Array, p. 360, a group of antennas operating to focus radiation in one or more directions.

Attenuation, p. 354, the decreasing of the amplitude of an EM wave due to losses in the medium.

Available power, p. 345, the maximum power that a circuit is capable of delivering into a matched load.

Base band, p. 343, the audio band.

Blackbody radiation, p. 344, thermal radiation, which creates noise in electronic systems. For resistors, the blackbody radiation is called Johnson noise.

Carrier, p. 330, a high frequency signal that is modulated to bear information.

Characteristic impedance, p. 351, the ratio of the voltage to the current waves that move along a transmission line.

Communication system, p. 328, a communication system consists of a source of information with some sort of encoder, a recipient of information with the appropriate decoder, and a medium connecting the two.

Demodulation, p. 338, the process of converting the information in a modulated signal to a usable form.

Depolarization, p. 354, the tendency of an EM wave to lose pure polarization when reflected from an irregular surface.

Diffraction, p. 354, the spreading of an EM wave due to self-interference.

Dipole, p. 355, a simple antenna consisting of two flared wires or a signal wire above a ground plane.

Dispersion, p. 351, the tendency of waves at different frequencies to travel at different velocities, causing distortion in the spectra of communication signals.

Effective area, p. 357, a measure of the ability produced by an antenna to withdraw power from a passing EM wave.

Flicker or 1/f noise, p. 344, noise due to slow changes in device properties from aging, temperature changes, and other physical changes.

Image band, p. 342, the undesired frequency band that is mixed into the IF system, usually eliminated by RF filtering.

Interference, p. 344, man-made noise.

Intermediate frequency (IF), p. 340, the frequency band in a radio circuit, between the RF and base band, where most of the gain and filtering are accomplished.

Local oscillator (LO), p. 340, the part of a radio circuit that produces a frequency that mixes with the RF signal to produce the IF signal.

Mixing, p. 338, the combining of two frequencies in a nonlinear device to produce sum and different frequencies, usually used to shift signals to lower frequencies.

Modulation, p. 330, the process by which information at a low frequency is encoded by some scheme as part of a high-frequency signal.

Noise, p. 344, any undesired signal in the system, whether natural or man-made.

Operating noise figure, p. 346, the input signal-to-noise ratio divided by the output signal-to-noise ratio.

Polarization, p. 353, the direction in space of the electric field for an EM wave.

Radar cross section, p. 363, a measure of an object to reflect EM waves in the direction from which they arrive.

Radar equation, p. 363, an equation that relates the available power at a receiving antenna to the transmitted power, the radar cross section of the target, distance to the target, and the gain of the antennas.

Reflection, p. 354, the bouncing of an EM wave off an object.

Reflection, p. 351, wave reflection occurs when the impedance of the load at the receiving end of a transmission line differs from the characteristic impedance of the line.

Refraction, p. 354, the bending of an EM wave due to variation in the medium of propagation.

RF, p. 335, radio frequency, the carrier and its sidebands.

Shot noise, p. 344, noise due to the discrete carriers that constitute a current.

Sidebands, p. 334, information bearing bands of frequency produced in amplitude modulation.

Single sideband, p. 335, a form of amplitude modulation that transmits only one sideband and no carrier.

Space loss, p. 358, the loss in energy of a wave due to its spherical spreading; also the term in the communication equation that represents this loss.

Standard noise figure, p. 347, a measure of the noise generated in a radio receiver, in which the internally generated noise is compared to the thermal noise of a matched resistor at a temperature of 290 K.

Superheterodyne receiver, p. 343, the most common radio circuit, which uses a local oscillator to mix the radio-frequency signal to an intermediate frequency for amplification and filtering.

System temperature, p. 345, a measure of the internally generated noise in a radio receiver, equal to the temperature of a matched input resistor producing as much noise as is generated internally.

Transmission line, p. 350, a means for guiding waves of EM energy from source to load.

Wave velocity, p. 351, the speed of wave motion in a transmission line; not always the energy velocity.

White noise, p. 345, broadband noise, usually of thermal origin.

PROBLEMS

Section 6.1: Radio Principles

6.1. A nonlinear circuit has input frequencies at 300 and 200 Hz. How many output frequencies are there below 1000 Hz, and what are these frequencies?

6.2. The input of a *linear* system contains frequencies of 40 and 100 Hz. Which of the following can be in the output: 0, 70, 100, 120, 210, and 270 Hz? Which of these can be in the output for a *nonlinear* system of the same input frequencies?

6.3. In a superheterodyne radio receiver, the receiver bandwidth is 100 to 101 MHz and the image bandwidth is 120 to 121 MHz. What is the LO frequency? What is the IF?

6.4. The AM radio band of carrier frequencies is 540 to 1600 kHz and the standard IF is 455 kHz. Calculate the image frequency when the radio is tuned to the bottom of the band to receive a station broadcasting with a carrier of 540 kHz. Is this image in the AM band?

6.5. An AM receiver has 10 channels, each 100 kHz wide, covering the bandwidth from 16 to 17 MHz. The IF is 2.6 MHz, with a 100-kHz bandwidth. Assume the LO frequency is below the RF. The LO is not continuously tuned, but each LO frequency is generated from a harmonic of a lower frequency, so the LO frequency is changed by selecting and amplifying the appropriate harmonic of a low-frequency oscillator.
  (a) What is the minimum LO frequency?
  (b) What is the maximum frequency of the low-frequency oscillator from which the LO frequencies are derived?
  (c) What are the harmonic numbers that must be selected to tune the receiver to all 10 channels?
  (d) If the receiver receives digital pulses, what is the approximate baud rate it can receive?

6.6. For the simplest design of the local oscillator (LO) in a radio, the ratio of the maximum to minimum LO frequencies should be as low as possible (that is, the percent tuning range of the LO should be minimum). To show why the LO is placed above the RF band, calculate the ratio f_{max}/f_{min} for the LO both above and below the RF in the standard AM radio.

6.7. For a standard AM superheterodyne radio receiver tuned to the station with a carrier of 1600 kHz find the following:

  (a) What is the IF frequency?
  (b) What is the LO frequency?
  (c) What is the IF bandwidth?
  (d) What is the maximum RF bandwidth?
  (e) Where does most of the gain occur?
  (f) Where does most of the filtering occur?
  (g) How is the receiver tuned to another station?

6.8. A superheterodyne radio receiver has the following characteristics: the antenna receives an RF spectrum of 10.160 MHz $\pm$ 32 kHz, with a voltage level of 4 μV; the RF amplifier has a gain of $+15$ dB; the mixer has a gain of -12 dB; the IF amplifier has a center frequency of 1.5 MHz and a gain of $+120$ dB; the second detector has a gain of -16 dB (based on audio voltage vs. IF voltage); and the audio amplifier has a gain of $+20$ dB. Find the following:
  (a) the LO frequency, assuming that the LO is above the RF frequency.
  (b) the IF bandwidth required.
  (c) the audio outuput voltage.
  (d) the maximum bandwidth of the RF amplifier.

6.9. The FCC has allocated the band from 88 to 108 MHz for FM broadcasting, with 200 kHz for each station. The intermediate-frequency amplifier in an FM receiver has a center frequency of 10.7 MHz and a bandwidth of 200 kHz.
  (a) What is the number of FM stations that can be assigned different carrier frequencies in the total FM band?
  (b) What range is required for the frequency of the local oscillator in an FM receiver, assuming that the LO is located above the RF frequency?
  (c) What is the maximum frequency that might be mixed into the IF passband from the image band of the mixer?

6.10. An FM radio receiver receives carriers at frequencies of 88.1, 88.3, . . . , 107.7, and 107.9 MHz. The RF bandwidth required by each station for the FM information is 200 kHz, the IF is 10.7 MHz, and the audio bandwidth is 50 to 15,000 Hz.
  (a) Find the lowest and highest LO frequency. State assumptions.
  (b) Find the image frequency of the first detector when the LO is tuned to its lowest frequency.

(c) What would be an appropriate bandwidth for the IF amplifier?

(d) What would be the highest frequency amplified by the audio amplifier?

6.11. An amplifier with a 290-K matched resistor at its input produces a certain amount of noise at its output. When the resistor is cooled to 77 K, the output noise power is observed to diminish by 15%. What is the system temperature of the amplifier?

6.12. A communication system has a bandwidth of 5 kHz and an output power spectrum of 2×10^{-4} (V)^2/Hz

Section 6.2: Electromagnetic Waves

6.14. The distributed inductance and capacitance of a lossless vacuum-filled coaxial transmission line are

$$ L = \frac{\mu_0}{2\pi} \ln\left(\frac{b}{a}\right) \text{ H/m} \quad \text{and} \quad C = \frac{2\pi\varepsilon_0}{\ln(b/a)} \text{ F/m} $$

where $b =$ inner radius of outer conductor and $a =$ outer radius of the inner conductor; $\mu_0 = 4\pi \times 10^{-7}$ H/m, and $\varepsilon_0 = 8.854 \times 10^{-12}$ F/m.
(a) Find the wave velocity in the cable.
(b) Determine the characteristic impedance if $b = 4a$.

6.15. A certain 60-Hz power transmission line has an inductive reactance of 1.4 Ω/mile and a capacitive susceptance of $3.0\,\mu\text{℧}$/mile.
(a) Convert these to H/m and F/m, as required for the calculation of the wave velocity and characteristic impedance.
(b) Determine the wave velocity and characteristic impedance of this transmission line.

6.16. Consider the Sun as an isotropic radiator of energy. The power density at the surface of the Earth is about 1000 W/m^2. Determine the total power radiated by the Sun, assuming 92.8 million miles distance between Sun and Earth.

6.17. An antenna with a gain of 12 dB radiates 10 W of power. What is the power density at 1-km distance?

6.18. The effective area of an antenna at 3000 MHz is 5 m^2. What is its gain in dB at that frequency?

6.19. Two half-wave dipoles, each having a gain of 1.3, are separated by 100 m. A 10-W signal at 100 MHz is transmitted by one antenna.
(a) What is the available power at the receiving antenna?

into a matched 100-Ω resistor. What is the rms current in the resistor?

6.13. A satellite receiver has a standard noise figure of 1.2. Its antenna looks at the cold sky and provides an input temperature, T_{in}, of 30 K.
(a) Find the operating noise figure of the receiver.
(b) If the satellite transmits with a signal-to-noise ratio of 70 dB, find the output signal-to-noise ratio of the receiver. The received power from the satellite is 6×10^{-8} watts in the 4-MHz bandwidth.

(b) What is the open-circuit rms voltage at the receiving antenna, assuming a 50-Ω output impedance?

6.20. A 10-ft-diameter satellite TV receiving antenna operates over a frequency range of 3700 to 4200 MHz. What is the maximum and minimum gain in decibels over this range of frequency?

6.21. An antenna produces at a certain point a wave intensity of 10^{-7} W/m^2, whereas an isotropic antenna gives a wave intensity of 10^{-8} W/m^2. What is the gain of the antenna?

6.22. The beam width of a reflector-type antenna describes the angular width of the region in space into which it focuses its radiated energy. Using Eqs. (6.32) and (6.42) and conservation of energy, show that the conical beam width (θ) of a reflector-type antenna is approximately

$$ \theta = \frac{4}{\pi}\sqrt{2}\left(\frac{\lambda}{D}\right) \text{ radians} $$

where D is the antenna diameter and λ is the wavelength.

6.23. Using Eq. (6.42), find the gain of a 6-ft-diameter reflector antenna at 500 MHz in dB.

6.24. A communication satellite transmitter radiates 10 watts of power. The satellite antenna focuses the power to cover the entire USA, exclusive of Alaska and Hawaii, more or less uniformly. (Assume a size of 1000 miles × 3000 miles, and a satellite height of 22,000 miles.)
(a) Estimate the average power density on the surface.
(b) The frequency is 6 GHz. Estimate the size of the satellite antenna.

Section 6.3: Examples of Communication Systems

6.25. For the FM station example, calculate the received power at the radio horizon (32 miles) and confirm the S/N of 90.3 dB, as stated in the text, page 363.

6.26. If a ship-borne radar must detect any target out to 10 miles, how high does the radar antenna have to be relative to the level of the sea?

6.27. It is desired to increase the range of an air-control radar by a factor of 50%.

    (a) If only the transmitted power is increased, what is the required percent of increase in power?

(b) If, rather than increasing the power, the reflector-type antenna is replaced by a larger antenna, what is the percent of increase in antenna diameter?

6.28. What would be the signal-to-noise ratio for the police radar if the device measured the speed for 0.3 s instead of 0.5 s?

6.29. Confirm that the gain of the police radar antenna described on page 364 is about 330.

Answers to Odd-Numbered Problems

6.1. $0 = dc$, $100 = 300 - 200$, 200, 300, $400 = 2 \times 200$, $500 = 200 + 300$, $600 = 2 \times 300$, $700 = 300 + 2 \times 200$, $800 = 4 \times 200$, $900 = 3 \times 300$.

6.3. LO = 110.5 MHz, IF = 10 MHz.

6.5. **(a)** 13.45 MHz; **(b)** 50 kHz; **(c)** all odd harmonics between 269th and 287th; **(d)** 50 kbaud.

6.7. **(a)** 455 kHz; **(b)** 2055 kHz; **(c)** 10 kHz; **(d)** 910 kHz; **(e)** in the IF amplifier; **(f)** in the IF amplifier; **(g)** by changing the LO frequency.

6.9. **(a)** 100; **(b)** 98.8 MHz < LO < 118.8 MHz; **(c)** 129.4 MHz.

6.11. 1130 K.

6.13. **(a)** 2.93 (4.67 dB); **(b)** 1.38 (1.40 dB).

6.15. **(a)** 2.31 μH/m and 4.95 pF/m; **(b)** 296,000 km/s and 683 Ω.

6.17. 12.6 μW/m^2.

6.19. **(a)** 9.63×10^{-5} W; **(b)** 0.139 V, rms.

6.21. 10 (10 dB).

6.23. 45.8 (16.6 dB).

6.25. 90.3 dB.

6.27. **(a)** 406%; **(b)** 50%.

6.29. 330.

7

Linear Systems

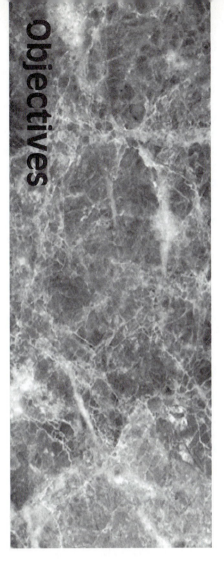
1. To understand how to recognize the class of functions that can be described by complex frequencies

2. To understand how to use generalized impedance to find the forced and natural response of a system

3. To understand how to recognize conditions for undamped, underdamped, critically damped, and overdamped responses in second-order systems

4. To understand how to use system notation to describe the properties of composite systems

5. To understand the role of loop gain in the dynamic response of a feedback system

This chapter on linear systems expands in several directions the tools for analysis of circuits presented in Chapter 1. The concept of frequency is generalized, and we expand frequency-domain techniques to a wide class of behavior. The results are then generalized to include systems with nonelectrical components and systems with feedback.

Introduction to Linear Systems

What is a system? A *system* consists of several components that together accomplish some purpose. For example, an automobile has a motor, steering mechanism, lights, padded seats, entertainment system, and more, operating together to give safe and pleasant transportation. Likewise, a stand-alone ac generator requires a control system to regulate the frequency and voltage of its output. Often, systems are modeled with linear equations. The analysis of such linear systems has furnished a powerful language for system description that builds on our earlier study of the frequency domain. This chapter introduces system models and explores basic techniques of linear system description and analysis.

Contents of this chapter. We begin by generalizing the concept of frequency. We then introduce the language of system notation by defining the generalized impedance of electrical circuits. From this impedance, we determine the natural frequencies and natural response of electrical circuits, and we then investigate the transient response of first- and second-order circuits. Next, this viewpoint is applied to a composite system involving electrical and thermal components, an oven. Finally, we regulate the oven with a feedback system and investigate the transient response and dynamic stability of the feedback system.

7.1 COMPLEX FREQUENCY

complex frequency

Definition of complex frequency. We resume our exploration of the frequency domain through the following definition of complex frequency. A time-domain variable, say, a voltage, is said to have a *complex frequency* $\underline{s}$ when it can be expressed in the form given in Eq. (7.1)

$$v(t) = \text{Re }\{\underline{V}e^{\underline{s}t}\} \tag{7.1}$$

IDEA 8
The Frequency Domain

where $\underline{V}$ is a complex number, a phasor, and

$$\underline{s} = \sigma + j\omega \quad \text{s}^{-1} \tag{7.2}$$

where $\underline{s}$, σ, and ω all have units of inverse seconds, s^{-1}. Equation (7.1) is a slightly modified version of, say, Eq. (1.98), except that frequency is now a complex number.

LEARNING OBJECTIVE 1.

To understand how to recognize the class of functions that can be described by complex frequencies

---

Functions that can be represented by complex frequencies. Before we explore the implications of complex frequency, we wish to relate Eq. (7.1) to familiar results. We have in previous chapters introduced several important functions that can be represented by complex frequencies.

When $\underline{s}$ is zero. When the complex frequency in Eq. (7.1) is zero, $\underline{s} = 0$, the voltage is a constant, or a dc voltage:

$$v(t) = \text{Re}\{\underline{V}e^{0t}\} = V_{dc} \qquad \text{for } \underline{s} = 0 \tag{7.3}$$

When $\underline{s}$ is real and negative. When the complex frequency $\underline{s}$ is real, $\underline{s} = \sigma$, the time-domain voltage is

$$v(t) = \text{Re}\{\underline{V}e^{\sigma t}\} = Ae^{\sigma t} \qquad \text{for } \underline{s} = \sigma \tag{7.4}$$

where A is a constant. For negative σ, the voltage is a decreasing exponential function, such as we encountered in first-order transient problems in Chap. 1. Specifically, the time constant is the negative of the reciprocal of σ:

$$\sigma = -\frac{1}{\tau} \quad \text{where} \quad \tau = R_{eq}C \quad \text{or} \quad \frac{L}{R_{eq}} \tag{7.5}$$

EXAMPLE 7.1 | **Transient problem**

Find the complex frequency representing the response of the circuit in Fig. 7.1.

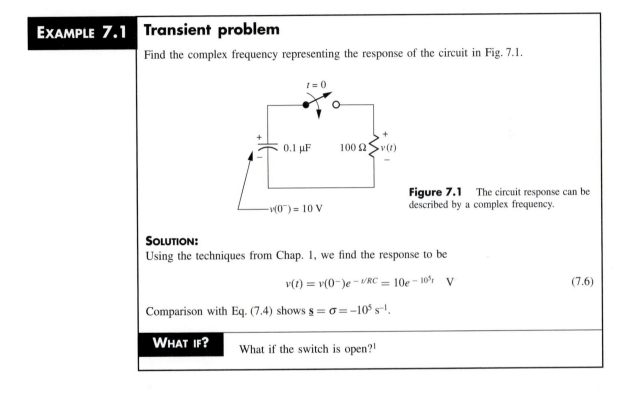

Figure 7.1 The circuit response can be described by a complex frequency.

SOLUTION:
Using the techniques from Chap. 1, we find the response to be

$$v(t) = v(0^-)e^{-t/RC} = 10e^{-10^5 t} \quad \text{V} \tag{7.6}$$

Comparison with Eq. (7.4) shows $\underline{s} = \sigma = -10^5 \text{ s}^{-1}$.

WHAT IF? | What if the switch is open?[1]

When $\underline{s}$ is real and positive. If σ is positive in Eq. (7.4), the voltage is an increasing exponential, which we have not encountered before. Figure 7.2 shows the time functions that result from zero and real complex frequencies. Thus, a complex frequency that is real includes the response that we studied as a transient solution in Chap. 1, except that we now include growing as well as decaying exponentials. Later we show that a growing exponential represents a possible response of an unstable system.

When $\underline{s}$ is pure imaginary. When the complex frequency is pure imaginary, Eq. (7.1) takes the form of a sinusoidal function, such as we studied in Chap. 1. In this case,

---

[1] Then the only possible response is $\underline{s} = 0$. In effect the resistance becomes infinite, but a constant voltage and a constant current (zero current) are still possible.

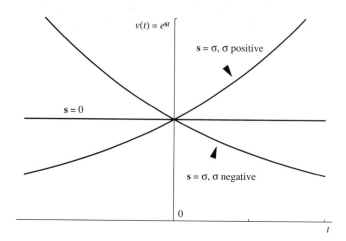

Figure 7.2 Time-domain responses for complex frequencies that are real.

$$v(t) = \text{Re}\{\underline{\mathbf{V}}e^{j\omega t}\} = V_p \cos(\omega t + \theta) \tag{7.7}$$

where $\underline{\mathbf{V}} = V_p e^{j\theta}$, with V_p the peak value and θ the phase of the sinusoidal function. Imaginary complex frequencies are used in the analysis of ac circuits.

Sinusoid = two complex frequencies. From another point of view, we may represent a sinusoidal function by two complex frequencies. This is implied by the "real part" operation in Eq. (7.1), because an alternate way to express the real part is through the identity in Eq. (7.8)

$$\text{Re}\{\underline{\mathbf{z}}\} = \tfrac{1}{2}(\underline{\mathbf{z}} + \underline{\mathbf{z}}^*) \tag{7.8}$$

where $\underline{\mathbf{z}}^*$ is the complex conjugate of $\underline{\mathbf{z}}$. Thus, Eq. (7.7) can be expressed in the form

$$V_p \cos(\omega t + \theta) = \text{Re}\{\underline{\mathbf{V}}e^{j\omega t}\} = \frac{\mathbf{V}}{2}e^{j\omega t} + \frac{\mathbf{V}^*}{2}e^{-j\omega t} \tag{7.9}$$

When we compare Eq. (7.9) with Eq. (7.1), we see that two complex frequencies, $\underline{\mathbf{s}} = j\omega$ and $\underline{\mathbf{s}} = -j\omega$, express a sinusoidal function. In our consideration of complex frequency, we often find this second point of view to be useful.

Frequency and complex frequency. We have a slight semantic problem in speaking of complex frequency. When the complex frequency $\underline{\mathbf{s}}$ is imaginary, the frequency, ω, is said to be real. Thus, a sinusoidal function has a real frequency but is described by a complex frequency that is imaginary. As shown before, a complex frequency that is real corresponds to a growing or decaying exponential function.

General interpretation of complex frequency. We now consider the meaning of $e^{\underline{\mathbf{s}}t}$, with $\underline{\mathbf{s}} = \sigma + j\omega$. In general

$$\begin{aligned}
v(t) &= \text{Re}\{\underline{\mathbf{V}}e^{\underline{\mathbf{s}}t}\} = \text{Re}\{\underline{\mathbf{V}}e^{(\sigma+j\omega)t}\} \\
&= e^{\sigma t}\text{Re}\{\underline{\mathbf{V}}e^{j\omega t}\} = V_p e^{\sigma t}\cos(\omega t + \theta)
\end{aligned} \tag{7.10}$$

Equation (7.10) expresses a time function that combines sinusoidal behavior with the

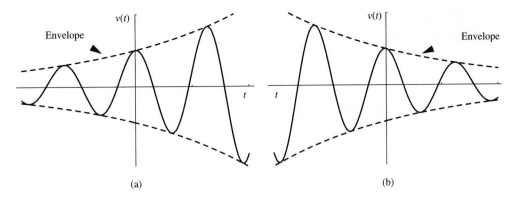

Figure 7.3 The function $\mathrm{Re}\{e^{(\sigma+j\omega)t}\}$ for (a) σ positive and (b) σ negative.

exponential behavior that we hitherto have associated with transients. Equation (7.10) also can be considered a sinusoidal function in which the peak value of the sinusoid changes exponentially with time. Figure 7.3 shows the character of the time function in Eq. (7.10).

The s-plane. A complex number, $\underline{s}$, can be represented by a point in the complex plane. In Fig. 7.4, we show such an $\underline{s}$-plane and identify the regions of complex frequency corresponding to associated time responses.

- The origin corresponds to a constant or dc function.
- Complex frequencies on the real axis correspond to growing and decaying exponential functions.
- Complex frequencies on the imaginary axis correspond to sinusoidal functions.
- The region to the right of the vertical axis, the right-half plane, corresponds to sinusoids that are growing exponentially.
- The left-half plane corresponds to sinusoids that are decreasing exponentially.

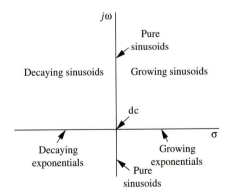

Figure 7.4 The $\underline{s}$-plane and associated time functions.

EXAMPLE 7.2 — Complex frequency

What complex frequency corresponds to the function

$$v(t) = 100e^{1000t} \sin(400\pi t + 75°) \text{ V} \tag{7.11}$$

SOLUTION:

There are two frequencies: $\sigma = 1000$ comes from the exponential part and $j\omega = \pm j400\pi$ comes from the sinusoidal part. Thus,

$$\underline{s}_1 = 1000 + j400\pi \quad \text{and} \quad \underline{s}_2 = 1000 - j400\pi \tag{7.12}$$

WHAT IF? What if the function were cosine instead of sine?[2]

IDEA 8 — The Time Domain

IDEA 7 — The Frequency Domain

eigenfunction

Time domain and frequency domain. The introduction of complex frequency expands our concept of the frequency domain to cover a wider class of time-domain behavior. We now have a frequency-domain representation of exponentially growing or decreasing time functions that may oscillate. As hinted before, this expansion will allow frequency-domain techniques to be applied to transient problems as well as sinusoidal steady-state problems such as we studied in Chap 1. Furthermore, these new functions allow the study of a wider class of transient problems.

Importance of $e^{\underline{s}t}$. The function $e^{\underline{s}t}$ is a mathematical probe[3] we use to investigate the properties of linear systems. At this point, we wish to look ahead and anticipate later results. We find that all linear systems are characterized by certain complex frequencies. Once we determine these frequencies for a given system, say, a feedback amplifier, we can determine from them the system response in the time domain or the frequency domain. We may thereby examine transient and steady-state responses, and we can determine if the system is stable or unstable. In the following section we show how to determine these characteristic frequencies and how to derive from them the time-domain response of the system.

Summary. In this section, we have generalized frequency to include growing or decreasing exponentials and exponentially growing or decreasing sinusoids. Such complex frequencies allow frequency-domain techniques to be applied to a wide class of systems. The characteristics of linear systems are often described in terms of complex frequency.

Check Your Understanding

1. What complex frequency or frequencies describe(s) an RC transient with a dc source if $R = 100 \ \Omega$ and $C = 10 \ \mu\text{F}$?

---

[2] The "sin/cos" and "75°" affect the phase but not the complex frequency.

[3] The mathematical term $e^{\underline{s}t}$ for is the *eigenfunction* for a linear DE.

2. What is the time between zero crossings for a function described by the complex frequency $\underline{s} = -2 + j10$?

3. Can pure sinusoid be described by a complex frequency that is pure imaginary, two complex frequencies that are pure imaginary, either, or neither.

4. Complex frequencies near the origin in the $\underline{s}$-plane describe functions that vary slowly with time. True or false?

Answers. **(1)** 0 (for the source) and -1000 s^{-1} (for the circuit); **(2)** 0.314 s; **(3)** either, depending on context; **(4)** true.

Generalized Impedance

generalized impedance

Impedance of R, L, and C. We may use complex frequency to generalize the concept of impedance. We use the technique presented in Chap. 1, where only real frequency was considered. As before, we argue that $e^{\underline{s}t}$ is a function that is indestructible to linear operations such as addition, differentiation, and integration. To determine the impedance, therefore, we excite a circuit with a voltage Re $\{\underline{V}(\underline{s})e^{\underline{s}t}\}$ and calculate a response, Re$\{\underline{I}(\underline{s})e^{\underline{s}t}\}$.

The *generalized impedance* is defined as

$$\underline{Z}(\underline{s}) = \frac{\underline{V}(\underline{s})e^{\underline{s}t}}{\underline{I}(\underline{s})e^{\underline{s}t}} \; \Omega \qquad (7.13)$$

We illustrate with an inductor, as shown in Fig. 7.5. The equations are

$$v(t) = L\frac{d}{dt}i(t) \Rightarrow \underline{V}(\underline{s})e^{\underline{s}t} = L\frac{d}{dt}\underline{I}(\underline{s})e^{\underline{s}t} = \underline{s}L\underline{I}(\underline{s})e^{\underline{s}t} \qquad (7.14)$$

where we have omitted the "real part of" for simplicity. The differentiation is performed only on the $e^{\underline{s}t}$ function because this is the only function of time. Thus the generalized impedance of an inductor is

$$\underline{Z}_L(\underline{s}) = \frac{\underline{V}(\underline{s})e^{\underline{s}t}}{\underline{I}(\underline{s})e^{\underline{s}t}} = \underline{s}L \; \Omega \qquad (7.15)$$

In like manner, we can establish the generalized impedances of resistors and capacitors:

$$\underline{Z}_R(\underline{s}) = R \; \Omega, \qquad \text{and} \qquad \underline{Z}_C(\underline{s}) = \frac{1}{\underline{s}C} \; \Omega \qquad (7.16)$$

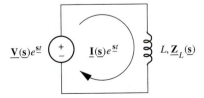

$\underline{V}(\underline{s})e^{\underline{s}t} \qquad \underline{I}(\underline{s})e^{\underline{s}t} \qquad L, \underline{Z}_L(\underline{s})$

Figure 7.5 An inductor. excited by $\underline{V}e^{\underline{s}t}$.

Application of generalized impedance. We now analyze the *RL* circuit shown in Fig. 7.6(a). In the frequency domain, Fig. 7.6(b), the impedances combine like resistors at dc:

$$\mathbf{Z}(\underline{s}) = 2\|(1 + \underline{s}/2) = \cfrac{1}{\cfrac{1}{2} + \cfrac{1}{1 + \underline{s}/2}} = \frac{2(\underline{s} + 2)}{(\underline{s} + 6)} \ \Omega \qquad (7.17)$$

We use the complex impedance function in Eq. (7.17) to explore the transient response of the circuit.

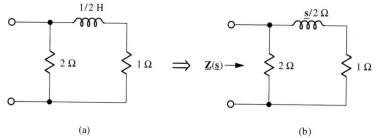

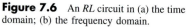

(a)   (b)

Figure 7.6 An *RL* circuit in (a) the time domain; (b) the frequency domain.

Transient Analysis

Forced and natural response.

When we excite a circuit with a voltage or current source, we force a certain response from the circuit. For example, if we apply a dc source, we expect a dc response. But, as we saw in Chap. 1, part of the transient response of the circuit takes a form that is natural to, and determined by, the circuit itself. Thus, the response is always of the form, say, for a current:

$$i(t) = i_f(t) + i_n(t) \qquad (7.18)$$

where $i_f(t)$ is the forced response and $i_n(t)$ is the natural response. We now explore the role of complex frequency and the impedance function in determining the forced and natural responses of a linear circuit.

Forced response from impedance.

The forced response of the circuit may be determined directly from the impedance function when the forcing function, input current or voltage, is of the form $e^{\underline{s}t}$. We evaluate the impedance function at the value of $\underline{s}$ of the forcing function. For example, if the voltage is a known function with a complex frequency $\underline{s} = \underline{s}_f$, the current must be of the same form and can be determined from Eq. (7.19)

$$\frac{\mathbf{V}_f e^{\underline{s}_f t}}{\mathbf{I}_f e^{\underline{s}_f t}} = \mathbf{Z}(\underline{s}_f) \ \Rightarrow \ \mathbf{I}_f e^{\underline{s}_f t} = \frac{\mathbf{V}_f}{\mathbf{Z}(\underline{s}_f)} \times e^{\underline{s}_f t} \qquad (7.19)$$

where $\mathbf{V}_f$ and $\mathbf{I}_f$ are phasors describing the source and response, respectively. The response in the time domain is thus

$$i_f(t) = \text{Re}\{\mathbf{I}_f e^{\mathbf{s}_f t}\} = \text{Re}\left\{ \frac{\mathbf{V}_f}{\mathbf{Z}(\mathbf{s}_f)} e^{\mathbf{s}_f t} \right\} \qquad (7.20)$$

When the exciting function is not of the form e^{st}, the forced response must be determined by other methods, such as Laplace transform theory.

EXAMPLE 7.3 **Forced response with a voltage source**

For the circuit in Fig. 7.6, a voltage source is connected to the input, $v_f(t) = 2e^{-t}$ for all time. Find the forced current response at the input.

SOLUTION:
The complex frequency of the input is $\mathbf{s}_f = -1 \text{ s}^{-1}$. The current must be of the form $I_f e^{-t}$, where I_f is a constant. From Eqs. (7.19) and (7.17),

$$\frac{2e^{-t}}{I_f e^{-t}} = \mathbf{Z}(-1) = \frac{2(-1+2)}{-1+6} = 0.4 \Rightarrow I_f = \frac{2}{0.4} = 5 \text{ A} \qquad (7.21)$$

Thus, $i_f(t) = 5e^{-t}$ A. This is the forced response of the circuit to the exponential voltage source.

WHAT IF?   What if $v(t) = 2e^{-2t}$?[4]

EXAMPLE 7.4 **Forced response with a current source**

Find the forced response of the circuit in Fig. 7.6 if excited by a current source with $i_f(t) = 1.5 \cos(2t + 42°)$.

SOLUTION:
The input can be represented by a phasor $i_f(t) = \text{Re}\{1.5 \angle 42° e^{j2t}\}$, so $\mathbf{s}_f = j2$. Thus, the phasor representing the forced voltage response is

$$\mathbf{V}_f = \mathbf{I}_f \times \mathbf{Z}(j2) = 1.5 \angle 42° \times \frac{2(j2+2)}{j2+6} = 1.34 \angle 68.6° \qquad (7.22)$$

Thus, the voltage produced in steady state is $v_f(t) = 1.34 \cos(2t + 68.6°)$ V. All this should look familiar to you because this is merely ac circuit analysis such as we presented in Chap. 1.

Finding the natural response. The impedance function can also assist us in finding the natural response of the circuit. Consider the case of a circuit excited by a voltage source and we wish to determine the current. The natural response of a linear circuit or system must be of the form

---

[4] Then $I_f = \infty$ by this method. More advanced methods must be used in that case.

$$i_n(t) = \text{Re}\{\mathbf{\underline{I}}_n e^{\mathbf{\underline{s}}_n t}\} \tag{7.23}$$

where $\mathbf{\underline{I}}_n$ is an unknown phasor and $\mathbf{\underline{s}}_n$ is the natural frequency of the system. We can determine the natural frequency from the impedance function. The natural frequency corresponds to the case where the exciting voltage is zero but the current is nonzero, being established solely by the circuit. The impedance function thus becomes

$$\mathbf{\underline{Z}}(\mathbf{\underline{s}}) = \frac{\mathbf{V}(\mathbf{\underline{s}})e^{\mathbf{\underline{s}}t}}{\mathbf{I}(\mathbf{\underline{s}})e^{\mathbf{\underline{s}}t}} = \frac{0}{\neq 0} \tag{7.24}$$

Equation (7.24) can be valid only if the natural frequency makes the impedance zero:

$$\mathbf{\underline{Z}}(\mathbf{\underline{s}}_n) = 0 \tag{7.25}$$

We illustrate with the circuit in Fig. 7.6 with the impedance function derived in Eq. (7.17). The circuit has only one natural frequency, which may be determined by setting the impedance to zero:

$$\mathbf{\underline{Z}}(\mathbf{\underline{s}}_n) = \frac{2(\mathbf{\underline{s}}_n + 2)}{(\mathbf{\underline{s}}_n + 6)} = 0 \implies \mathbf{\underline{s}}_n = -2 \tag{7.26}$$

The natural frequency in this first-order circuit corresponds to the negative of the reciprocal of the time constant. Because we are exciting the circuit with a voltage source, the time constant must be determined with the input shorted.

EXAMPLE 7.5 | **Transient with a voltage source**

Find the current input to the circuit in Fig. 7.6 if the voltage source input of Fig. 7.7 is applied as an input.

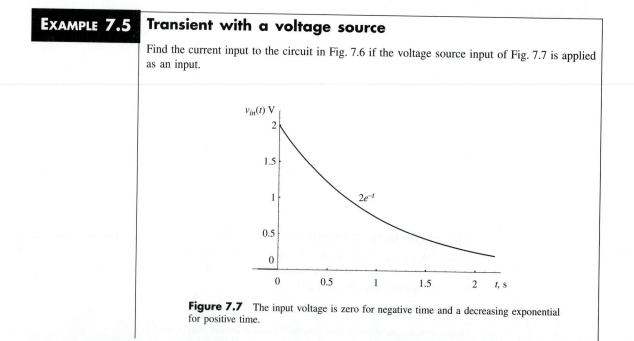

Figure 7.7 The input voltage is zero for negative time and a decreasing exponential for positive time.

SOLUTION:

The current consists of a forced and a natural component, as in Eq. (7.18). The forced component is derived in Eq. (7.21) and the natural response is of the form of Eq. (7.23) with $\underline{s}_n = -2$, as shown in Eq. (7.26). Thus, the current produced by the voltage in Fig. 7.7 must be of the form

$$i(t) = 5e^{-t} + Ae^{-2t} \tag{7.27}$$

where A is to be determined from the initial conditions. In this case, the initial current must be 1 A because the initial voltage is 2 V and the inductor acts as an open circuit.[5] Thus, A comes from Eq. (7.27) at $t = 0$:

$$1 = 5e^0 + Ae^0 \implies A = -4 \tag{7.28}$$

The input current is, therefore,

$$i(t) = 5e^{-t} - 4e^{-2t} \text{ A} \tag{7.29}$$

which is shown in Fig. 7.8.

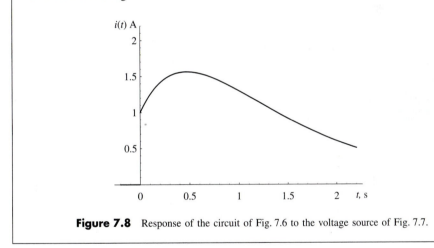

Figure 7.8   Response of the circuit of Fig. 7.6 to the voltage source of Fig. 7.7.

Open-circuit and short-circuit natural frequencies.

We showed that the value of $\underline{s}$ that makes the impedance function go to zero corresponds to the natural frequency with a turned-OFF voltage source at the input.[6] In Fig. 7.9, for example, the inductor sees an equivalent resistance of 1 Ω if the 2-Ω resistor is shorted; thus, the short-circuit time constant of the circuit is $L/R = 0.5$ s. This time corresponds to a natural frequency of $\underline{s}_n = -2$, Eq. (7.5).

Additional information can be derived from the value of $\underline{s}$ that makes the impedance function go to infinity. Infinite impedance corresponds to input current, $\underline{I}(\underline{s})$, going to zero, an open circuit, with input voltage, $\underline{V}(\underline{s})$, nonzero.

---

[5] See the discussion on p. 42.

[6] A turned-OFF voltage source is equivalent to a short circuit; see p. 19.

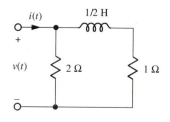

Figure 7.9 Circuit of Fig. 7.6(a) repeated.

$$\mathbf{Z}(\mathbf{s}) = \frac{\mathbf{V}(\mathbf{s})e^{\mathbf{s}t}}{\mathbf{I}(\mathbf{s})e^{\mathbf{s}t}} = \frac{\neq 0}{0} \qquad (7.30)$$

For Eq. (7.17), this requires

$$\frac{2(\mathbf{s} + 2)}{(\mathbf{s} + 6)} = \frac{\neq 0}{0} \quad \Rightarrow \quad \mathbf{s}_n = -6 \qquad (7.31)$$

This natural frequency corresponds to the negative of the reciprocal of the time constant of the circuit with the input open-circuited. The open circuit corresponds to exciting the circuit with a current source that is turned OFF for the natural response.

EXAMPLE 7.6 **Transient with a current source**

Find the voltage produced in the circuit of Fig. 7.6 if the current in Fig. 7.10 were applied.

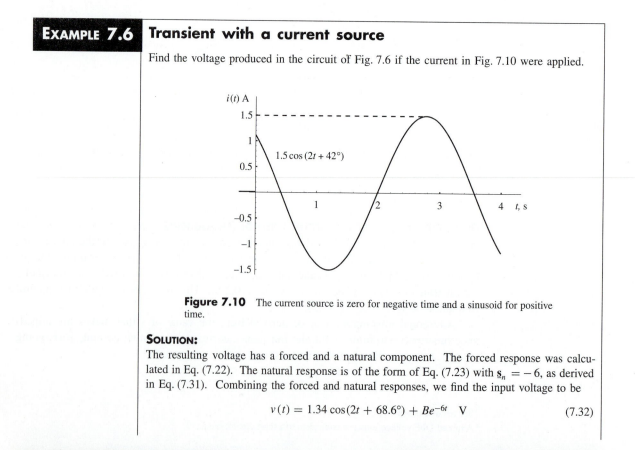

Figure 7.10 The current source is zero for negative time and a sinusoid for positive time.

SOLUTION:

The resulting voltage has a forced and a natural component. The forced response was calculated in Eq. (7.22). The natural response is of the form of Eq. (7.23) with $\mathbf{s}_n = -6$, as derived in Eq. (7.31). Combining the forced and natural responses, we find the input voltage to be

$$v(t) = 1.34 \cos(2t + 68.6°) + Be^{-6t} \quad \text{V} \qquad (7.32)$$

where B is a constant to be determined from the initial conditions. As argued before, the inductor acts initially as an open circuit, so the initial value of the voltage must be $1.5 \cos{(42°)}\text{A} \times 2\,\Omega = 2.23\,\text{V}$. From Eq. (7.32) at $t = 0$,

$$2.23 = 1.34 \cos{(0 + 68.6°)} + Be^0 \Rightarrow B = 1.74\,\text{V} \tag{7.33}$$

Hence, the voltage produced by the current in Fig. 7.10 in the circuit in Fig. 7.9 is

$$v(t) = 1.34 \cos{(2t + 68.6°)} + 1.74e^{-6t}\quad \text{V} \tag{7.34}$$

which is shown in Fig. 7.11.

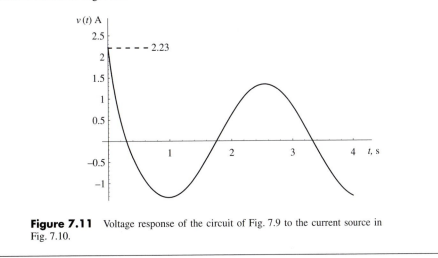

Figure 7.11 Voltage response of the circuit of Fig. 7.9 to the current source in Fig. 7.10.

Summary. The value of $\underline{s}_n$ that sets the impedance function to zero is called a *zero* of the impedance function and corresponds to the short-circuit natural frequency. The value of $\underline{s}_n$ that makes the impedance function go to infinity is a *pole* of the impedance function and corresponds to the open-circuit natural frequency of the circuit. These natural frequencies are normally called the zeros and poles of the impedance function.

zero,
pole

Poles and zeros. The impedance function of a complicated circuit generally takes the form of Eq. (7.17), except that the numerator and denominator are polynomials in $\underline{s}$. Thus, the function can always be factored into the form

$$\frac{\underline{V}(\underline{s})e^{\underline{s}t}}{\underline{I}(\underline{s})e^{\underline{s}t}} = \underline{Z}(\underline{s}) = K\frac{(\underline{s} - \underline{z}_1)(\underline{s} - \underline{z}_2)}{(\underline{s} - \underline{p}_1)(\underline{s} - \underline{p}_2)(\underline{s} - \underline{p}_3)} \tag{7.35}$$

where K is a constant, and we assumed a second-order polynomial in the numerator and a third-order polynomial in the denominator.[7] In Eq. (7.35), $\underline{z}_1$ and $\underline{z}_2$ are called the zeros of the function because these are the values of $\underline{s}$ at which the function goes to zero. Likewise, $\underline{p}_1$, $\underline{p}_2$, and $\underline{p}_3$ are called the poles of the function because these are the values of $\underline{s}$ at which the function goes to infinity.

---

[7] This indicates a circuit with three independent energy-storage elements.

Natural frequencies. As we have shown, the zeros correspond to the short-circuit natural frequencies and the poles correspond to the open-circuit natural frequencies of the circuit. Except for a multiplicative constant, the poles and zeros of the impedance function fully establish the behavior of the circuit. They give the natural frequencies, as we have illustrated before, and they also give directly the forced response to excitations of the form $e^{\underline{s}t}$, such as dc ($\underline{s} = 0$) or ac ($\underline{s} = j\omega$).

transfer function **Second-order circuit.** The circuit of Fig. 7.12 is identical to that in Fig. 7.6 except that a 2-H inductor is added at the input and the voltage across the 1-Ω resistor is taken as an output.[8] The ratio of the output and the input is the *transfer function*, $\underline{T}(\underline{s})$

$$\frac{\underline{V}_{out}(\underline{s})e^{\underline{s}t}}{\underline{V}_{in}(\underline{s})e^{\underline{s}t}} = \underline{T}(\underline{s}) \tag{7.36}$$

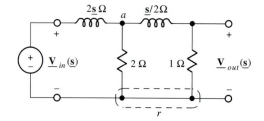

Figure 7.12 The same circuit as that in Fig. 7.6 with an added inductor. An output voltage is indicated because we will derive the transfer function.

The transfer function is like an impedance function except it relates circuit variables at different places in a system. The transfer function is a generalization of the filter functions presented in Chaps. 4 and 5.

Transfer-function derivation. We now derive the filter function with a nodal analysis. Because the output is open-circuited, we first determine the voltage at a and then find $\underline{V}_{out}(\underline{s})$ with a voltage divider. Kirchhoff's current law for node a in the frequency domain becomes

$$\frac{\underline{V}_a - \underline{V}_{in}}{2\underline{s}} + \frac{\underline{V}_a - (0)}{2} + \frac{\underline{V}_a - (0)}{\underline{s}/2 + 1} = 0 \tag{7.37}$$

Conservation of Charge which, after a bit of algebra, becomes

$$\underline{V}_a = \frac{\underline{s} + 2}{\underline{s}^2 + 7\underline{s} + 2} \times \underline{V}_{in} \tag{7.38}$$

Using a voltage divider relationship, we find the transfer function to be

$$\underline{T}(\underline{s}) = \frac{\underline{V}_a}{\underline{V}_{in}} \times \frac{\underline{V}_{out}}{\underline{V}_a} = \frac{\underline{s} + 2}{\underline{s}^2 + 7\underline{s} + 2} \times \frac{1}{\underline{s}/2 + 1} = \frac{2}{\underline{s}^2 + 7\underline{s} + 2} \tag{7.39}$$

---

[8] And we have labeled the circuit for nodal analysis.

The methods we developed for transient analysis work equally well for transfer functions, as shown by the following example.

<table>
<tr><td>**EXAMPLE 7.7**</td><td>**Transient analysis of a transfer function**</td></tr>
</table>

The circuit of Fig. 7.12 has a switched-battery input, as shown in Fig. 7.13. Find the output voltage.

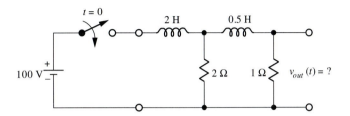

Figure 7.13 The circuit of Fig. 7.12 with a switched-battery input. We determine the output voltage.

SOLUTION:

The output voltage consists of a forced and natural component. The forced component has a frequency of $\underline{s}_f = 0$, because the input is dc. The output voltage in the frequency domain will therefore be from Eq. (7.36) at $\underline{s} = 0$

$$\frac{\mathbf{V}_f e^{0t}}{100 e^{0t}} = \mathbf{T}(0) = \left. \frac{2}{\underline{s}^2 + 7\underline{s} + 2} \right|_{\underline{s}=0} = 1 \tag{7.40}$$

Thus, the forced component of the output voltage is 100 V. This occurs because the inductors become short circuits at dc.

The natural response corresponds to having nonzero output with zero input:

$$\frac{\mathbf{V}_{out}}{\mathbf{V}_{in}} = \mathbf{T}(\underline{s}) = \frac{\neq 0}{0} \tag{7.41}$$

which can be valid only if the denominator of Eq. (7.39) is set to zero:

$$\underline{s}^2 + 7\underline{s} + 2 = 0 \implies \underline{s}_n = -0.298, -6.70 \quad \text{s}^{-1} \tag{7.42}$$

Thus, we have two natural frequencies that are real and negative. The output is, therefore, of the form

$$v_{out}(t) = 100 + Ae^{-0.298t} + Be^{-6.70t} \quad \text{V} \tag{7.43}$$

where A and B are constants. The two inductors block the initial voltage from the output; hence, the initial conditions on $v_{out}(t)$ are

$$v_{out}(0^+) = 0 \quad \text{and} \quad \left. \frac{dv_{out}(t)}{dt} \right|_{t=0^+} = 0 \tag{7.44}$$

We may substitute Eq. (7.43) into Eqs. (7.44) and solve simultaneously for A and B, with the result

$$v_{out}(t) = 100 - 104.66e^{-0.298t} + 4.66e^{-6.70t} \text{ V} \qquad (7.45)$$

which is plotted in Fig. 7.14.

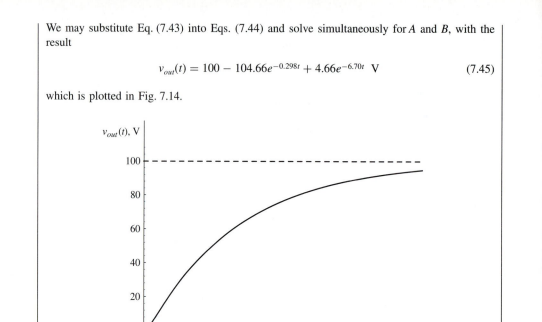

Figure 7.14 Response of the circuit of Fig. 7.13.

The time constants derived from Eq. (7.45) do not correspond to any L/R combination in the circuit, but are influenced by the interactions of all components. When the energy-storage elements in a circuit are all inductors or all capacitors, the natural frequencies are always real and negative.

Frequency Response

Poles, zeros, and frequency response.
The poles and zeros of a transfer function define the frequency response of the circuit in sinusoidal steady state. In this context, the system acts as a filter with $\underline{s} = j\omega$ and the poles and zeros correspond to the critical frequencies defining the Bode plot.

EXAMPLE 7.8 | **Bode plot of a filter function**

Give the filter function and Bode plot of the network in Fig. 7.12.

SOLUTION:
The filter function $\underline{F}(j\omega)$ is the transfer function given in Eq. 7.39 with $\underline{s} = j\omega$

$$\underline{F}(j\omega) = \frac{2}{(j\omega)^2 + 7(j\omega) + 2} = \frac{2}{(j\omega + 0.298)(j\omega + 6.70)} \qquad (7.46)$$

where we have factored the denominator using the results of the previous example. Equation (7.46) can be placed in the form

$$\underline{F}(j\omega) = \frac{1}{(1 + j\omega/0.298)(1 + j\omega/6.70)} \tag{7.47}$$

This is the response of a low-pass filter with critical frequencies at $\omega_{c1} = 0.298$ and $\omega_{c2} = 6.70$ rad/s. The asymptotic Bode plot of this filter function is shown in Fig. 7.15. Thus, we see that the poles of the transfer function correspond to the critical frequencies of the filter response; indeed, these critical frequencies are often called the "poles" of the filter.

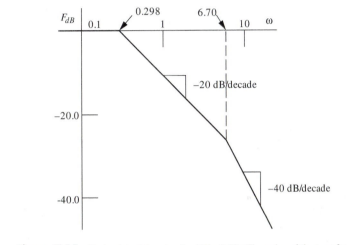

Figure 7.15 Bode plot of the circuit of Fig. 7.12. The poles of the transfer function correspond to the critical frequencies of the filter, as indicated in the Bode plot.

 ### Check Your Understanding

1. What is the impedance of a 100-µF capacitor at a complex frequency of $\underline{s} = -100$ s^{-1}?

2. An impedance consists of $R = 100\ \Omega$ and $C = 10\ \mu F$ in series. What impedance does this present to an input voltage $v(t) = 10e^{-500t}$?

3. A 10-Ω resistor is connected in series with a 100-µF capacitor. At what complex frequency does this combination look like a short circuit?

4. An impedance with a zero at $\underline{s} = 0$ passes a dc current. True or false?

5. The open-circuit natural frequencies of a circuit correspond to the zeros of the impedance function. True or false?

6. A circuit with two inductors must have natural frequencies that are real. True or false?

Answers. (1) –100 Ω; (2) –100 Ω; (3) –1000 s^{-1}; (4) true; (5) false; (6) true.

Types of Natural Responses in *RLC* Circuits

Transfer function of an *RLC* circuit. The circuit of Fig. 7.16 can exhibit an oscillatory response because the circuit contains an inductor and capacitor. This circuit can be analyzed as a voltage divider:

$$\mathbf{T}(\underline{s}) = \frac{\mathbf{V}_{out}(\underline{s})e^{\underline{s}t}}{\mathbf{V}_{in}(\underline{s})e^{\underline{s}t}} = \frac{R\|1/\underline{s}C}{\underline{s}L + R\|1/\underline{s}C} = \frac{1}{LC}\frac{1}{\underline{s}^2 + (\underline{s}/RC) + 1/(LC)} \tag{7.48}$$

where $\mathbf{T}(\underline{s})$ is the transfer function.

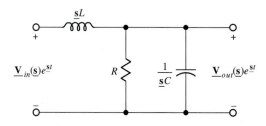

Figure 7.16 An *RLC* circuit in the frequency domain. The transfer function is the ratio of output to input voltage.

Natural frequencies of the circuit. The poles of the transfer function correspond to an output voltage with no input voltage, and hence are the natural frequencies of the circuit with the input short-circuited. These poles are investigated in detail because they reveal the various types of behavior that can result with an *RLC* circuit. We may determine the poles with the quadratic formula

$$\underline{s}^2 + \frac{\underline{s}}{RC} + \frac{1}{LC} = 0 \;\Rightarrow\; \underline{s}_n = -\frac{1}{2RC} \pm \sqrt{\left(\frac{1}{2RC}\right)^2 - \frac{1}{LC}} \;\; s^{-1} \tag{7.49}$$

There are four possibilities, depending on the relative values of R, L, and C. The possibilities are undamped, underdamped, critically damped, and overdamped responses.

undamped response

Undamped behavior. First, we consider the case where the resistance is infinite, which is equivalent to removing the resistor from the circuit. Equation (7.49) gives imaginary roots:

$$\underline{s}_1 = +j\omega_0 \quad \text{and} \quad \underline{s}_2 = -j\omega_0 \; s^{-1} \tag{7.50}$$

where $\omega_0 = 1/\sqrt{LC}$ rad/s. Thus, with no resistance, the short-circuit natural frequencies are imaginary, meaning that the natural response of the circuit is a sinusoid of constant amplitude. This may be understood as a lossless resonance, with ω_0 the resonant frequency. With no resistor, any energy imparted to the circuit alternates between the inductor and capacitor and produces a sinusoidal output. This is called an *undamped response* because the oscillations do not diminish with time.

damped oscillation, damping constant

Underdamped behavior. For large but finite resistance, the square root remains imaginary, and the roots of the quadratic are complex:

$$\mathbf{s}_{1,2} = -\alpha \pm \sqrt{\alpha^2 - \omega_0^2} \qquad \text{or} \qquad \mathbf{s}_{1,2} = -\alpha \pm j\omega \quad \text{s}^{-1} \tag{7.51}$$

where $\alpha = 1/2RC$ s^{-1} and $\omega = \sqrt{\omega_0^2 - \alpha^2}$ s^{-1}. The two roots are complex conjugates and correspond to an exponentially decreasing sinusoid, a *damped oscillation*. Thus, the natural response of the circuit is of the form

$$v_{out}(t) = Ae^{-\alpha t} \cos(\omega t + \theta) \tag{7.52}$$

where A and θ are constants. The constant α is called the *damping constant* and is the reciprocal of the time constant of the dying oscillation. The frequency of the oscillation, ω, is less than the resonant frequency, ω_0, because the exchanges of energy between inductor and capacitor are slowed down by loss in the resistor.

underdamped response

Condition for underdamped response.
The exponentially decreasing oscillation is called an *underdamped response*. The condition for an underdamped response is

$$\alpha < \omega_0 \qquad \text{or} \qquad R > \frac{1}{2}\sqrt{\frac{L}{C}} \tag{7.53}$$

Thus, for large values of resistance, the response of the circuit is dominated by the resonance of the inductor and capacitor, but the oscillations die out in time due to the loss in the resistor.

critical damping

Critical damping.
Critical damping occurs when $\alpha = \omega_0$, and the roots of Eq. (7.49) become real and equal. This is interesting mathematically but unimportant in practice because it exists only for one exact value of resistance, given by Eq. (7.53) with an equality sign. Even if we wished to produce this type of response in a physical circuit, we would be able to achieve the required resistance only with great care or good fortune. For us, critical damping is important as the boundary between underdamped oscillations and overdamped behavior.

overdamped response

Overdamped behavior.
When $\alpha > \omega_0$, the roots of Eq. (7.49) are real, negative, and unequal.

$$\mathbf{s}_1 = -\alpha + \sqrt{\alpha^2 - \omega_0^2} \qquad \text{and} \qquad \mathbf{s}_2 = -\alpha - \sqrt{\alpha^2 - \omega_0^2} \tag{7.54}$$

Thus, the *overdamped response* has the form

$$v_{out}(t) = Ae^{\mathbf{s}_1 t} + Be^{\mathbf{s}_2 t} \tag{7.55}$$

where $\mathbf{s}_1$ and $\mathbf{s}_2$ are real, negative numbers. This response consists of two time constants, which are the negatives of the reciprocals of $\mathbf{s}_1$ and $\mathbf{s}_2$. Thus, for small resistance, the increased loss eliminates the resonance between the inductor and capacitor.

Summary.
The natural frequencies of the *RLC* circuit are derived from the roots of a quadratic equation. The roots may be imaginary, complex, real and equal, or real and unequal. Imaginary roots indicate an undamped oscillation. Complex roots indicate a damped oscillation. Real, unequal roots indicate two time constants and no oscillation. Real, equal roots represent the mathematical boundary between oscillatory and non-oscillatory behavior. These time-domain responses are derived through the application of frequency-domain techniques.

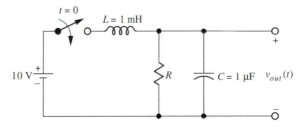

Figure 7.17 An *RLC* circuit with a source in the time domain.

RLC Circuit Transient Behavior

Forced response. We now investigate the time-domain response of the circuit shown in Fig. 7.17, in which a dc voltage is applied to the *RLC* circuit of the previous section. The output voltage consists of two components, a natural response and a forced response due to the dc input. We may determine the forced response by the methods presented in Chap. 1, by treating the capacitor as an open circuit and the inductor as a short circuit. Alternately, we may consider the input as $10e^{\mathbf{s}_f t}$, with $\mathbf{s}_f = 0$, and derive the forced response from the transfer function given in Eq. (7.48) with $\mathbf{s} = 0$. Using the latter method, we find the forced component of the output voltage to be

$$\frac{V_{out}e^{0t}}{V_{in}e^{0t}} = \mathbf{T}(0) \Rightarrow V_{out} = T(0) \times 10e^{0t} = 10 \text{ V} \tag{7.56}$$

This indicates a dc output of 10 V due to the battery. The total response is, therefore,

$$v_{out}(t) = 10 + v_n(t) \text{ V} \tag{7.57}$$

where $v_n(t)$ is the natural response. Several forms are possible for the natural response, depending on the resistance.

Natural response: undamped case. We consider first the response for $R = \infty$:

$$v_{out}(t) = 10 + A \cos(\omega_0 t + \theta) \tag{7.58}$$

where A and θ must be determined from the initial conditions.

Initial conditions. We may establish the initial conditions of the output voltage through the techniques presented in Chap. 1. The capacitor acts initially as a short circuit; hence, the output voltage must be zero at $t = 0^+$:

$$0 = 10 + A \cos \theta \tag{7.59}$$

The inductor acts initially like an open circuit. Consequently, the initial current through the capacitor is also zero, and the derivative of the output voltage must also be zero at $t = 0^+$:

$$\left. \frac{dv_{out}(t)}{dt} \right|_{t=0^+} = 0 \Rightarrow 0 = -\omega_0 A \sin\theta \tag{7.60}$$

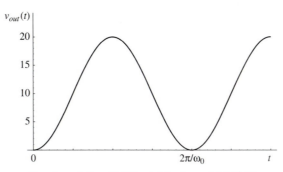

Figure 7.18 Response of the circuit o Fig. 7.17 with $R = \infty$. The peak voltage i twice the input voltage, a property of the circuit that finds many applications.

so $\theta = 0$ and Eq. (7.59) yields $A = -10$ V.[9] Hence, the response is

$$v_{out}(t) = 10 - 10\cos(\omega_0 t) \tag{7.61}$$

This response is shown in Fig. 7.18.

resonant charging

Resonant charging. The voltage on the capacitor reaches twice that of the input because the inductor gives a momentum to the current to continue charging past the equilibrium point. If a diode is placed in series with the inductor to prevent the discharge of the capacitor, the addition of the diode and inductor results in a fourfold increase in stored energy in the capacitor, compared to charging through a resistor. This technique is known as *resonant charging* of the capacitor, and applications are found in radar and power electronics.

Underdamped response. For the inductance and capacitance values given in Fig. 7.17, the value of the resistance for critical damping is given by Eq. (7.53) with an equality sign:

$$R_c = \frac{1}{2}\sqrt{\frac{L}{C}} = \frac{1}{2}\sqrt{\frac{10^{-3}}{10^{-6}}} = 15.8 \ \Omega \tag{7.62}$$

where R_c is the resistance for critical damping. We consider first the underdamped response, which requires a resistance larger than the critical value. We assume $R = 5R_c = 79.1 \ \Omega$. The resonant frequency and damping coefficient, derived from Eq. (7.51):

$$\omega_0 = \frac{1}{\sqrt{LC}} = 31{,}600 \ \text{s}^{-1} \quad \text{and} \quad \alpha = \frac{1}{2RC} = 6320 \ \text{s}^{-1} \tag{7.63}$$

and thus the natural frequencies, determined from Eq. (7.51), are

$$\underline{s}_1 = -6320 + j31{,}000 \quad \text{and} \quad \underline{s}_2 = -6320 - j31{,}000 \tag{7.64}$$

The total response is

$$v_{out}(t) = 10 + Ae^{-6320t}\cos(31{,}000t + \theta) \tag{7.65}$$

---

[9] An identical answer results from $\theta = \pi$ and $A = +10$.

The initial conditions are the same as for the undamped response. Hence, at $t = 0^+$,

$$0 = 10 + A \cos \theta \tag{7.66}$$

and

$$\frac{dv_{out}(t)}{dt}\bigg|_{t=0^+} = 0 \Rightarrow 0 = A(-6320 \cos\theta - 31{,}000 \sin\theta) \tag{7.67}$$

Equations (7.66) and (7.67) yield $\theta = -11.5°$ and $A = -10.21$ V. The solution is, therefore,

$$v_{out}(t) = 10 - 10.21e^{-6320t} \cos(31{,}000t - 11.5°) \text{ V} \tag{7.68}$$

which is shown in Fig. 7.19.

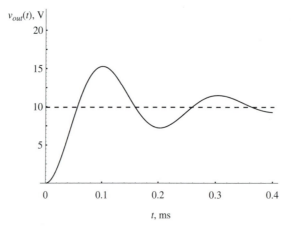

Figure 7.19 Underdamped response for the circuit in Fig 7.17 with $R = 79.1 \ \Omega$.

Overdamped response. To obtain an overdamped response, we use a resistance one-fifth the critical value given by Eq. (7.62): $R = R_c/5 = 3.16 \ \Omega$. With this resistance, $\alpha = 158{,}000$, and Eq. (7.54) gives the natural frequencies:

$$\underline{s}_1 = -3190 \text{ s}^{-1} \quad \text{and} \quad \underline{s}_2 = -313{,}000 \text{ s}^{-1} \tag{7.69}$$

Thus, the output voltage is

$$v_{out}(t) = 10 + Ae^{-3190t} + Be^{-313{,}000t} \text{ V} \tag{7.70}$$

where A and B are constants. The first term is the forced response, which is the same as before. The second term has a time constant of 0.313 ms and the third term a time constant of 3.19 μs. The initial conditions are the same as before; hence, we may solve for A and B from the equations

$$0 = 10 + A + B \quad \text{and} \quad 0 = -3190A - 313{,}000B \tag{7.71}$$

which yield $A = -10.10$ V and $B = 0.10$ V. Thus, the overdamped response is

$$v_{out}(t) = 10 - 10.10e^{-3190t} + 0.10e^{-313{,}000t} \text{ V} \tag{7.72}$$

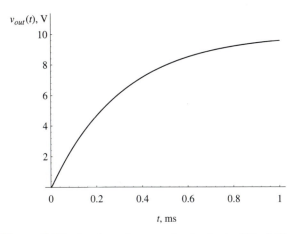

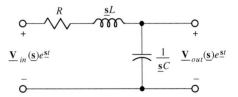

Figure 7.20 Overdamped response for the circuit of Fig. 7.17.

Figure 7.21 This is a series *RLC* circuit when driven by a voltage source.

This response is shown in Fig. 7.20. The third term in Eq. (7.72) has a brief influence to give the zero derivative at the origin, but otherwise the response is that of a first-order transient of the *RL* part of the circuit.

Parallel and series *RLC* circuits. The circuit of Fig. 7.17, when excited by a voltage source, is a parallel *RLC* circuit. With the source turned OFF, the resistor, inductor, and capacitor are connected in parallel. In the circuit shown in Fig. 7.21, the resistor, inductor, and capacitor are connected in series. The analysis of the series *RLC* circuit for the transfer function, natural frequencies, and transient response follows the same lines as for the parallel circuit, but the results differ in detail. We leave the analysis of the series *RLC* circuit for the reader.

Relationship with Laplace transform theory. The Laplace transform gives a strong mathematical foundation to the techniques used in this chapter and is closely associated with linear system theory. Although the Laplace transform yields the total response of a circuit or system, including initial conditions, its primary importance lies in furnishing the tools for describing linear systems that we have developed in this chapter.

The Frequency Domain

IDEA 7

Frequency domain. Functions such as $\underline{V}(\underline{s})$ and $\underline{I}(\underline{s})$, which we have been treating as constants, can be considered *transforms* of the time-domain voltage and current. These transforms can be interpreted as generalized spectra of the time-domain functions. Similarly, generalized impedances and transfer functions are frequency-domain transforms of the linear differential equations that describe the circuit or system component in the time domain.

Summary. In this section, we used complex frequency to investigate transient behavior of first- and second-order circuits. Natural responses have been decreasing exponentials or exponentially decreasing sinusoids. The natural frequencies and the character of the natural response may be determined from the poles and zeros of the impedance or transfer functions. In the next section, we illustrate how the transfer function describes the properties of a linear-system component.

Check Your Understanding

1. Describe the largest number and the character of the natural frequencies of the following circuits: (a) two resistors and a capacitor; (b) one resistor and two capacitors; (c) three resistors and two inductors; (d) and two resistors, a capacitor, and an inductor.

2. Determine the natural frequencies of the circuit in Fig. 7.17 if the resistor has the value for critical damping.

Answers. **(1)** (a) One natural frequency, real and negative; (b) one natural frequency, real and negative; (c) two natural frequencies, real and negative; and (d) two natural frequencies, either complex with negative real part or both real and negative; **(2)** $\underline{s} = -31,623 \text{ s}^{-1}$ (repeated).

7.4 SYSTEM ANALYSIS

> **LEARNING OBJECTIVE 4.**
>
> To understand how to use system notation to describe the properties of composite systems

Introduction. In this section, we explore the notation and techniques commonly used to describe linear systems. Our example is a thermostatically controlled oven. We determine its response when controlled by a feedback control system. Although heating of the oven is inherently nonlinear, we linearize the problem to use the techniques of the frequency domain. We show that a second-order system, two independent modes of energy storage, is unconditionally stable, although the system response may be unacceptable. We then show that a third-order system can become unstable for large loop gain.

Problem description. In Chap. 4, p. 237, we discussed a system for controlling an oven with a feedback control system. We described the system in general terms and mentioned briefly the dynamic stability of the system. We now possess the tools to investigate the system stability. Figure 7.22 shows the oven control system. The oven is heated by a resistive element driven by an amplifier. The oven temperature is monitored by a sensor, which produces a voltage proportional to its temperature. The op amp is driven by the difference between an input set voltage and a feedback signal from the sensor.

System Functions

system function

We now express the properties of the sensor in the frequency domain through a system function. A *system function*, $\underline{F}(\underline{s})$, relates the transforms of input and output in the equation

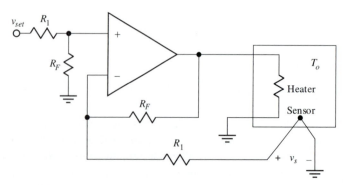

Figure 7.22 Oven control system.

$$\frac{\underline{\mathbf{X}}_{out}(\underline{\mathbf{s}})}{\underline{\mathbf{X}}_{in}(\underline{\mathbf{s}})} = \underline{\mathbf{F}}(\underline{\mathbf{s}}) \qquad (7.73)$$

where the $\underline{\mathbf{X}}$'s are frequency-domain transforms of the time-domain input and output variables. The system function is similar to the transfer function except that the input and output variables may be any of a wide class of variables. In this case, the input is the oven temperature and the output is the sensor voltage.

Sensor-system function. We assume the sensor produces a voltage proportional to the oven temperature, $v_s = K_s T_o$, and responds to changes in oven temperature with a time constant τ_s. The specification of a time constant implies a first-order system. To satisfy both properties, the output voltage of the sensor must satisfy the DE

$$\tau_s \frac{dv_s}{dt} + v_s = K_s T_o(t) \qquad (7.74)$$

where $v_s(t)$ is the sensor voltage, τ_s is the time constant of the sensor, and $T_o(t)$ is the oven temperature. We may transform Eq. (7.74) into the frequency domain by assuming input temperature and output voltage to be of the form $e^{\underline{s}t}$, with the results

$$(\tau_s \underline{\mathbf{s}} + 1)\underline{\mathbf{V}}_s(\underline{\mathbf{s}}) = K_s \underline{\mathbf{T}}_o(\underline{\mathbf{s}}) \;\Rightarrow\; \underline{\mathbf{V}}_s(\underline{\mathbf{s}}) = \underbrace{\frac{K_s/\tau_s}{\underline{\mathbf{s}} + (1/\tau_s)}}_{\substack{\text{system function} \\ \text{for sensor}}} \times \underline{\mathbf{T}}_o(\underline{\mathbf{s}}) \qquad (7.75)$$

where $\underline{\mathbf{V}}_s(\underline{\mathbf{s}})$ is the transform of the sensor voltage and $\underline{\mathbf{T}}_o(\underline{\mathbf{s}})$ is the transform of the oven temperature. Thus, we may represent the sensor in the frequency domain by the system component shown in Fig. 7.23.

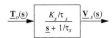

Figure 7.23
System component representing the sensor. The input and output are frequency-domain transforms of the oven temperature and sensor voltage, respectively.

Oven-system function. The oven temperature will be controlled by the temperature of the heater element. We will assume a time constant of τ_o for this process. Thus the heater temperature and oven temperature are related by a system function similar to the sensor

$$\underline{\mathbf{T}}_o(\underline{\mathbf{s}}) = \underbrace{\frac{1/\tau_o}{\underline{\mathbf{s}} + (1/\tau_o)}}_{\substack{\text{system function} \\ \text{for oven}}} \times \underline{\mathbf{T}}_H(\underline{\mathbf{s}}) \qquad (7.76)$$

where τ_o represents the oven time constant and $\mathbf{T}_H(\underline{\mathbf{s}})$ is the transform of the heater temperature.

Heater voltage. Although the heater is nonlinear, we consider only small changes in the signal levels and assume that a linear increment in voltage produces a linear increment in power in the heater, and hence a linear increment in the heater temperature. There is a time constant associated with the thermal mass of the heater element, but we ignore this effect in the present analysis. Thus, we assume for now that changes in heater voltage produce proportional and instantaneous changes in the temperature of the heater element. These changes are represented in the time domain and frequency domain by

$$T_H(t) = K_H v_H(t) \implies \underline{\mathbf{T}}_H(\underline{s}) = K_H \underline{\mathbf{V}}(\underline{s}) \tag{7.77}$$

where $\underline{\mathbf{V}}_H(\underline{s})$ is the transform of the heater voltage and K_H deg/V is a constant relating heater temperature to changes in heater voltage.

System diagram. We may represent the interaction of the various system variables by the system diagram shown in Fig. 7.24. The amplifier is represented by a summer (differencer) and a gain of R_F/R_1.

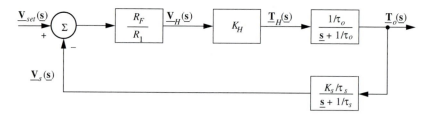

Figure 7.24 System diagram for the oven control system.

Summary. In this section, we introduced the system functions of the various components of a feedback control system. Each variable is represented by its frequency-domain transform. Each component is represented by its system function. We may now use feedback theory to determine the overall system function of the controller. We determine the natural frequencies of the system and, from these, the transient response of the system in the time domain.

Dynamic Stability of Feedback Systems

Feedback analysis. The gain with feedback of an amplifier with feedback is given in Eq. (4.77) as

$$A_f = \frac{A}{1 - L} \tag{7.78}$$

where A_f is the gain with feedback, A is the gain without feedback, and L is the loop gain. Here we are expressing the system properties in the frequency domain, so our signals and gains are complex functions of $\underline{s}$. The gain without feedback in Fig. 7.24 is the product of the system functions between input and output:

$$\underline{\mathbf{A}}(\underline{s}) = \frac{R_F}{R_1} \times K_H \times \frac{1/\tau_o}{\underline{s} + 1/\tau_o} \tag{7.79}$$

and the loop gain is the product of the system functions around the loop:

$$\underline{\mathbf{L}}(\underline{s}) = (-1)\frac{R_F}{R_1} \times K_H \times \frac{1/\tau_o}{\underline{s} + 1/\tau_o} \times \frac{K_s/\tau_s}{\underline{s} + 1/\tau_s} = +\frac{L_0/\tau_o \tau_s}{(\underline{s} + 1/\tau_o)(\underline{s} + 1/\tau_s)} \tag{7.80}$$

where $L_0 = -(R_F/R_1)K_H K_s$ is the dc loop gain at $\underline{s} = 0$, a dimensionless quantity.

System natural frequencies. Our investigation of the dynamic behavior of the system does not require that we substitute Eqs. (7.79) and (7.80) into Eq. (7.78). We are interested in the natural response of the system, when the input voltage is zero but the output is nonzero. This can occur only at the poles of the overall system function with feedback, which requires $1 - \underline{L}(\underline{s}) = 0$. Thus we may determine the natural frequencies from the quadratic equation

$$1 - \frac{L_0/\tau_s\tau_o}{(\underline{s} + 1/\tau_s)(\underline{s} + 1/\tau_o)} = 0 \implies \underline{s}^2 + \left(\frac{1}{\tau_o} + \frac{1}{\tau_s}\right)\underline{s} + \frac{1 - L_0}{\tau_o\tau_s} = 0 \qquad (7.81)$$

We continue our analysis assuming a time constant of 300 s for the oven and a time constant of 30 s for the sensor. The roots of Eq. (7.81) for dc loop gains of $L_0 = -1$, -10, and -50 are given in Fig. 7.25(a). The associated dynamic responses shown in Fig. 7.25(b) are derived in the next section.

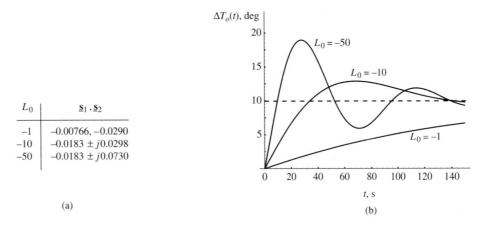

L_0	$\underline{s}_1, \underline{s}_2$
-1	$-0.00766, -0.0290$
-10	$-0.0183 \pm j0.0298$
-50	$-0.0183 \pm j0.0730$

(a)

(b)

Figure 7.25 (a) Natural frequencies for dc loop gains of -1, -10, and -50; (b) the output temperature response to a sudden increase in input voltage for the various dc loop gains.

Dynamic response: overdamped case. We may determine the dynamic response of the circuit from the natural frequencies, which are the poles of the overall system function with feedback, and from the initial and final conditions of the output. We assume in this section that the input voltage is increased suddenly to produce an eventual 10°F increase in the oven temperature. For $L_0 = -1$, the natural frequencies of the system are given in Fig. 7.25(a) as $\underline{s}_1 = -0.00766$ and $\underline{s}_2 = -0.0290$. The response is overdamped, and the total response is of the form

$$\Delta T_o(t) = A + Be^{-0.00766t} + Ce^{-0.0290t} \qquad (7.82)$$

where $\Delta T_o(t)$ is the change in the oven temperature and A, B, and C are constants to be determined from the initial and final conditions. The final condition is that $\Delta T_o(t)$ should approach $+10°F$ as $t \to \infty$; hence, $A = +10$.

Initial conditions. The initial conditions can be deduced from the time delays in the system. Because it takes time for the oven to heat up, the initial value of $\Delta T_o(t)$ is zero,

$\Delta T_o(0^+) = 0$. The initial value of the derivative is not affected by the feedback effects because of the time delay in the sensor. Because the feedback reduces the final value by the factor $1 - L_0$, the initial derivative is that of the function

$$T_o(t) = \underbrace{\Delta T_o(\infty)(1 - L_0)}_{\substack{\text{final temperature} \\ \text{with no feedback}}}(1 - e^{-t/\tau_o}) \tag{7.83}$$

Therefore, the initial value of the derivative of $\Delta T_o(t)$ is

$$\left. \frac{d\Delta T_o(t)}{dt} \right|_{t=0^+} = \frac{\Delta T_o(\infty)(1 - L_o)}{\tau_o} \tag{7.84}$$

With these two initial conditions, we can readily determine the constants B and C in Eq. (7.82). The initial condition on the temperature increment is

$$0 = 10 + B + C \tag{7.85}$$

and the initial condition on the derivative of $\Delta T_o(t)$ for $L_0 = -1$ follows from Eqs. (7.82) and (7.84):

$$\frac{10}{300} \times [1 - (-1)] = -0.00766B - 0.0290C \tag{7.86}$$

Simultaneous solution of Eqs. (7.85) and (7.86) yields $B = -10.466$ and $C = +0.466$. Hence, the response of the system with a loop gain of -1 to a sudden increase in input voltage is

$$\Delta T_o(t) = 10 - 10.466e^{-0.00766t} + 0.466e^{-0.0290t} \tag{7.87}$$

which is shown in Fig. 7.25(b). We note that the transient is dominated by the longer time constant of $1/0.00766 = 130.5$ s. This is the time constant of the oven, sped up by a factor of approximately 2 by the feedback.

Dynamic response: underdamped case. For dc loop gains greater than 2.025, the roots of Eq. (7.81) are complex, indicating underdamped response. For $L_0 = -10$, the roots are given in Fig. 7.25(a) as $\mathbf{s}_{1,2} = -0.0183 \pm j0.0298$. The form of the response can be written in several ways, but the form given in Eq. (7.88) is most convenient for our analysis.

$$\Delta T_o(t) = A + e^{-0.0183t}[B \cos(0.0298t) + C \sin(0.0298t)] \tag{7.88}$$

where, again, A, B, and C are constants to be determined from the initial and final conditions. The final value, as before, is $A = +10$. The initial condition is $\Delta T_o(t) = 0$, as before, which allows B to be determined by inspection as -10 because the sine term in Eq. (7.88) vanishes at $t = 0$. The initial value of the derivative is given by Eq. (7.84):

$$\frac{d\Delta T_o(t)}{dt} = \frac{10}{300} \times [1 - (-10)] = 0 - 0.0183B + 0.0298C \tag{7.89}$$

which yields $C = +6.159$. Hence, the response for a dc loop gain of -10 is

$$\Delta T_o(t) = 10 + e^{-0.0183t}[-10\cos(0.0298t) + 6.159\sin(0.0298t)] \qquad (7.90)$$

This response is shown in Fig. 7.25(b), where we have also shown the response for a dc loop gain of -50.

stable

Summary of feedback effects for second-order systems.   Figure 7.25(b)

reveals the effects of feedback on the second-order system, which has delays due to thermal energy storage in both oven and sensor. As the magnitude of the dc loop gain is increased, the response goes from overdamped to underdamped. The character of the various responses is shown in Fig. 7.25(b), which indicates that the system speeds up with increasing loop gain. However, the use of excessive feedback causes a severe overshoot problem.

In a feedback system, instability is caused by excessive phase shift around the feedback loop. If at some frequency the phase shift approaches $-180°$, negative feedback becomes positive feedback and oscillations can occur. A second-order system is unconditionally *stable*, meaning that oscillations always die out, regardless of the amount of feedback, because the $-180°$ phase shift is never reached. As we see in the next section, a third-order system can have growing oscillations if the loop gain is high. Such unstable behavior renders the system useless and can make it dangerous to life and property.

Control theory.   Figure 7.25(b) shows that large loop gain degrades the dynamic re-

sponse of the system by causing a "hunting" type of response. But with low loop gain, we lose the benefits of feedback. One goal of control theory is to achieve an acceptable compromise between these two effects by modifying the system properties with electrical filters.

Dynamic analysis of third-order system.   We now repeat the analysis of the

oven-control system considering the delay in the heater element. When the power is applied to the heater element, its temperature increases with a time constant τ_H. This changes the system function of the heater element in Fig. 7.24 to that shown in Fig. 7.26. The analysis of the feedback system proceeds as before, and the natural frequencies are the roots of $1 - \mathbf{L}(\mathbf{s}) = 0$. This results in the cubic equation

$$\left(\mathbf{s} + \frac{1}{\tau_H}\right)\left(\mathbf{s} + \frac{1}{\tau_o}\right)\left(\mathbf{s} + \frac{1}{\tau_s}\right) + \frac{L_0}{\tau_H \tau_o \tau_s} = 0 \qquad (7.91)$$

We determined the roots of Eq. (7.91) assuming time constants of 15 s, 300 s, and 30 s

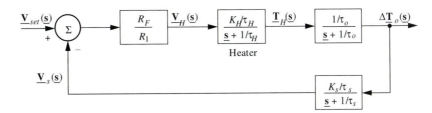

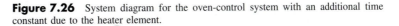

Figure 7.26   System diagram for the oven-control system with an additional time constant due to the heater element.

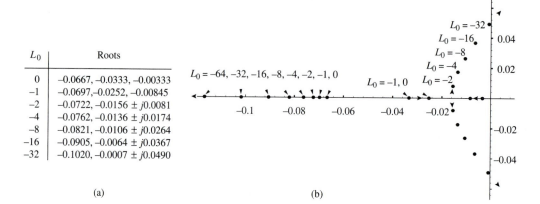

L_0	Roots
0	−0.0667, −0.0333, −0.00333
−1	−0.0697, −0.0252, −0.00845
−2	−0.0722, −0.0156 ± j0.0081
−4	−0.0762, −0.0136 ± j0.0174
−8	−0.0821, −0.0106 ± j0.0264
−16	−0.0905, −0.0064 ± j0.0367
−32	−0.1020, −0.0007 ± j0.0490

(a)   (b)

Figure 7.27   (a) Natural frequencies for the system in Fig. 7.26; (b) root plot for third-order system with increasing loop gain.

for the heater, oven, and sensor, respectively. Figure 7.27(a) tabulates the roots for various values of dc loop gain.

Root-locus plot.   Figure 7.27(b) shows the motion of the natural frequencies in the complex plane as the magnitude of the dc loop gain is increased. For a loop gain of −1, all three roots are real, indicating exponential behavior in the time domain. For loop-gain magnitudes greater than about 1.3, two roots become complex and begin to move toward the real axis as the oscillation frequency increases. For loop-gain magnitudes greater than about 25, the roots move into the right half of the complex plane, indicating oscillations that increase with time, similar to the response shown in Fig. 7.3(a). This type of response indicates dynamically unstable behavior. The three poles cause excessive phase shift at high frequency, the feedback becomes positive, and oscillations grow until limited by nonlinear effects.

Summary.   In this section, we explored the dynamic response of second- and third-order feedback systems. We modeled system components and derived the system function with feedback. The dynamic response of the second-order system was found to be unconditionally stable but can exhibit an unacceptable hunting response if the loop gain is too high. The third-order system has similar behavior but can go unstable as loop gain is increased.

CHAPTER SUMMARY

This chapter introduces the tools and concerns of linear system analysis. We generalize the concept of frequency to include exponential growth and decay, plus damped or growing oscillations. We demonstrate the significance of the poles and zeros of the impedance and system functions in establishing the transient and steady-state response of a system. We explore the various transient responses of which second-order systems are capable. Finally, we investigate the effect of loop gain on the response and dynamic stability of a feedback system.

Objective 1. To understand how to organize the class of functions that can be described by complex frequencies. Complex frequencies can represent constant signals, growing and decaying exponentials, sinusoidal functions, and growing and decaying sinusoidal functions. These functions have a broad practical significance, but also are useful as a mathematical probe to explore the behavior of linear systems.

Objective 2. To understand how to use generalized impedance to find the forced and natural response of a system. We generalize the impedance expressions from Chap. 4 using complex frequency to include a broader class of behavior than simple sinusoids. The poles and zeros of an impedance or transfer function determine both the natural and forced responses when the circuit is excited at a complex frequency.

Objective 3. To understand how to recognize conditions for undamped, underdamped, critically damped, and overdamped responses in second-order systems. We study *RLC* circuits as an example of a second-order system. The four types of response are named and described.

Objective 4. To understand how to use system notation to describe the properties of composite systems. Hybrid and nonelectrical linear systems can be described by complex-frequency techniques. The idea of system analysis is to focus on the dynamic interaction of system components without excessive attention to physical details.

Objective 5. To understand the role of loop gain in the dynamic response of a feedback system. We analyze second- and third-order systems with feedback. We show that second-order systems are unconditionally stable, but response may be unacceptable for high loop gain. Third- and higher-order systems can become unstable, with growing oscillations driving the system beyond linear operation, if loop gain is sufficiently high. We use the root-locus description in the stability analysis.

Throughout this chapter, frequency-domain concepts are emphasized. Input and output signals have been the frequency-domain transforms of time-domain signals and system functions replace differential equations representing dynamic behavior.

GLOSSARY

Complex frequency, $\underline{s}$, p. 374, a time-domain variable has a complex frequency, $\underline{s} = \sigma + j\omega$, when it can be expressed in the form $v(t) = Re\{\underline{V}e^{\underline{s}t}\}$.

Critical damping, p. 391, when the natural frequencies of a circuit are real and equal. Critical damping is the boundary between oscillatory and nonoscillatory behavior.

Damping constant, p. 390, the reciprocal of the time constant of the amplitude of a damped oscillation.

Generalized impedance, p. 379, the phasor voltage divided by phasor current when phasors are functions of complex frequency.

Overdamped response, p. 391, a response consisting of two or more time constants, but no oscillation.

Pole, p. 385, a frequency at which an impedance or transfer function goes to infinity.

Resonant charging, p. 393, charging a capacitor through an inductor and a diode in series.

Stable system, p. 401, a system in which oscillations die out.

System, p. 374, several components that together accomplish some purpose.

System function, p. 396, the ratio of the transforms of input and output as a function of complex frequency.

Undamped response, p. 390, a transient response in which oscillations do not diminish with time.

Underdamped response, p. 391, a response when circuit natural frequencies are complex conjugates and correspond to an exponentially decreasing sinusoid, also termed a damped oscillation.

Zero, p. 385, a frequency at which an impedance or transfer function is zero.

PROBLEMS

Section 7.1: Complex Frequency

7.1. The half-life of the exponential decay of carbon 14 is 3730 years. What complex frequency in s^{-1} describes this process?

7.2. A decaying sinusoid has a complex frequency $\underline{s} = -2 + j1$ and has its maximum value at $t = 0$. What is the angle of the phasor representing this signal in the frequency domain?

7.3. A decaying sinusoid crosses zero every 10 ms and each positive peak is 90% of the previous positive peak. What complex frequency describes this function?

7.4. A time-domain function is shown in Fig. P7.4. The function is of the form $v(t) = A + Be^{\sigma t}\cos(\omega t)$.
  (a) From the graph, determine A, B, σ, and ω.
  (b) Estimate the complex frequencies that can be used to describe this function.

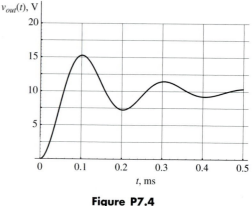

Figure P7.4

Section 7.2: Impedance and the Transient Behavior of Linear Systems

7.5. Consider a resistor, R, in series with a capacitor, C.
  (a) Determine the generalized input impedance as a function of complex frequency and, from that, the pole and zero of the circuit.
  (b) What is the interpretation of the pole?
  (c) What is the interpretation of the zero?

7.6. The circuit shown in Fig. P7.6 is excited by a voltage source that is zero for negative time and exponential for positive time, as indicated.
  (a) Determine the impedance of the circuit, $\underline{Z}(\underline{s})$.
  (b) What is the complex frequency of the source for $t > 0$?
  (c) What is the natural frequency of the circuit?

  (d) Determine the forced response of the current.
  (e) Determine the total response of the current for $t > 0$, including the initial condition.

7.7. A circuit has the impedance function

$$\underline{Z}(\underline{s}) = 10^4 \frac{\underline{s}^2 + 4\underline{s} + 1}{\underline{s}(\underline{s} + 2)} \ \Omega$$

  (a) What is the impedance at dc?
  (b) What are the natural frequencies of the circuit if the input is shorted?
  (c) What are the natural frequencies if the input is open-circuited?
  (d) In addition to resistors, this circuit has one

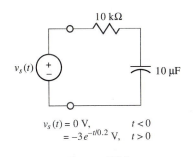

$v_s(t) = 0$ V, $t < 0$
       $= -3e^{-t/0.2}$ V, $t > 0$

Figure P7.6

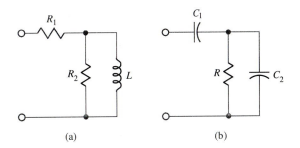

Figure P7.9

inductor, or one capacitor; two inductors, or two capacitors; or one capacitor and one inductor? Which combinations are possible (may be more than one)? Explain your answer.

(e) If the circuit is excited with a current source of value $-2e^{-t}$ A that is suddenly turned on at $t = 0$, what is the forced response of the voltage at the input to the circuit?

(f) What is the total response for $t > 0$ for this input voltage if the initial value of the voltage and its rate of change are zero?

7.8. A system has a transfer function

$$\mathbf{T}(\mathbf{s}) = \frac{K}{(\mathbf{s} - \mathbf{s}_1)(\mathbf{s} - \mathbf{s}_2)}.$$

The input voltage is zero for negative time and $+10$ for positive time. With this input the output is

$$v_{out}(t) = 5 + 10e^{-t} - 5e^{-2t} \text{ V}.$$

(a) Find K, $\mathbf{s}_1$, and $\mathbf{s}_2$.
(b) If the input voltage is $3e^{-t/2}$ for $t > 0$, find the output voltage for $t > 0$, assuming the same initial conditions.

7.9. Determine the open-circuit and short-circuit natural frequencies for the circuits shown in Fig. P7.9 and interpret these in terms of the time constants of the circuits.

7.10. Figure P7.10(a) shows the input function to the low-pass filter shown in Fig. P7.10(b). Determine the equation of the output voltage in the time domain.

7.11. The circuit shown in Fig. P7.11 is excited by a voltage source that is zero for negative time, but $v(t) = 100e^{-10,000t}$ V for positive time. Determine the current for positive time, assuming zero current at $t = 0$.

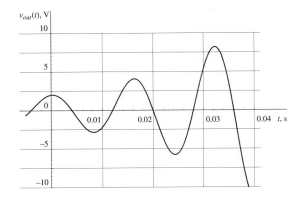

(a)

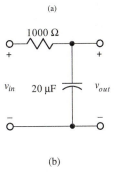

(b)

Figure P7.10

7.12. The impedance of the circuit in Fig. P7.12 has one zero at $\mathbf{s} = 0$ and one pole at $\mathbf{s} = -1$ s^{-1}. At $\mathbf{s} = +1$ s^{-1}, the impedance into the circuit has a value of 15 Ω. The circuit is excited by a current source that is zero for negative time and has a value of $i(t) = 0.5e^{-2t}$ A for positive time. Determine the input voltage, as shown, assuming $v(0+) = +2$ V.

7.13. An RC network has the impedance function

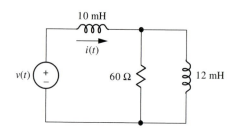

Figure P7.11

$$\mathbf{Z(\underline{s})} = 100 \, \frac{\underline{s}^2 + 25\underline{s} + 100}{\underline{s}(\underline{s} + 8)} \, \Omega$$

(a) Does the circuit allow a dc current?
(b) Find the impedance of the circuit at a frequency of 5 Hz.
(c) If the circuit is excited with a current source, what is (are) the natural frequency (frequencies) of the circuit?
(d) If the circuit source is $3e^{-10t}$ A, what input voltage results?
(e) What are the short-circuit natural frequencies of the network?

7.14. The system diagram shown in Fig. P7.14 gives the relationship between input and output voltage in the

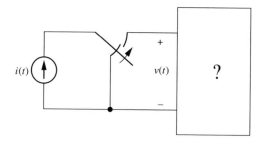

Figure P7.12

frequency domain. No voltage is applied prior to $t = 0$.

(a) Does this system pass a dc signal? How do you know?
(b) If the input is $v_{in}(t) = 2e^{-10t}$ for $t > 0$, and $v_{out}(0^+) = 0$, find the output for $t > 0$.
(c) If the input is $v_{in}(t) = 2\cos(2t)$ for $t > 0$, and $v_{out}(0^+) = 0$, find the output for $t > 0$.

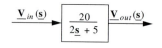

Figure P7.14

Section 7.3: Transient Response of *RLC* Circuits

7.15. For the *RLC* circuit of Fig. 7.16, what would be the natural frequencies of the network if R and L were exchanged? What would be the initial value and initial derivative of the output response to the sudden application of an input voltage?

7.16. For the series *RLC* circuit of Fig. 7.21, with $L = 1$ mH and $C = 1$ μF, find the following:
(a) Determine the transfer function, $\mathbf{T(\underline{s})}$.
(b) From the transfer function, determine the natural frequencies of the circuit if excited by a voltage source.
(c) What is the value of resistance for critical damping of the circuit? Is this different from the critical value given in Eq. (7.62) for the parallel *RLC* circuit?
(d) For resistances one-half and twice the critical value established in part (*c*), determine the natural frequencies of the circuit and write the corresponding time responses with unknown constants.
(e) Consider that a 10-V source is suddenly applied to the input in the manner shown in Fig. 7.17.

What are the initial value and initial derivative of the output voltage? Work out the complete response for the overdamped case calculated in part (d).

7.17. For the circuit shown in Fig. P7.17, find the following:
(a) Find the input impedance, $\mathbf{Z(\underline{s})}$.
(b) Determine the open-circuit natural frequencies.
(c) What value (range) of resistance corresponds to behavior that is (1) undamped, (2) underdamped, (3) critically damped, and (4) overdamped?

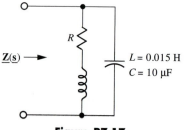

Figure P7.17

7.18. A circuit impedance has the pole–zero pattern shown in Fig. P7.18 and has an impedance magnitude of 30 Ω at $\underline{s} = +1$ s^{-1}. The circuit is excited by a voltage source that is zero for negative time and has a value of $v(t) = 5\cos(2t)$ for positive time. Determine the current into the circuit for positive time, assuming zero for the initial value and initial derivative of the current.

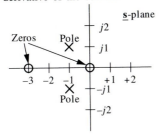

Figure P7.18

7.19. Figure P7.19 shows a circuit with a disconnected source.
  (a) Find the generalized impedance of the circuit, $\underline{Z}(\underline{s})$.
  (b) What is the impedance at $\underline{s} = 0$?
  (c) What is the impedance at $\underline{s} = \infty$?
  (d) If the switch is closed, what are the natural frequencies of the circuit?
  (e) What is the character of the response (undamped, underdamped, critically damped, overdamped)?
  (f) What is the initial value of the current in the resistor?
  (g) What is the final value of the current in the resistor?

7.20. Figure P7.20 shows an undamped LC circuit that resonant charges the capacitor to twice the power-supply voltage, and the diode holds the voltage until the capacitor is discharged. Determine the inductance and capacitance to store 2 J of energy in the capacitor in 1 ms from the time of switch closure. Assume an ideal diode.

Section 7.4: System Analysis

7.22. A simplified analysis of an electric-oven control system includes the following effects:
  • When the heater element in turned on, it heats with a time constant of 15 s.
  • If the heater element were fully hot, the air in the oven would heat with a time constant of 3 min.

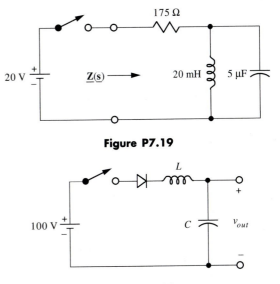

Figure P7.19

Figure P7.20

7.21. A circuit has the impedance function

$$\underline{Z}(\underline{s}) = 12\,\frac{\underline{s}}{\underline{s}^2 + 9\underline{s} + 6}\ \Omega$$

  (a) What is the impedance at dc?
  (b) What are the natural frequencies of the circuit if the input is shorted?
  (c) What are the natural frequencies if the input is open-circuited?
  (d) In addition to resistors, this circuit has one inductor, or one capacitor; two inductors, or two capacitors; or one capacitor and one inductor? Which combinations are possible (may be more than one)? Explain your answer.
  (e) If the circuit is excited with a current source of value $-2e^{-2t}$ A that is suddenly turned on at $t = 0$, what is the forced response of the voltage at the input to the circuit?
  (f) What is the total response for the input voltage if the initial value of the voltage and its rate of change are zero?

  • The time constant of the thermostat is negligible.
  • If the heater element were left on a long time, the oven would heat to 600°F relative to ambient temperature.

Figure P7.22 shows a block diagram of the system.

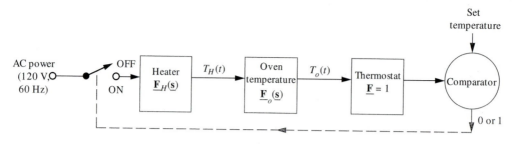

Figure P7.22

The output of the thermostat is compared to the oven-temperature setting, and the output of the comparator is represented as a dashed line to the switch.

(a) Determine the form of the transfer functions for the heater and oven-temperature blocks.

(b) Derive the system function from the input of the heater to the output of the thermostat.

(c) What is the form of the output signal in the time domain, given that the oven is excited with a constant voltage at the heater element?

(d) Determine the time required to heat the oven to 350°F and compare that time with the time required if there were no heat-up delay in the heater element. Assume an ambient temperature of 75°F.

7.23. A feedback system is shown in Fig. P7.23.

(a) What is the system function with feedback?

(b) Find A for critical damping, that is, to put the system on the boundary between under- and overdamped behavior.

(c) For a value of A 10 times that computed in part (b), what is the form of the natural response of the system?

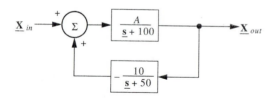

Figure P7.23

7.24. Show that in the absence of feedback, the final oven temperature used as an example in Sec. 7.4 reaches $10(1 - L_0)$, where L_0 is the dc loop gain and the

final change in the oven temperature is 10°F. *Hint:* Determine the increment in input temperature to produce a 10°F change in the oven temperature with feedback operating.

7.25. Determine from Eq. (7.81) the value of dc loop gain to give critical damping. Show that this is −2.025 for $\tau_o = 300$ s and $\tau_s = 30$ s, as stated in the text.

7.26. Show from Eq. (7.81) that the second-order system cannot become dynamically unstable, regardless of the dc loop gain.

7.27. A linear system is described by the differential equation

$$\frac{d^2 x_{out}}{dt^2} + 5\frac{dx_{out}}{dt} + 16x_{out} = 5\frac{dx_{in}}{dt}$$

where x_{out} is the output signal and x_{in} is the input signal. A fraction of the output, call it β, is fed back to the input and subtracted from the input in a negative feedback system.

(a) Draw a block diagram of the system in the frequency domain.

(b) What range of β gives transient behavior that is overdamped?

7.28. The block diagram for a linear process is shown in Fig. P7.28.

(a) Determine the maximum value of B for an overdamped response.

(b) Find the transfer function for $B = 0.1$.

(c) What value(s) of $\underline{s}$ allow output with no input?

7.29. The circuit shown in Fig. P7.29 has a main amplifier and a feedback network. The main amplifier has a gain of 500, infinite input impedance, zero output impedance, and a −3-dB point at 5 kHz (one pole only). Find C such that the poles are real and negative and differ by a factor of 2.

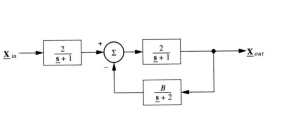

Figure P7.28

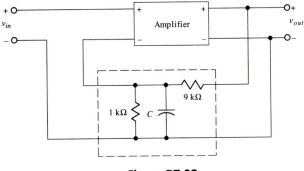

Figure P7.29

Answers to Odd-Numbered Problems

7.1. 5.89×10^{-12} s^{-1}.

7.3. $-5.27 + j314$.

7.5. (a) $\underline{Z}_{eq} = (\underline{s}RC + 1)/\underline{s}C$; (b) at dc, we can have voltage, but no current, open-circuit natural frequency; (c) at $\underline{s} = -1/RC$, we can have current but no voltage, short-circuit natural frequency.

7.7. (a) ∞; (b) -0.268, -3.73 s^{-1}; (c) 0, -2 s^{-1}; (d) must have two energy-storage elements and block dc: 2 Cs or LC is possible; (e) $-4 \times 10^4 e^{-t}$ V; (f) $v(t) = 2 \times 10^4(1 + e^{-2t} - 2e^{-t})$.

7.9. (a) short circuit: $-(R_1 \| R_2)/L$, open circuit: $-R_2/L$; (b) $-1/R(C_1 + C_2)$, open circuit: 0, $-1/RC_2$;

7.11. $i(t) = 5(e^{-10^4 t} + e^{-1.1 \times 10^4 t})$ A.

7.13. (a) No dc current; (b) $116\angle -27.2°$; (c) 0, -8 s^{-1}; (d) $-750e^{-10t}$ V; (e) -5, -20 s^{-1}.

7.15. Same as before, 0 initial value but V/RC for initial derivative.

7.17. (a) $\dfrac{R + \underline{s}L}{\underline{s}^2 LC + \underline{s}RC + 1}$;

(b) $\underline{s} = -\dfrac{R}{2L} \pm \sqrt{\left(\dfrac{R}{2L}\right)^2 - \dfrac{1}{LC}}$;

(c) 77.5 Ω is critically damped, undamped is 0, underdamped up to 77.5 Ω, overdamped over 77.5 Ω.

7.19. (a) $\dfrac{175(\underline{s}^2 + 10^7) + 2 \times 10^5 \underline{s}}{\underline{s}^2 + 10^7}$ Ω;

(b) 175 Ω, by inspection of the circuit; (c) 175 Ω, by inspection of the circuit; (d) $\underline{s} = -571 \pm j3110$ s^{-1}; (e) underdamped; (f) 20/175 A; (g) 20/175 A.

7.21. (a) 0 Ω; (b) 0 s^{-1}; (c) -0.725, -8.27 s^{-1}; (d) two inductors because it is second-order and it approaches a short at dc, or could be LC; (e) $-6e^{-2t}$ V; (f) $v(t) = -6e^{-2t} + 4.99e^{-0.725t} + 1.01e^{-8.27t}$ V.

7.23. (a) $\dfrac{A(\underline{s} + 50)}{(\underline{s} + 100)(\underline{s} + 50) + 10A}$; (b) 62.5;

(c) $-75.0 \pm j75.0$ s^{-1}.

7.25. For critical damping, $\dfrac{1}{4}\left(\dfrac{1}{\tau_o} + \dfrac{1}{\tau_s}\right)^2 = \dfrac{1 - L_0}{\tau_o \tau_s}$.

7.27. (a)

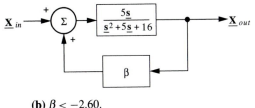

(b) $\beta < -2.60$.

7.29. 8.05 μF.

Index

1/f noise, 296
3 dB frequency, 223

A

analog-to-digital converter (A/D, ADC), 304
ac component, 211
acceptor dopant, 82
acoustic spectra, 200
active filter, 294
  high-pass, 299
  log-pass, 298
  Butterworth, 300
adder, BCD, 156
aliasing, 311
alpha of transistor, 93
ALU, Arithmetic Logic Unit, 173
AM, Amplitude Modulation, 330
  broadcast band, 335
  radio, 214
  signal spectrum, 334
ammeter, 28
Ampere's circuital law, 349
amplifier
  feedback, 230
  instrumentation, 290
  inverting, 247, 314
  JFET, 109
  noninverting, 290
  transistor, 94, 100, 242
analog, 196
analog computers, 245
Analog-to-Digital (A/D) conversion, 280, 304
  precision, 304
  range, 304
  resolution, 304
AND, 139, 147
antenna, 24
  cornucopia, 351
  gain, 355
  receiving, 356
  reflector-type, 359
  transmitting, 355
  wire, 355
antilog amplifier, 260
arithmetic circuits, 156
arithmetic-logic unit, 173
array, antenna, 360, 361
ASCII, 170

assembler, 175
assembly language, 175
asymptotic Bode plot, 225
  high-pass filter, 226, 299
  low-pass filter, 227, 299, 389
attenuation, wave, 355
attenuator, 294
available power, 26, 244
  antenna, 357

B

balanced bridge, 281
bandpass filter, 228
bandwidth, 209, 311
  information rate, 214
  pulse width, 209
base, transistor
  breakdown, 95
  transistor, 89
base 10 numbers, 154
base 16 numbers, 155
base 2 numbers, 153
base band, 344
battery, 14
  properties, 70
baud, 215
bell, 209
beta of transistor, 93
Binary Coded Decimal (BCD), 154, 170
  adder, 156
binary numbers, 153
binary variables, 137
bipolar junction transistors (BJT), 89, 93, 98,
      100, 141, 143
bit, 136, 154
blackbody radiation, 344
Bode, H. W., 224
Bode plot, 221, 223, 388
  high-pass filter, 226, 299
  low-pass filter, 225, 227, 299, 389
bonds, covalent, 80
Boole George, 147
Boolean theorems, 147
Boolean algebra, 147
bridge circuit, 281
bridge rectifier, 73
buffer, 252, 308
bus, data, 170

Butterworth filter, 300
byte, 170

C

cabling, 302
calculator charger, 136
calculator, 18
capacitor, capacitance, 33, 34
  applications, 310
  circuit symbol, 34
  coupling, 100
  distributed, 350, 351
  energy, 393
  filter, 75
  final values, 42
  generalized impedance, 379
  KCL, 35
carrier, 330
carriers, 81
cascaded systems, 223
  dB gain, 228
cathode ray tubes, 68
center-tapped transformer full-wave rectifier, 75
central processing unit (CPU), 173
channel, 170
characteristic impedance, 351
charge, 4
chopped signal, 296
circuit symbol
  resistance, 12
  FET, 105
circuits
  linear, 69, 328
  nonlinear, 328
circular polarization, 362
CMOS, 146, 152
CMRR, 291
coaxial cables, 350, 351
collector, transistor, 89
combinational logic, 161
common, voltmeter, 30
common collector, 242
common mode signal, 291
common-emitter connection, 94
common-mode rejection ratio, 291
communication equation, 357
communication link, 358
communication systems, 328